Horticulture: Ecology and Agriculture

Horticulture: Ecology and Agriculture

Edited by Wendel Mason

SYRAWOOD
PUBLISHING HOUSE

New York

Published by Syrawood Publishing House,
750 Third Avenue, 9th Floor,
New York, NY 10017, USA
www.syrawoodpublishinghouse.com

Horticulture: Ecology and Agriculture
Edited by Wendel Mason

International Standard Book Number: 978-1-68286-382-4 (Hardback)

Cataloging-in-publication Data

Horticulture : ecology and agriculture / edited by Wendel Mason.
    p. cm.
Includes bibliographical references and index.
ISBN 978-1-68286-382-4
1. Horticulture. 2. Agricultural ecology. 3. Agricultural biotechnology. 4. Agriculture--Environmental aspects.
I. Mason, Wendel.
SB318 .H67 2017
635--dc23

Printed in the United States of America.

TABLE OF CONTENTS

Preface...VII

Chapter 1   **Breeding without Breeding: Is a Complete Pedigree Necessary for Efficient Breeding?**...1
Yousry A. El-Kassaby, Eduardo P. Cappa, Cherdsak Liewlaksaneeyanawin, Jaroslav Klápště, Milan Lstibůrek

Chapter 2   **Multiple Origins of the Sodium Channel *kdr* Mutations in Codling Moth Populations**...12
Pierre Franck, Myriam Siegwart, Jerome Olivares, Jean-François Toubon, Claire Lavigne

Chapter 3   **Carnivore use of Avocado Orchards across an Agricultural-Wildland Gradient**...........22
Theresa M. Nogeire, Frank W. Davis, Jennifer M. Duggan, Kevin R. Crooks, Erin E. Boydston

Chapter 4   **The Effect of Urbanization on Ant Abundance and Diversity: A Temporal Examination of Factors Affecting Biodiversity**...28
Grzegorz Buczkowski, Douglas S. Richmond

Chapter 5   **Quasi-Double-Blind Screening of Semiochemicals for Reducing Navel Orangeworm Oviposition on Almonds**...37
Kevin Cloonan, Robert H. Bedoukian, Walter Leal

Chapter 6   **Comparative Toxicities and Synergism of Apple Orchard Pesticides to *Apis mellifera* (L.) and *Osmia cornifrons* (Radoszkowski)**...43
David J. Biddinger, Jacqueline L. Robertson, Chris Mullin, James Frazier, Sara A. Ashcraft, Edwin G. Rajotte, Neelendra K. Joshi, Mace Vaughn

Chapter 7   **Identification of Land-Cover Characteristics using MODIS Time Series Data: An Application in the Yangtze River Estuary**...49
Mo-Qian Zhang, Hai-Qiang Guo, Xiao Xie, Ting-Ting Zhang, Zu-Tao Ouyang, Bin Zhao

Chapter 8   **Ecological and Genetic Differences between *Cacopsylla melanoneura* (Hemiptera, Psyllidae) Populations Reveal Species Host Plant Preference**.........................60
Valeria Malagnini, Federico Pedrazzoli, Chiara Papetti, Christian Cainelli, Rosaly Zasso, Valeria Gualandri, Alberto Pozzebon, Claudio Ioriatti

Chapter 9   **Arbuscular Mycorhizal Fungi Associated with the Olive Crop across the Andalusian Landscape: Factors Driving Community Differentiation**.........................69
Miguel Montes-Borrego, Madis Metsis, Blanca B. Landa

Chapter 10   **Manure Refinement Affects Apple Rhizosphere Bacterial Community Structure: A Study in Sandy Soil**...81
Qiang Zhang, Jian Sun, Songzhong Liu, Qinping Wei

Chapter 11   **A Simple Model to Predict the Probability of a Peach (*Prunus persicae*) Tree Bud to Develop as a Long or Short Shoot as a Consequence of Winter Pruning Intensity and Previous Year Growth**...89
Daniele Bevacqua, Michel Génard, Françoise Lescourret

Chapter 12   **Spatial and Temporal Variations of Ecosystem Service Values in Relation to Land Use Pattern in the Loess Plateau of China at Town Scale**.. 95
Xuan Fang, Guoan Tang, Bicheng Li, Ruiming Han

Chapter 13   **Development and Validation of a Weather-Based Model for Predicting Infection of Loquat Fruit by *Fusicladium eriobotryae***.. 108
Elisa González-Domínguez, Josep Armengol, Vittorio Rossi

Chapter 14   **Determination of 17 Organophosphate Pesticide Residues in Mango by Modified QuEChERS Extraction Method Using GC-NPD/GC-MS and Hazard Index Estimation in Lucknow, India**..120
Ashutosh K. Srivastava, Satyajeet Rai, M. K. Srivastava, M. Lohani, M. K. R. Mudiam, L. P. Srivastava

Chapter 15   **Spatial and Temporal Patterns of Carbon Storage in Forest Ecosystems on Hainan Island, Southern China**..130
Hai Ren, Linjun Li, Qiang Liu, Xu Wang, Yide Li, Dafeng Hui, Shuguang Jian, Jun Wang, Huai Yang, Hongfang Lu, Guoyi Zhou, Xuli Tang, Qianmei Zhang, Dong Wang, Lianlian Yuan, Xubing Chen

Chapter 16   **A Melting Pot of Old World Begomoviruses and their Satellites Infecting a Collection of *Gossypium* Species in Pakistan**..141
Muhammad Shah Nawaz-ul-Rehman, Rob W. Briddon, Claude M. Fauquet

Chapter 17   **Aerial Application of Pheromones for Mating Disruption of an Invasive Moth as a Potential Eradication Tool**..164
Eckehard G. Brockerhoff, David M. Suckling, Mark Kimberley, Brian Richardson, Graham Coker, Stefan Gous, Jessica L. Kerr, David M. Cowan, David R. Lance, Tara Strand, Aijun Zhang

Chapter 18   **Migrating Giant Honey Bees (*Apis dorsata*) Congregate Annually at Stopover Site in Thailand**..172
Willard S. Robinson

Permissions

List of Contributors

Index

PREFACE

Horticulture is the growth of plants for self-sustenance as well for as commercial purposes. Horticultural practices dominate the practical aspect of agriculture and horticultural theory too, plays an important aspect in agronomic studies. This book will provide interesting topics for research which interested readers can take up. This book studies, analysis and upholds the pillars of horticulture and its utmost significance in modern times. It attempts to understand the multiple branches that fall under the discipline of horticulture and now such concepts have practical applications primarily with respect to ecology and agriculture. It will help the readers in keeping pace with the rapid changes in this field. Those with an interest in the horticulture field would find this book helpful.

Various studies have approached the subject by analyzing it with a single perspective, but the present book provides diverse methodologies and techniques to address this field. This book contains theories and applications needed for understanding the subject from different perspectives. The aim is to keep the readers informed about the progress in the field; therefore, the contributions were carefully examined to compile novel researches by specialists from across the globe.

Indeed, the job of the editor is the most crucial and challenging in compiling all chapters into a single book. In the end, I would extend my sincere thanks to the chapter authors for their profound work. I am also thankful for the support provided by my family and colleagues during the compilation of this book.

Editor

Breeding without Breeding: Is a Complete Pedigree Necessary for Efficient Breeding?

Yousry A. El-Kassaby[1]*, Eduardo P. Cappa[2], Cherdsak Liewlaksaneeyanawin[1¤], Jaroslav Klápště[1], Milan Lstibůrek[3]

1 Department of Forest Sciences, Faculty of Forestry, University of British Columbia, Vancouver, British Columbia, Canada, 2 Instituto Nacional de Tecnología Agropecuaria (INTA), Instituto de Recursos Biológicos, Hurlingham, Buenos Aires, Argentina, 3 Department of Dendrology and Forest Tree Breeding, Faculty of Forestry and Wood Sciences, Czech University of Life Sciences Prague, Praha, Czech Republic

Abstract

Complete pedigree information is a prerequisite for modern breeding and the ranking of parents and offspring for selection and deployment decisions. DNA fingerprinting and pedigree reconstruction can substitute for artificial matings, by allowing parentage delineation of naturally produced offspring. Here, we report on the efficacy of a breeding concept called "Breeding without Breeding" (BwB) that circumvents artificial matings, focusing instead on a subset of randomly sampled, maternally known but paternally unknown offspring to delineate their paternal parentage. We then generate the information needed to rank those offspring and their paternal parents, using a combination of complete (full-sib: FS) and incomplete (half-sib: HS) analyses of the constructed pedigrees. Using a random sample of wind-pollinated offspring from 15 females (seed donors), growing in a 41-parent western larch population, BwB is evaluated and compared to two commonly used testing methods that rely on either incomplete (maternal half-sib, open-pollinated: OP) or complete (FS) pedigree designs. BwB produced results superior to those from the incomplete design and virtually identical to those from the complete pedigree methods. The combined use of complete and incomplete pedigree information permitted evaluating all parents, both maternal and paternal, as well as all offspring, a result that could not have been accomplished with either the OP or FS methods alone. We also discuss the optimum experimental setting, in terms of the proportion of fingerprinted offspring, the size of the assembled maternal and paternal half-sib families, the role of external gene flow, and selfing, as well as the number of parents that could be realistically tested with BwB.

Editor: Pär K. Ingvarsson, University of Umeå, Sweden

Funding: This study is supported by the Natural Sciences and Engineering Research Council of Canada - Discovery and IRC Grants and the Johnson's Family Forest Biotechnology Endowment to YAE. The funders had no role in study design, data collection and analysis, decision to publish, or preparation of the manuscript.

Competing Interests: The authors have declared that no competing interests exist.

* E-mail: y.el-kassaby@ubc.ca

¤ Current address: SCG Paper PLC, Bangkok, Thailand

Introduction

Plant breeding, including tree improvement, typically follows the classical recurrent selection scheme, which is characterized by systematic and repetitive cycles of breeding, testing, and selection [1], [2]. These programs deal with multiple populations (e.g., base, breeding, and deployment) and large numbers of parents and offspring, planted over multiple sites and years, and requiring extensive monitoring and maintenance. Selection of elite genotypes for either further breeding and/or inclusion in production populations is commonly performed based on their breeding values, determined from the intra-class correlation among relatives produced from elaborate mating designs [3]. As breeding programs advance, the number of parents' increases and their genealogy overlaps, and mating designs become more elaborate and the time required for their completion become real breeding programs' limiting factors [4]. To alleviate the efforts associated with generating offspring with complete pedigree information, specifically for early generation testing, forest geneticists have adopted simplified protocols, ranging from those not requiring a pedigree (e.g., bulk samples from natural populations known as

provenance testing [5] to those with incomplete pedigrees (e.g., open-pollinated [6] or polycross mating [7]). Data analyses with incomplete pedigrees often require invoking and/or accepting untestable assumptions related to the genetic constitution of the tested families and the numbers of male parents involved in their formation, as well as their proportionate contributions. Since these assumptions are not inordinately realistic in practice, the resulting genetic parameters and their associated inferences are often biased, ultimately leading to various degrees of inaccuracy and inefficiency [8]–[10].

The availability of affordable, highly informative DNA markers, coupled with the development of sophisticated pedigree reconstruction methods, has enhanced their utility in converting incomplete pedigree trials into (effectively) complete trials, thus eliminating the pitfalls associated with the invocation of unfulfilled assumptions [11]. Lambeth *et al.* [11] initiative of converting the polycross mating design's incomplete pedigree to complete made proper quantitative genetic analyses possible and the method was repeatedly evaluated for several species [12]–[16]. El-Kassaby *et al.* [13] and El-Kassaby and Lstibůrek [17] capitalized on the restricted maximum likelihood-based "animal model" [18] capability of analysing unbalanced and incomplete pedigree data, along with pedigree reconstruction (tantamount to paternity assignment), to introduce the concept of "Breeding without Breeding (BwB)." The basic idea of BwB is to combine the use of offspring with incomplete pedigree information (an entire open-pollinated test) with a subset of offspring with complete pedigree information, to construct both parental and offspring breeding values, thus incorporating backwards, forwards, and combined selection into an efficient breeding framework [13], [17]. Most of the DNA fingerprinting effort is dedicated to a subset of the offspring from a small number of known maternal parents (seed donors) to generate information about the entire population (maternal and paternal parents, as well as offspring) after reassembling paternal half-sib families from the pedigree reconstruction of the fingerprinted subset. Pedigree reconstruction permits connecting the entire parental population (sampled or not) through their shared offspring thus allowing the implementation of classical quantitative genetics analyses [18].

Here we experimentally demonstrate the utility, the increased precision of genetic parameters estimation, and increased accuracy of predicted breeding values, hence the effectiveness of the "Breeding without Breeding" concept, using open-pollinated offspring from 15 of 41 parents in a western larch (*Larix occidentalis* Nutt.) "breeding population." We compared the performance of the combined incomplete (half-sib: HS) + complete (full-sib: FS) analysis to that of both the incomplete and complete pedigree designs. Finally, we illustrate the optimum experimental efforts needed for the successful implementation of BwB and discuss the role of factors such as external gene flow, expansion of the test population (i.e., the number of tested parents), and the size of half- or full-sib family needed for accurate genetic parameter determination.

Results

Pedigree Reconstruction/Mating Design Assembly

The partial pedigree reconstruction allowed direct estimation of gene flow, selfing rate, male reproductive success, and the number and/or size of maternal and paternal half-sib families on the individual as well as the population level (Figure 1). With 95% confidence, 1,419 out of 1,538 (92.3%) fingerprinted offspring were assigned to male parents within the orchard (Figure 1). The remaining 119 paternally unassigned offspring were identified as the product of introgression from an adjacent orchard, suggesting a pollen immigration rate of 7.7%. In addition, a total of 113 individual offspring resulted from selfing (average: 7.4%), ranging from 0.0 to 26.8% among seed donors, reflecting the 15 maternal parents propensity variation to selfing. This variability could be caused by maternal parents' pollen shed and receptivity period synchrony differences.

Pedigree reconstruction resulted in the formation of 349 full-sib families, nested within the 15 maternal and 38 paternal half-sib families, respectively, indicating that three out of the orchard's potential 41 male parents did not participate in pollination, at least of these 15 maternal parents, most likely due to their recent introduction to the seed orchard population (pers. observation). The 15 maternal half-sib families had an average size of 283.9 (range: 222–397) and the 38 paternal half-sib families had an average size of 37.3 (range: 1–193 among the 38 recovered paternal sibships), the latter evidently reflecting male fecundity variation within the orchard. There was an apparently high correlation between the difficult to assess male reproductive investment (male strobili production) and male reproductive success (determined by paternity analysis [19] ($r = 0.87$; $P < 0.001$).

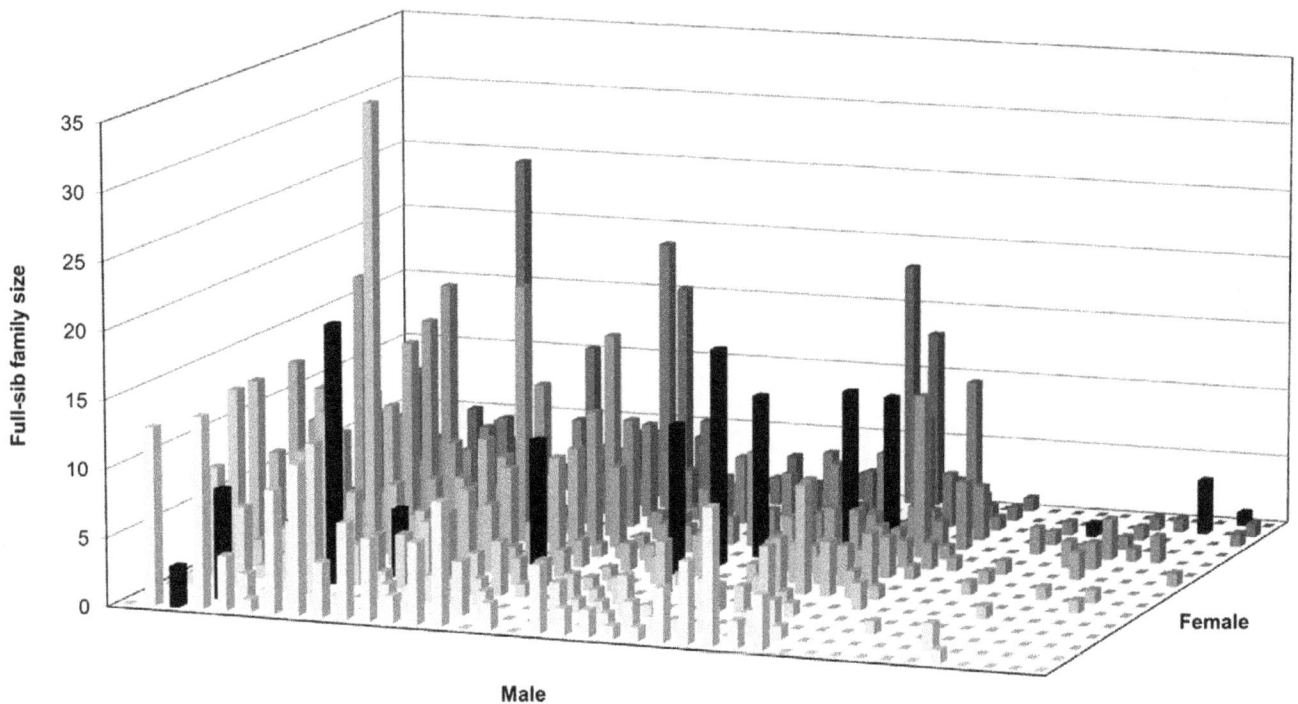

Figure 1. Pedigree reconstruction results showing the formation of full-sib families nested within the maternal and paternal half-sib families (black bars represent selfing).

Table 1. Forth-year height variance components and narrow sense heritability values (h^2_{ns}) and their standard errors for the half-sib (HS), combined half-sib+full-sib (HS+FS) and full-sib (FS) models.

Source of variation	Variance component		
	Incomplete pedigree		Complete pedigree
	HS	HS+FS	FS
Additive	156.8±80.0	69.3±26.9	55.93±25.42
Plot	48.7±11.2	80.7±17.2	101.95±23.93
Error	266.4±60.5	332.4±20.1	315.99±19.52
Total	471.9	482.5	473.9
h^2_{ns}	0.33±0.16	0.14±0.05	0.12±0.05

The reconstructed pedigree formed a structured mating design, which we used to generate quantitative genetic parameters for the complete pedigree model (FS), and was used in concert with the non-fingerprinted individuals within each of the 15 HS families to form a combined pedigree model, consisting of half- and full-sib families (HS+FS) (see below). A minimum paternal half-sib family size threshold of six individuals was established for inclusion in quantitative genetic analyses. Seven male parents did not meet this threshold, but two were retained, because they were also represented as seed-donors, thus far exceeding the established minimum family size threshold.

Estimation of Quantitative Genetic Parameters

Following the classical individual-tree additive model, three analyses were conducted. The first is for the 15 open-pollinated families (HS) with sample size of $N = 5,796$ individuals (i.e., incomplete pedigree). The second is also for the same 15 HS families ($N = 5,796$) but after the inclusion of the male parent for 1,419 individuals (i.e., a combination of half- and full-sib families (HS+FS) and also represents an incomplete pedigree). While the third representing full pedigree ($N = 1,419$) and was solely based on full-sib families formed by the pedigree reconstruction (FS) (Figure 1; Table 1). Relative to the combined HS+FS model, the HS model grossly overestimated the additive genetic variance (156.8 vs. 69.3), which more than doubled the height heritability estimate (0.33 vs. 0.14) (Table 1). The precision of the additive genetic variance (80.0 vs. 26.9) and heritability (0.16 vs. 0.05) estimates for these two models produced higher standard error for the HS as compared to the combined HS+FS model (Table 1). Additionally, the inclusion of more genetic information in the combined HS+FS model (i.e., those from FS families) increased the sensitivity of the analysis, as subtle plot-to-plot variation was detected, resulting in a more realistic assessment of the residual error term (Table 1). Parental breeding values' comparisons was limited to only the 15 maternal parents in the HS analysis with their corresponding 15 estimates from the HS+FS analysis and produced non-significant product-moment ($r = 0.44$ (CI: -0.099, 0.775); $p = 0.105$, Figure 2) and rank ($\rho = 0.44$ (CI: -0.099, 0.775); $p = 0.105$) correlations. The corresponding comparison of HS with HS+FS breeding values for the offspring yielded significant product-moment ($r = 0.69$ (CI: 0.672, 0.700); $p = 0.0001$, Figure 3) and rank ($\rho = 0.67$ (CI: 0.656, 0.686); $p = 0.0001$) correlations. Both results

$r = 0.44$ ($p < 0.105$)

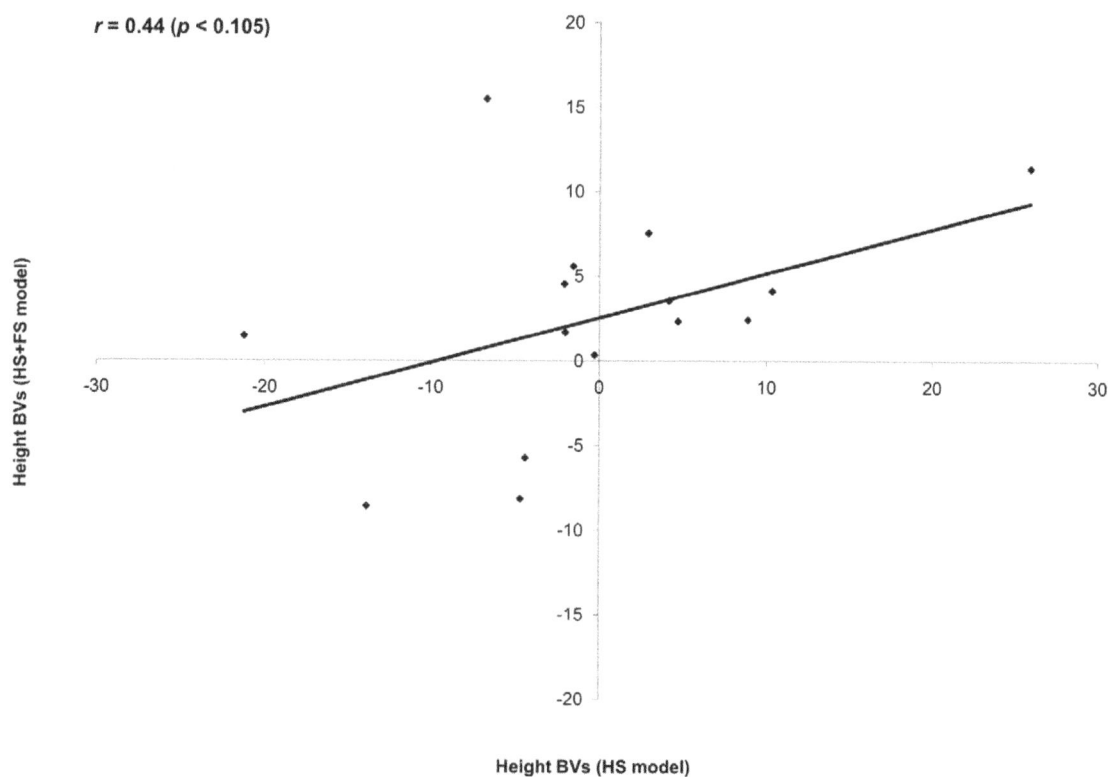

Figure 2. Scatter plot of predicted breeding values for parents from the two incomplete pedigree models (HS and combined FS+HS). Pearson correlation (r) is in the left corner of the graph.

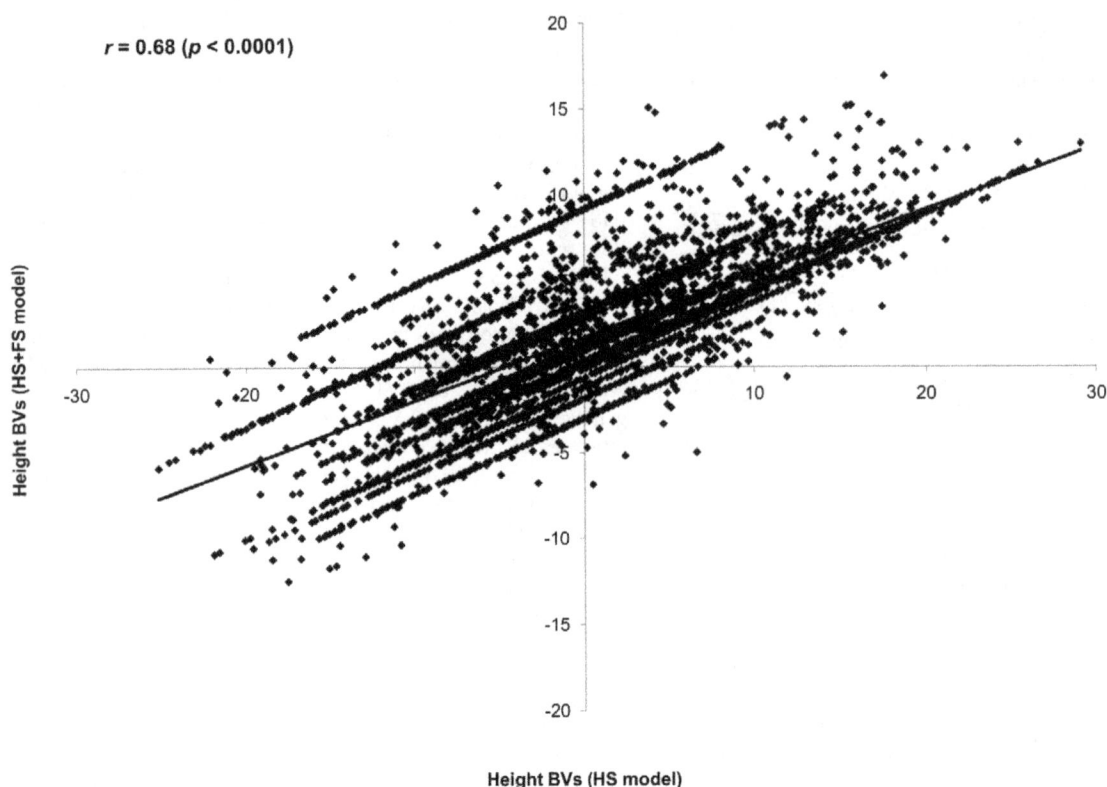

Figure 3. Scatter plot of predicted breeding values for offspring from the two incomplete pedigree models (HS and combined HS+FS). Pearson correlation (*r*) is in the left corner of the graph (note the greater extent of variation between the two models).

clearly demonstrate the reduced utility of the HS model's estimates for forward selection, relative to the results from the HS+FS treatment as indicated by both product-moment and rank correlations. Finally, the average accuracy of predicted breeding values, calculated from the combined HS+FS model was higher for parents (0.81) and offspring (0.55), than their corresponding values (0.56 and 0.45, respectively), calculated from HS model.

The full (FS) and combined HS+FS pedigree models produced comparable additive and heritability estimates, with similar precision (Table 1). Predictions of parental breeding values extracted from both models were comparable and highly correlated (product-moment ($r = 0.96$ (CI: 0.928, 0.982); $p = 0.0001$, Figure 4) and rank ($\rho = 0.94$ (CI: 0.875, 0.968); $p = 0.0001$) correlations). The same was true for offspring breeding values (product-moment ($r = 0.97$ (CI: 0.971, 0.976); $p = 0.0001$, Figure 5) and rank ($\rho = 0.97$ (CI: 0.967, 0.973); $p = 0.0001$) correlations). The results from the combined HS+FS pedigree approach are robust and reliable. Moreover, the average accuracy of breeding values from parents and offspring calculated from the FS model (0.78 and 0.69, respectively) were very similar to those estimated from the combined HS+FS model (0.76 and 0.64, respectively). It is interesting to note that predicted parental breeding values were produced for the entire parental population (i.e., all seed and pollen donors), even when only 15 maternal parents were used and these estimates were based on the entire population ($N = 5,796$) for the combined HS+FS model as opposed to $N = 1,419$ for the FS model.

Production Population Selection

We implemented three selection options; namely, forwards, backwards, and combined (combination of backwards and forwards), utilizing either the parental (backwards) and/or offspring

(forwards) "Best Linear Unbiased Predictors" (BLUPs) generated from the HS or the combined HS+FS models. The backwards selection option was applied exclusively to the combined HS+FS model as parental breeding values were determined from both maternal and paternal information. The limited number of maternal parents (15 seed donors) precluded the application of the backwards selection option under the HS model; however, maternal breeding values along with offspring was used in the HS combined selection. Additionally, the limited number of maternal parents minimized the response to selection's differences between the forwards and combined selections resulting in somewhat identical results (Figure 6). Without exception and across the range of effective population size tested, the HS model overestimated the response to selection as compared to that from the combined HS+FS model, reflecting the observed additive genetic variance overestimation (Figure 6). For example, compared to the combined HS+FS model, the HS combined selection overestimated the response to selection by a range of 15 and 25% for effective population size of 10 and 40, respectively (Figure 6). The combined HS+FS model's forward and/or combined selections were superior to their backward with response to selection differences ranging between 7 and 12% for effective population size of 10 and 30, respectively (the paternal HS family size restriction of n = 6 limited the effective population size range for backward) (Figure 6). Finally, as expected and for all selection methods and both HS and the combined HS+FS models, the response to selection decreased with increased in effective population size (Figure 6).

Estimating Offspring Optimum Sample Size

Drastic difference in the additive genetic variance magnitude and its standard error was observed with increasing the number of

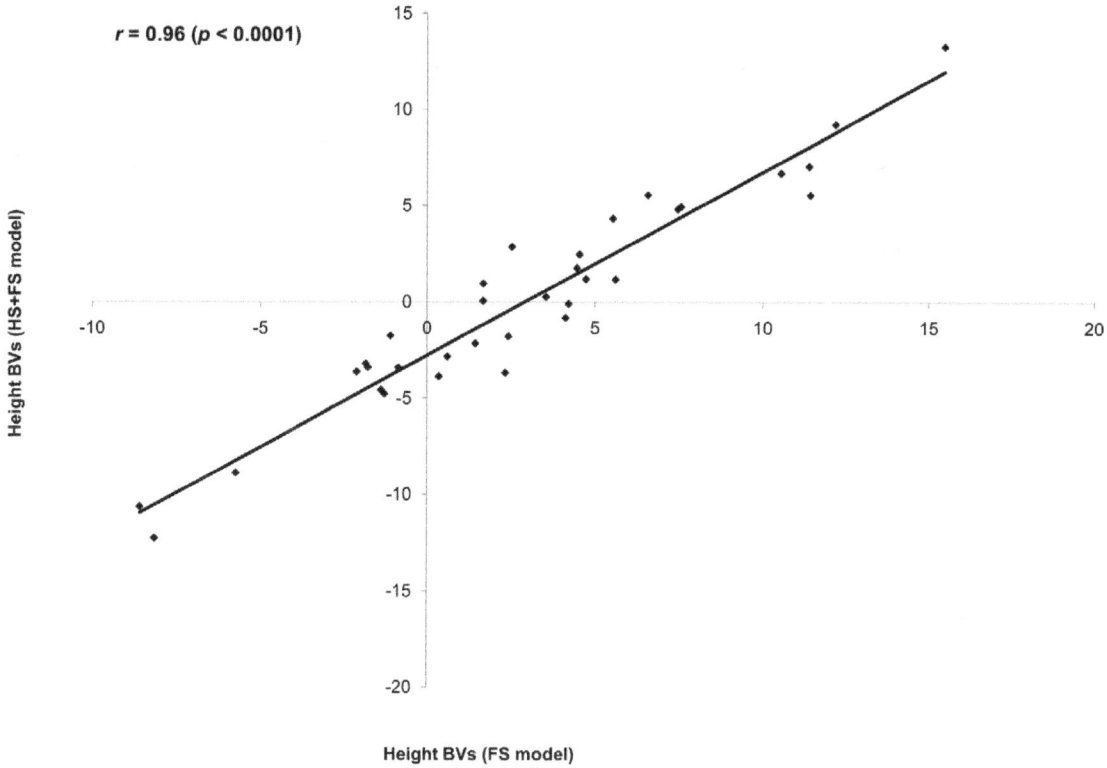

Figure 4. Scatter plot of predicted breeding values for parents from the incomplete (combined HS+FS) and complete (FS) pedigree models. Pearson correlation (*r*) is in the left corner of the graph.

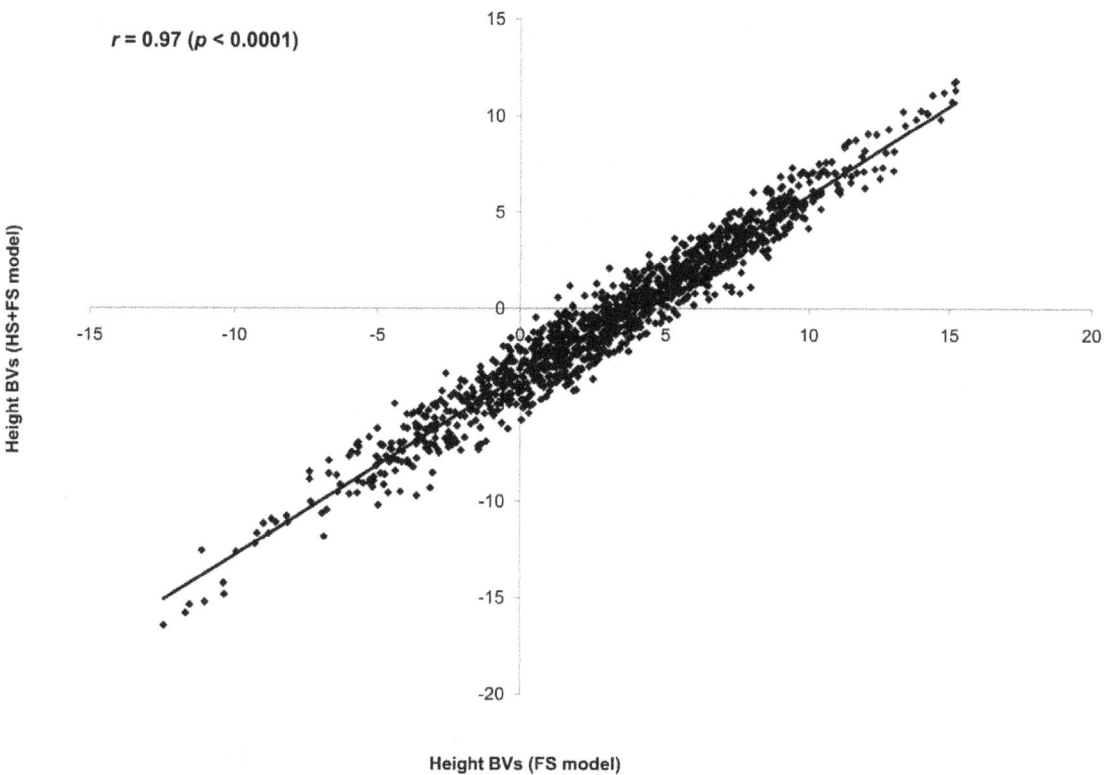

Figure 5. Scatter plot of predicted breeding values for offspring from the incomplete (combined HS+FS) and complete (FS) pedigree models. Pearson correlation (*r*) is in the left corner of the graph.

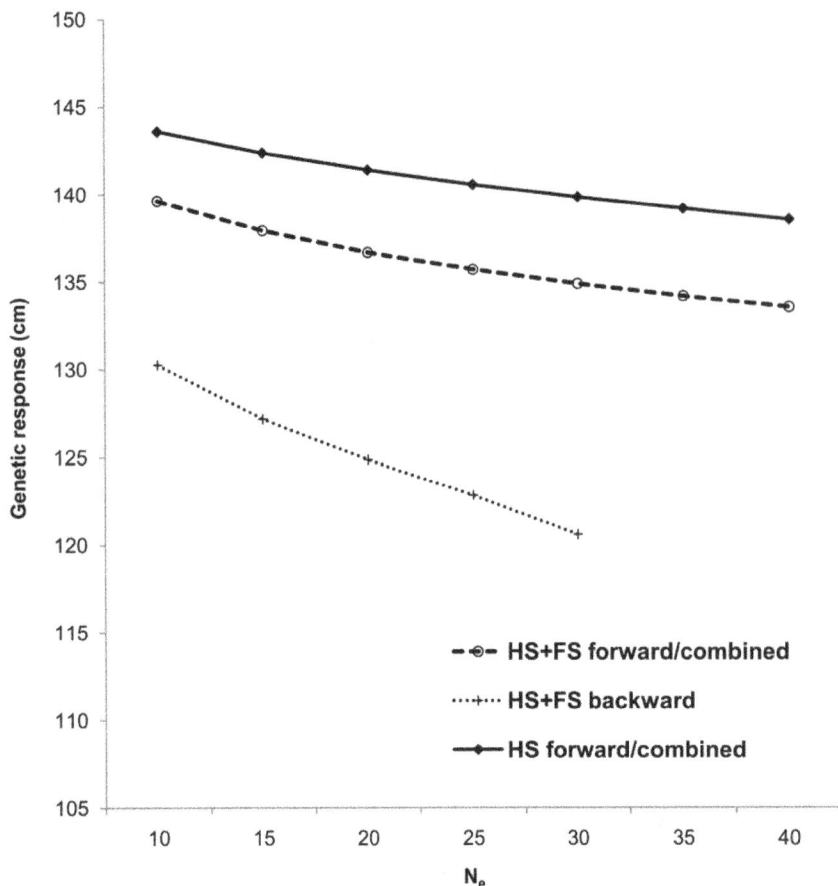

Figure 6. Response to selection comparison between the half-sib (HS) (forward and combined) and combined half- and full-sib (HS+FS) (backward, forward and combined) models assessed across various effective population sizes (10 to 40). (The small number of tested parents resulted in identical results for forward and combined selection methods under the combined HS+FS and HS scenarios).

trees with known paternal information (Figure 7). Increasing the number of trees with known fathers (i.e., those from the pedigree reconstruction) to those already with known mothers improved the direct and/or indirect connectedness among parents and thus permitted their unbiased comparison as well as their genetic parameters' estimation. The observed improvement in the additive genetic variance precision leveled after the inclusion of 600 individuals and no substantial fluctuations were observed beyond this point, indicating that a threshold was reached and the inclusion of any additional offspring would not substantially affect the results (Figure 7). Based on the observed trend and in this particular case, it appears that the inclusion of paternal information for 10% of the evaluated offspring population is adequate to create the direct and/or indirect connectedness among parents is sufficient to achieve the available precision.

Discussion

The concept of marker-assisted estimation of quantitative genetic parameters was introduced by Ritland [20], whereby traits' heritabilities and the magnitude and direction of their genetic correlations are derived from regressing pair-wise phenotypic similarity on their corresponding pair-wise genetic relatedness. This concept is appealing, because of its obvious simplicity, *in situ* nature (i.e., no experiments or mating designs), and most of all its suitability to long-lived organisms such as trees

or wildlife that require long-term experiments or extensive field observations. The distribution of relatedness among the studied individuals is assumption-free, thus it is applicable to natural populations where a vast array of genetic relationships can occur [21], [22]. In situations where offspring are derived from random mating among a set of known parents and more specifically when their number is somewhat limited, the no *a priori* assumption about the expected distribution of genetic relationship becomes inappropriate for a network of full-sibs, half-sibs, and selfs (albeit absence of spatial autocorrelation in relationship coefficients as well as in trait performance in the wild are assumed). It should be stated; however, that the regression approach does not permit the prediction of parents and/or offspring breeding values, thus its application to selection and breeding is somewhat limited.

Conventional tree breeding programs are structured around three main activities: breeding, testing and selection [23]. These activities are long-term endeavours, based on structured pedigree produced from one or a combination of different mating designs [23]; they also require extensive testing in large experimental settings, distributed throughout vast territories [4], and (most important of all), they require sustained organizational and financial commitment. Obviously, simplified breeding schemes that reduce time and cost would be of great value. The generation of complete pedigreed offspring for testing and selection is an obvious target for simplification, fostering incomplete pedigree methods such as open-pollinated family testing [6] and polycross

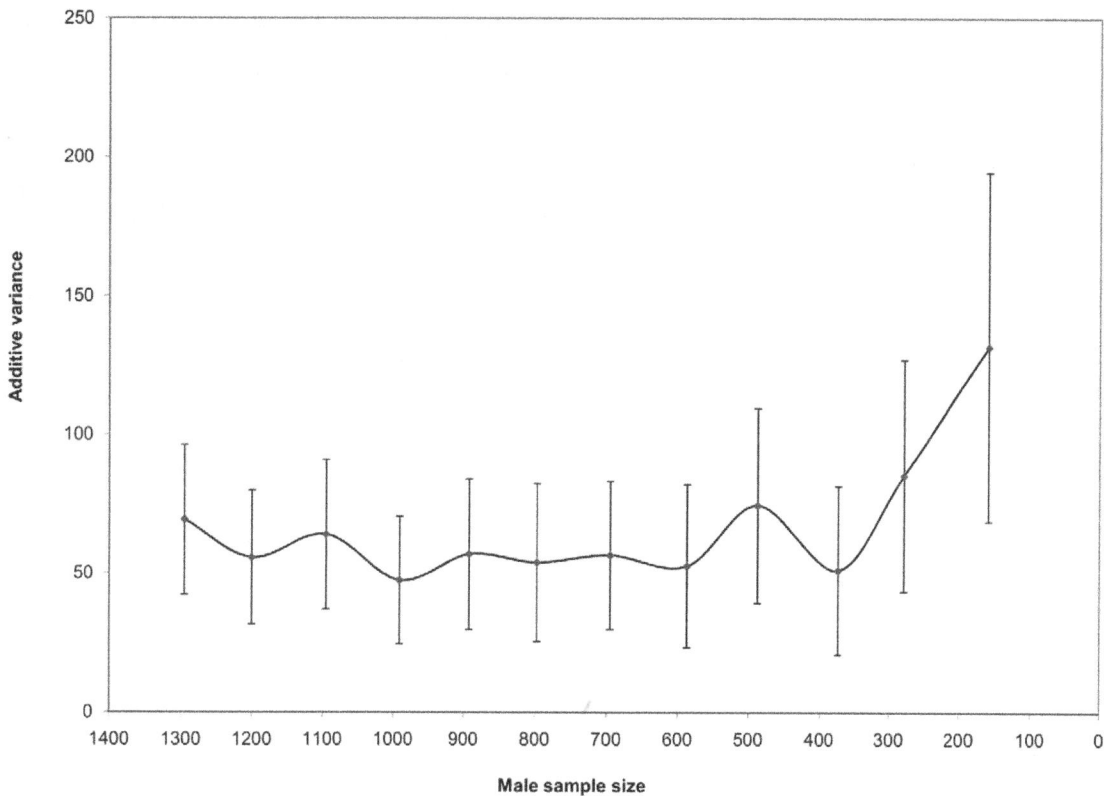

Figure 7. Additive genetic variance estimates as affected by variation in the number of offspring with known male parents. Vertical lines represent the standard error bars for additive genetic variance estimates.

designs [7]. Incomplete pedigree methods, however, are not without their limitations. In particular, open-pollinated testing of the offspring of each maternal parent (seed-donor) assumes they are half-sibs (i.e., sired by different fathers) and non-inbred, and the covariance among half-sib families is assumed to be equal to one-fourth of the additive genetic variance [3]. Both theoretical and empirical studies indicate that this assumption is often violated; as a practical consequence, additive genetic variance is typically overestimated [8]–[10]. The extent of the half-sib assumption violation is expected to be greater if the number of male parents is restricted, as it typically is in a confined breeding population, the usual strategy in breeding arboreta.

To avoid the inaccuracies associated with quantitative genetic parameters assessment from incomplete pedigrees, Lambeth *et al.* [11] proposed the use of molecular genetic markers for paternity assignment, thus converting the incomplete to a complete pedigree, allowing proper genetic parameters estimation and reliable parental and offspring ranking. The same approach was also introduced to open-pollinated testing by Grattapaglia *et al.* [12], who reconstructed the complete pedigrees.

El-Kassaby *et al.* [13] and El-Kassaby and Lstibůrek [17] introduced the concept of "Breeding without Breeding" as a simple, alternative scheme to conventional tree breeding. The method uses: 1) large open-pollinated (i.e., incomplete pedigree) as a primary mean to simplify testing, 2) informative DNA fingerprinting and pedigree reconstruction for a randomly selected subset of the tested individuals to determine their genetic relationship (i.e., complete pedigree) and hence provide adequate bridges between all parents (female and male), 3) the animal model [18] to concurrently analyse the combination of complete (FS:

subset) and incomplete (HS: open-pollinated families) pedigree to generate the quantitative genetic parameters needed for selection, and 4) application of an optimization protocol [17] that maximizes the genetic gain at any desired genetic diversity level in a selection scheme. The method capitalizes on the animal model's [18] capabilities of analysing unbalanced and incomplete pedigree to generate the genetic parameters using the "Best Liner Unbiased Prediction" procedure (BLUP [24] needed for parental and offspring evaluation thus facilitating backwards, forwards, or a combined (backwards and forwards) selection in a breeding framework. Therefore, the fundamental difference between the assembled genetic relationship among individuals in the BwB scheme (present study) and those from either the polycross [11], [16] or open-pollinated testing [12]–[16] is that the former does not require complete pedigree for the tested population (a combination of large half-sibs and several smaller full-sibs families) while the latter explicitly stipulates the availability of complete pedigree information for every individual for quantitative genetic parameters estimation.

Quantitative genetic parameters comparison between the two incomplete-pedigree models (i.e., HS and the combined HS+FS) indicated that the HS model over-estimated the additive genetic variance and its surrogate heritability and under-estimated the environmental effects (Table 1). As expected, the genetic relationships (half-sib, and full-sib; Figure 1) within the studied 15 half-sib families should have reduced the average covariance among relatives within the HS model, thus the resulting additive genetic variance is unrealistically inflated. Furthermore, the HS model failed to detect the subtle site heterogeneity present in the experimental site [25], hence the observed under-estimation of the plot effect (Table 1). This is due to the fact that the 15 half-sib families were present in 4 large, 10×10 replications which made it difficult to definitively separate the genetic and environmental effects within experimental units (i.e., plots). In multiple-tree and contiguous plots designs, substantial environmental covariance among family members is confounded with genetic covariance of a given plot [25]. The degree of confounding depends on the size of the plots and the patterns of environment variability. In general, the larger the plot, the more difficult it is to cleanly separate genetic from environmental effects. On the other hand, site heterogeneity was clearly detected after the inclusion of more genetic information in the combined HS+FS model (i.e., those resulting from the pedigree reconstruction of 1,419 individuals which resulted in a better site heterogeneity detection due to their presence across all half-sib families and their respective replications). It is noteworthy to mention that the changes in variance components apportionment over the HS and the combined HS+FS models' sources of variation, collectively affected the resulting heritability estimate (Table 1). While it is only for a subset of the offspring, the inclusion of additional paternal information in the combined HS+FS model permitted covariance among relatives adjustment and hence the observed improvement in the generated parameters, a situation cannot be attained under the HS model (i.e., open-pollinated test). The discrepancy between the two models is further demonstrated by the low to moderate correlations between either paternal or offspring breeding values (Figs. 2–3) and their different average accuracy of prediction (0.56 vs. 0.81 for parents and 0.45 vs. 0.55 for offspring), highlighting the reduced reliability of the open-pollinated testing for either backwards or forwards selection. Furthermore, the combined HS+FS model allowed predicting the breeding value for the entire parental population (38 vs. 15) as it utilized all offspring information irrespective of parental gender (i.e., as pollen and/or seed donors) while the HS model was restricted only to the maternal population (i.e., seed donors).

The observed differences between the two incomplete pedigree models (HS and the combined HS+FS) support the beneficial role of including the pedigree reconstruction information even though it is only from a subset of the studied population. The inclusion of additional genetic information allowed the creation of linkages among the 15 half-sib families (known seed donors) with all parents participated in mating (pollen donors), thus increasing the sample size (i.e., higher genetic parameters' precision and breeding values' accuracy) and maximizing the BLUP-method utilization for breeding values prediction (see Ronningen and Van Vleck [24], for detailed explanation). The comparison between the combined HS+FS and full pedigree (FS) models is also needed to illustrate the advantages of partial pedigree inclusion. The full pedigree (FS) model is based on the assembled mating design from the pedigree reconstruction that is based on 1,419 offspring. Variance components and their precision and parental and offspring breeding values comparison between the two models produced similar estimates (Table 1; Figs. 4–5) and accuracies for parents (0.76 vs. 0.78) and offspring (0.64 vs. 0.69) were virtually identical. Heritability estimates are known to be population-specific [3]; however, the two models produced comparable 4-year height heritability estimates (HS+FS: 0.14±0.05; FS: 0.12±0.05), indicating similar magnitude/trajectory. This is not surprising since the two populations share 1,419 individuals in common and the combined HS+FS model included additional 4,258 individuals with known maternal parents. More importantly, the striking similarity between parental and offspring breeding values between the two models are indicative of similar ranking even though different number of individuals and genetic information were used.

The observed high correspondence between the suggested combined HS+FS and complete pedigree models highlights the superiority of the proposed BwB [17] indicating that a mixture of incomplete (half-sibs) and complete (full-sibs) pedigree is an efficient approach for acquiring reliable quantitative genetic parameters. The fingerprinting of a subset of the testing population is expected to substantially reduce the cost associated with pedigree reconstruction without any parameters' precision penalties.

The advantage of the combined HS+FS model over the HS and/or FS models is clearly demonstrated at the selection stage (Figure 6). Notwithstanding the overestimation of the additive genetic variance, the HS model is restricted to backward selection from the studied female parents as no BLUP values are generated for their male counterparts (i.e., 15 out of 38). The FS model is better than the HS as it allows the generation of accurate BLUP values for the 38 parents participated in mating as well as their offspring (N = 1,419) which is a subset of the tested population (N = 5,796), thus limiting forward selection to the fingerprinted offspring and thus does not consider any of the non-fingerprinted offspring which represent a substantial part of the tested population (57%). The combined HS+FS model, on the other hand, provides BLUP values for the parents and their offspring, irrespective of their family status, thus increasing the efficiency of forwards selection and improving the precision of backwards selection as well as combined selection. Additionally, the establishment of open-pollinated vs. those based of full pedigree field tests is more simplistic and can be effectively done with reduced efforts and cost.

The large number of parents commonly tested in traditional tree improvement programs requires the use of "efficient" mating designs so manageable number of crosses are made (e.g., disconnected partial diallel [4], [23]). In these mating designs, the parental population is divided to multiple subsets of parents with crosses are often restricted to within parental subsets with minimal or no matings among members of the different subset, thus creating opportunities for genetic sampling (i.e., no opportunity for cross referencing across set). The present study has demonstrated that paternity assignment of wind-pollinated half-sib families from know seed-donors provided adequate linkage across parents, hence we propose the implementation of similar approach concurrently with the selected traditional mating schemes to provide means for cross referencing and the avoidance of genetic sampling.

If BwB is to be considered as a viable option for tree breeding, then several additional questions must to be answered, among them: 1) what is the proportion of the population needed for pedigree reconstruction? 2) What is the minimal HS and/or FS family size required for proper BLUP analysis? 3) What is the role of elevated gene flow or selfing in the breeding population? 4) How many parents can be realistically tested? 5) How are we to expand testing beyond those parents present in the breeding population? The observed changes in the additive genetic variance estimates and their associated precision that accompany changes in the number of genotyped individual with known male parents (i.e., those resulting from the pedigree reconstruction) suggest that the inclusion of approximately 10% of the tested population is adequate to reach stable parameter estimates (Figure 7). The main function of these individuals is to create enough connections between parents, thus permitting direct and/or indirect comparison among the parental population members, a fundamental prerequisite for the BLUP analysis [24]. Increasing the number of offspring with known fathers to those already with known mothers increased the direct and/or indirect connectedness among parents and thus permitted their unbiased comparison as well as the estimated genetic parameters. Rönningen and Van Vleck [24] explicitly stated that a minimum of two offspring between any two males is needed for proper parameters estimation. In the present analysis, we imposed a minimum half-sib family size of six for any parent to be included and the observed correspondence between parents and offspring breeding values between the combined HS+FS and FS models is a reflection of this practice. The number of offspring designated for fingerprinting will also be affected by the degree of gene flow. As gene flow increases, more genotyped individuals will not provide any paternal information for connecting the different parents, but those individuals will remain in the analysis if they are among the maternal parents evaluated. Additionally, as the selection differential between the gene flow's source and the parental breeding population increases, the greater the difference in their offspring performance. A simple offspring phenotypic ranking followed by truncation selection theoretically could eliminate a substantial amount of the inferior offspring [17]. Offspring produced through selfing, while limited, remained in the data analysis through the inclusion of the pedigree information, and thus the estimated genetic parameters should be minimally affected. The rate of selfing among the tested parents is expected to provide an idea of the selfing propensity variation, which may shed some light on the relationship between selfing rate and general combining ability. As the number of parents' increases, the number of informative genetic markers must increase to allow for the exclusion power needed for pedigree reconstruction. The use of paternally inherited markers such as cpDNA could aid in differentiating among males with similar autosomal multilocus genotypes. Increasing the number of marker loci and including paternally inherited markers is expected to increase the experimental efforts; however, the increased efforts should be evaluated in light of the number of parents tested. Finally, increasing the number of tested parents beyond what is present in the breeding population could be accomplished through the use of supplemental-mass-pollination, a technique known to successfully incorporate pollen from specific parents in natural wind pollination of unprotected receptive females [26].

Materials and Methods

Plant material

In 2005, wind-pollinated seed samples from 15 unrelated parents were collected from a 41-parent western larch (*Larix occidentalis* Nutt.) seed orchard. The sampled orchard is one of two genetically distinct (41 and 62 parents) orchards established by British Columbia Ministry of Forests, Lands and Natural Resource Operations to provide genetically improved seed to the Nelson (<1,300 m) and East Kootenay (800–1,500 m) seed production units. These orchards are located near Vernon, B.C., Canada (altitude 480 m, latitude 50°14′N, longitude 119°16′E), in an area devoid of western larch background pollen. The orchards are separated by an 8 m wide road and a row of black cottonwood (*Populus* trichocarpa Torr. & Gray) trees, acting as a partial pollen barrier. Seed samples and orchard's reproductive survey data were provided by British Columbia Ministry of Forests, Lands and Natural Resource Operations as the orchard is part of a co-operative arrangements among government-private industry-academia. Seed were sown (February, 2006) by individual maternal family in a commercial nursery in growing blocks (80 cavities/block), soil mixes, irrigation, heating, and fertilization regimes similar to those operationally applied for reforestation seedling production. Seed pre-treatment (i.e., pre-chilling to break dormancy) prior to sowing followed International Seed testing

Association procedures [27]. At the end of the growing season (September, 2006), seedlings were extracted, by family, and used to establish a common garden trial.

Common garden trial

The trial was established at the University of British Columbia's Research Facility (latitude 49° 15′N, longitude 123° 15′W, elevation 79 m), laid out as a randomized complete design with four replications. Each replication consisted of 10×10 square plots at a spacing of 0.3×0.3 m (100 seedlings/family). At the end of the third field growing season (fall of 2009, 4 years from germination), total seedling heights (HT in cm) were measured on all surviving trees (5,306). The trial was watered and weeded when needed, and survival was 88% at the time of height measurement.

DNA fingerprinting and paternity assignment

The two orchards (studied and neighbouring, with their 41 and 62 parents, respectively) represent the possible paternal parents for a randomly selected 1,538 offspring that were genotyped with 16 microsatellite (SSR) markers. The SSR markers used were: 1) seven developed for *Larix occidentalis* [28], 2) two developed for *L. lyalli* [29], with one primer (UAKLly13) amplifying two loci (UAKLly13-1 and UAKLly13-2) in *L. occidentalis*, and [3] seven developed for *L. kaempferi* [30] (Table S1). Touchdown PCR was performed according to the protocol used by Isoda and Watanabe [30]: 94°C for 1 min followed by 10 cycles/30 s at 94°C, 30 s/ 63°−53°C (−1°C at each cycle) for 1 min, followed by 25 cycles of 30 s/94°C, 30 s/53°C and 1 min/72°C followed by 10 min/ 72°C. The CERVUS program ver. 3.0.3 was used to estimate null allele frequencies in the studied orchard's parental population [31–32], as null alleles introduce errors in parentage analysis by leading to high frequencies of false parentage exclusions [33]. PCR multiplexing was developed for four sets of loci sharing the same annealing temperature: 1) UBCLXdi-16, UBCLX1-10, and UBCLXtet-21, 2) UBCLXtet_2-12, UAKLly10, and UAKLly13, 3) bcLK33, bcLK66, bcLK211, and bcLK258, and 4) bcLK232 and bcLK263 (Table S1). Our preliminary paternity analysis, showed a 10% increase of paternity assignment after removing SSR loci with high null allele frequencies, but we included UBCLXtet-21 in spite of its high null allele frequency, because it was easy to multiplex and score as tetra-nucleotide SSR. Additionally, our results showed that the inclusion of this locus did not introduce serious parentage assignment bias. In total, 10 SSR loci were used for parentage assignment (Table S1). After paternity assignment (below), 98% of genotyped offspring were sired by members of the two orchards' panel of fathers. The CERVUS program [31], [32] provides likelihood based paternity inference with a known level of statistical confidence that accounts for genotyping error; we used it to to assign the pollen donor for 1,538 offspring. A genotyping error rate of 0.03 across the 10 loci was estimated from the known mother-offspring genotypes (Table S1). The paternity assignment was based on 10,000 simulations, with the 41-parents as candidate fathers. The log-likelihood (LOD) score, the likelihood that the candidate parent is the true parent divided by the likelihood that the candidate parent is not the true parent, was calculated for each putative parent. The delta score, the difference in LOD scores of the two most likely candidate parents, was used as a criterion for assignment of parentage at the 95% level of confidence in our analysis.

Quantitative genetics analyses

A classical individual-tree additive model, assuming no dominance and epistatic effects, was used. The model included a fixed effect of overall mean (β), a normally distributed random additive genetic effect (a, breeding values), with covariance matrix $A = \{\sigma_a^2\}$ where A is the additive relationship matrix (see below [34]) among all trees: parents without records, plus offspring with data, and σ_a^2 the additive genetic variance. The model also included a normally distributed random plot effect term (p) with mean zero and variance σ_p^2. Finally, a normally distributed random error (e) with mean zero and variance σ_e^2 were included. Let y be a vector containing the tree individual observations for height. Then, in matrix notation, the classical individual-tree additive model can be described as:

$$y = X\beta + Z_p p + Z_a a + e \qquad (1)$$

Let A be the additive relationship matrix based on pedigree. The A matrix has diagonal elements equal to $1 + F_i$, where F_i is the inbreeding coefficient for the ith individual and off diagonals equal to the additive relationships A_{ij} between tree i and j. Three individual-tree additive mixed models (model 91)) were evaluated using different pedigree files. Assuming that parent trees were unrelated, the first model, half-sib (HS model), was used with the known female parent of each individual, where all individuals are assumed not inbred (i.e., $F_i = 0$), and the additive genetic relationship are 0.25 or 0.0 for both trees with different fathers (with unrelated pollen), thus being maternal half-sibs and unrelated trees, respectively. This model is commonly used by forest geneticists and is called the open-pollinated test, where individuals within an open-pollinated family are assumed to be half-sibs [8]. The pedigree reconstruction created two more scenarios, one includes known female parents for all individuals in the common garden (the sampled 15 seed donors) and the male parents (any one of the orchard's 41 parents) for those used in the pedigree reconstruction (1,419 seedlings) (combined HS+FS model). The second includes only the 1,419 seedlings with their known maternal and paternal parents (known as the FS model). When male parents are known, correct inbreeding coefficient (i.e., $F_i = 0.5$) and additive relationship between trees ranging from selfs to half-sibs (e.g., $A_{ij} = 1$ if two individuals are generated by self-pollination or $A_{ij} = 0.5$ if two individuals are full-sibs through a common father) were considered in the A matrix.

Variance components

Restricted Maximum Likelihood (REML [34]) was used to estimate variances for the random effects of the classical individual-tree additive model (model (1)) and was obtained with the ASReml program [35], which uses the average information algorithm described by Gilmour et al. [36]. The narrow-sense individual heritability (h^2) was calculated as $h^2 = \sigma_\alpha^2 / \left(\sigma_\alpha^2 + \sigma_p^2 + \sigma_e^2\right)$, where σ_i^2 with $i = a$, p, and e are the values of the additive, plot, and error variance of the individual-tree model (1). Additionally, the inclusion of male information in the pedigree matrix allowed expanding model (1) to estimate the additive genetic variance after considering the additional genetic relationships generated by pedigree reconstruction. This was done to allow comparing the classical individual-tree additive models used. An important limitation of the REML (co)variance estimates is that their distribution is unknown. Only an approximate measure of precision of the estimates based on asymptotic or large sample theory can be calculated. Approximate standard errors (s.e.) of the σ_i^2 and h^2 were computed with the "delta method" based on the Taylor expansion [18] using ASREML [35].

Prediction of the breeding values and response to selection

The analysis of a progeny test normally involves two steps: first the estimation of variance components and second the prediction of breeding values for the individuals, using the variance components estimated in the first step. In the three models, the "Best Linear Unbiased Predictors" (BLUPs) of parent and offspring breeding values were computed with ASReml from the estimated variance components. The accuracy of the predicted breeding values was calculated using the following expression: $r = \sqrt{1 - \dfrac{PEV}{(1 + F_i)\sigma_\alpha^2}}$. The acronym PEV stands for 'prediction error variance' [36] of predicted breeding values, using the BLUPs of parent and offspring and F_i is the inbreeding coefficient for the ith individual. The PEV is calculated as the diagonal elements of the inverse of the coefficient matrix from the mixed model equations [36]. To make the accuracies comparable across models (i.e., HS, combined HS+FS and FS), the variance components required to set up the mixed model equations were those estimated from the combined HS+FS. Pearson product-moment correlation and Spearman rank-order correlation were also calculated to compare whether the strength of linear dependence and the ranking of predicated breeding values differed among models. Additionally we have included confidence intervals of all correlation estimates to evaluate jointly the variance and sample size under the alpha value of 0.05. Individual tree BLUP values were used to compare the response to selection under the HS (forward and combined) and combined HS+FS (backward, forward and combined) models, as affected by effective population size, using the optimization protocol outlined in El-Kassaby and Lstibůrek [17].

Estimating offspring optimum sample size

To determinate the optimum number of individuals with known fathers needed for obtaining reliable genetic parameters and thus reducing the DNA fingerprinting efforts, the classical individual-tree additive model (1) was fitted with several pedigree files, where the male information was randomly and progressively deleted, thus increasing percentage of omitted male data from 7 to 92% (i.e., reducing the number of individuals with known male parents). These pedigrees with randomly deleted males provided us with a range of values and standard errors associated with them that the different parameters may take and permitted us to investigate the robustness of results under reduced fingerprinting efforts (i.e., reduce the number of offspring with known paternal parents). For this data set, we set the minimum paternal HS family to n = 6 for inclusion in the analyses and hence the generation of precise genetic parameters and their respective predicted breeding values.

Supporting Information

Table S1 Annealing temperature in °C, number of alleles, observed (H_o) and expected (H_e) heterozygosities, and estimated frequencies of null alleles and genotyping error of the seed orchard population used in the present study (41-Parents).

Acknowledgments

We are most grateful to D. Reid and C. Walsh for providing seed and reproductive survey data; J. Halusiak for seedling production; UBC graduate students N. Massah, B. Lai, M. Ismail and R. Soolanayakanahally for assistance with the common garden establishment; I. Fundova, T. Funda and B. Lai for trial maintenance and measurements; I. Fundova and C.N. Takuathung for DNA extraction, L. Bouffier, R.J. Peti, M. Stoehr and P.E. Smouse for critical and constructive review of earlier draft.

Author Contributions

Conceived and designed the experiments: YAE. Performed the experiments: YAE CL. Analyzed the data: YAE EPC CL JK ML. Wrote the paper: YAE.

References

1. Allard RW Principles of plant breeding (John Wiley and Sons, YN).
2. Namkoong G, Kang HC, Brouard JS Tree breeding: principles and strategies (Springer-Verlag, NY, Monograph, Theor Appl Genet 11).
3. Falconer DS, Mackay TFC Introduction to quantitative genetics (Longman, NY).
4. White TL, Adams WT, Neale DB Forest genetics (CABI Publishing, Cambridge, MA).
5. Stern K, Roche L Genetics of forest ecosystems (Chapman and Hall, London).
6. Cotterill PP (1986) Genetic gains expected from alternative breeding strategies including simple low cost options. Silvae Genet 35: 212–223.
7. Burdon RD, Shelbourne CJA (1971) Breeding populations for recurrent selection: conflicts and possible solutions. NZ J For Sci 1: 174–193.
8. Namkoong G (1966) Inbreeding effects on estimation of genetic additive variance. For Sci 12: 8–13.
9. Squillace AE (1974) Average genetic correlations among offspring from open-pollinated forest trees. Silvae Genet 23: 149–156.
10. Askew GR, El-Kassaby YA (1994) Estimation of relationship coefficients among progeny derived from wind-pollinated orchard seeds. Theor Appl Genet 88: 267–272.
11. Lambeth C, Lee BC, O'Malley D, Wheeler N (2001) Polymix breeding with parental analysis of progeny: an alternative to full-sib breeding and testing. Theor Appl Genet 103: 930–943.
12. Grattapaglia D, Ribeiro VJ, Rezende GDSP (2004) Retrospective selection of elite parent trees using paternity testing with microsatellite markers: an alternative short term breeding tactic for Eucalyptus. Theor Appl Genet 109: 192–199.
13. El-Kassaby YA, Lstibůrek M, Liewlaksaneeyanawin C, Slavov GT, Howe GT (2006) Breeding without breeding: approach, example, and proof of concept. In: Low input breeding and genetic conservation of forest tree species (IUFRO, Antalya, Turkey). pp 43–54.
14. Gaspar MJ, de-Lucas AI, Alia R, Paiva JAP, Hidalgo E, et al. (2009) Use of molecular markers for estimating breeding parameters: a case study in a *Pinus pinster* Ait. progeny trial. Tree Genet Genomes 5: 609–616.
15. Hansen OK, McKinney LV (2010) Establishment of a quasi-field trial in *Abies nordmanniana* – test of a new approach to forest tree breeding. Tree Genet Genomes 6: 345–355.
16. Doreksen TK, Herbinger CM (2010) Impact of reconstructed pedigrees on progeny-test breeding values in red spruce. Tree Genet Genomes DOI 10.1007/s11295-010-0274-1.
17. El-Kassaby YA, Lstibůrek M (2009) Breeding without breeding. Genet Res 91: 111–120.
18. Lynch M, Walsh B Genetics and analysis of quantitative traits (Sinauer Associates, Sunderland, MA).
19. Funda T, Liewlaksaneeyanawin C, Fundova I, Lai BSK, Walsh C, et al. (2011) Congruence between clonal reproductive investment and success as revealed by DNA-based pedigree reconstruction in seed orchards of lodgepole pine, Douglas-fir, and western larch. Can J For Res 41: 380–389.
20. Ritland K (1996) A marker-based method for inferences about quantitative inheritance in natural populations. Evolution 50: 1062–1073.
21. Ritland K, Ritland C (1996) Inferences about quantitative inheritance based upon natural population structure in the common monkeyflower, *Mimulus guttatus*. Evolution 50: 1074–1082.
22. Ritland K, Travis S (2004) Inferences involving individual coefficients of relatedness and inbreeding in natural populations of Abies. Fort Ecol Manage 197: 171–180.
23. Namkoong G Introduction to quantitative genetics in forestry (US Depart Agriculture, Forest Service, Washington, DC, Tech Bulletin No 1588).
24. Rönningen K, Van Vleck LD General and quantitative genetics (World Animal Science, Elsevier, NY). pp 187–225.
25. Cappa EP, Lstiburek M, Yanchuk AD, El-Kassaby YA (2011) Two-dimensional penalized splines via Gibbs sampling to account for spatial variability in forest genetic trials with small amount of information available. Silvae Genet 60: 25–35.
26. El-Kassaby YA, Barnes S, Cook C, MacLeod DA (1993) Supplemental-mass-pollination success rate in a mature Douglas-fir seed orchard. Can J For Res 23: 1096–1099.
27. International Seed Testing Association (1993) International rules for seed testing. Seed Sci Technol 21S. 284 p.
28. Chen CC, Liewlaksaneeyanawin C, Funda T, Kenawy AMA, Newton CH, et al. (2008) Development and characterization of microsatellite loci in western larch (*Larix occidentalis* Nutt.). Mol Ecol Res 9: 843–845.
29. Khasa PD, Newton C, Rahman M, Jaquish B, Dancik BO (2000) Isolation, characterization and inheritance of microsatellite loci in alpine larch and western larch. Genome 43: 439–448.
30. Isoda K, Watanabe A (2006) Isolation and characterization of microsatellite loci from *Larix kaempferi*. Mol Ecol Notes 6: 664–666.
31. Marshall TC, Slate J, Kruuk LEB, Pemberton JM (1998) Statistical confidence for likelihood-based paternity inference in natural populations. Mol Ecol 7: 639–655.
32. Kalinowski ST, Taper ML, Marshall TC (2007) Revising how the computer program CERVUS accommodates genotyping error increases success in paternity assignment. Mol Ecol 16: 1099–1106.
33. Dakin EE, Avise JC (2004) Microsatellite null alleles in parentage analysis. Heredity 93: 504–509.
34. Henderson CR Applications of linear models in animal breeding (University of Guelph, Guelph, ON, Canada).
35. Gilmour AR, Gogel BJ, Cullis BR, Thompson R ASReml user guide (Release 2.0 VSN International, Hemel Hempstead, UK).
36. Gilmour AR, Thompson R, Cullis BR (1995) Average information REML, an efficient algorithm for variance parameter estimation in linear mixed models. Biometrics 51: 1440–1450.

2

Multiple Origins of the Sodium Channel *kdr* Mutations in Codling Moth Populations

Pierre Franck*, Myriam Siegwart, Jerome Olivares, Jean-François Toubon, Claire Lavigne

INRA, UR1115 Plantes et Systèmes de culture Horticoles, Avignon, France

Abstract

Resistance to insecticides is one interesting example of a rapid current evolutionary change. DNA variability in the voltage-gated sodium channel gene (trans-membrane segments 5 and 6 in domain II) was investigated in order to estimate resistance evolution to pyrethroid in codling moth populations at the World level. DNA variation among 38 sequences revealed a unique *kdr* mutation (L1014F) involved in pyrethroid resistance in this gene region, which likely resulted from several convergent substitutions. The analysis of codling moth samples from 52 apple orchards in 19 countries using a simple PCR-RFLP confirmed that this *kdr* mutation is almost worldwide distributed. The proportions of *kdr* mutation were negatively correlated with the annual temperatures in the sampled regions. Homozygous *kdr* genotypes in the French apple orchards showed lower P450 cytochrome oxidase activities than other genotypes. The most plausible interpretation of the geographic distribution of *kdr* in codling moth populations is that it has both multiple independent origins and a spreading limited by low temperature and negative interaction with the presence of alternative resistance mechanisms to pyrethroid in the populations.

Editor: João Pinto, Instituto de Higiene e Medicina Tropical, Portugal

Funding: This study was partially funded by the French program ECOGER "Ecco des vergers". The funders had no role in study design, data collection and analysis, decision to publish, or preparation of the manuscript. No additional external funding was received for this study.

Competing Interests: The authors have declared that no competing interests exist.

* E-mail: Pierre.Franck@avignon.inra.fr

Introduction

The codling moth, *Cydia pomonella* (L.) (Lepidoptera: Tortricidae), is one of the major insect pests in the orchards (mainly apple, pear and walnut orchards), worldwide distributed in the temperate regions [1]. Chemical insecticides remain the major means used to maintain populations of this pest at a low level. As a consequence of these treatments, *C. pomonella* developed resistance to numerous insecticides in Australia [2], Americas [3,4,5], and Eurasia [6,7] including resistance to synthetic pyrethroids [8,9].

Resistance to pyrethroids is mainly conferred by modification of their primary target site: the voltage-gated sodium channel [10]. Computer-generated 3D models characterized a small number of mutations linked to insecticide-binding sites in the voltage-gated sodium channel, most of them being in the trans-membrane segments 4 to 6 of the domain II region of the protein [11]. L1014F and M918T originally found in the housefly and respectively referred to as *kdr* and *super-kdr* mutations [12] are two of the most common of these mutations in insects [13]. The *kdr* mutation is associated to moderate resistance to DTT and pyrethroids. The *super-kdr* mutation is usually linked to the *kdr* mutation and increases by tenfold primary pyrethroid resistance due to *kdr* [13,14]. The L1014F mutation is the only voltage-gated sodium channel mutation reported so far in the codling moth [15]. It was detected in few populations over the World [8,16,17]. A resistance ratio of about 80-fold to the pyrethroid insecticide deltamethrin is conferred by this recessive mutation in first-instar codling moth larvae [18,19]. A low level of pyrethroid resistance in the codling moth is also attributed to enhanced detoxification activity notably due to the P450 cytochrome oxidases [18,20].

The evolution of resistance in insect pest populations depends on both historic and current selective processes that should be understood to manage resistance [21]. To shed light on the evolutionary processes linked to the evolution of pyrethroid resistance in *C. pomonella* populations, we analysed genetic variations at the *para* sodium channel gene. We first report countries over the World where *kdr* resistance has been observed to establish origins of resistance alleles and identify factors that may affect their global spreading. Secondly we present a more detailed analysis on populations from South-eastern France to document the impact of local pyrethroid treatments on resistance evolution at the population level.

Materials and Methods

Sampling

The evolution of pyrethroid resistance conferred by the sodium channel gene was investigated on codling moth samples from 52 different apple commercial orchards in 19 countries (Table 1). Codling moths were collected as diapausing larvae using corrugated cardboard traps wrapped around the trunks of apple tree. Among these 52 codling moth population samples, seven populations were previously studied [16,17]. Codling moth populations from 21 orchards showing high larva density in South-eastern France [22] were further analysed to determine the impact of current pyrethroid treatments and interaction between resistance mechanisms on genetic variation in the *para* sodium channel gene (Table 2). Pyrethroid treatments in these French

Table 1. Origins of the codling moth samples, meteorology characteristic (annual mean of the daily minimal temperature in celsius degree, annual number of freezing days) and proportions in each country of alleles (77 and 112) and homozygous genotypes (77/77) detected with the PCR-RFLP test [16].

Country	Year	Minimal Temperature	Freezing days	N	n	112	77	77/77
Armenia[a]	2002	n.a.	n.a.	30	1	0.00	1.00	1.00
Argentina[b]	2005	7.5	45	24	1	0.23	0.54	0.25
France	2006	9.1	56	771	21	0.40	0.24	0.08
New Zealand[b]	2005	8.7	5	18	1	0.61	0.17	0.00
Turkey	2010	11.8	0	42	1	0.56	0.14	0.05
USA	2008	1.3	177	217	5	0.54	0.09	0.01
Bulgaria	2007	5.3	100	60	3	0.54	0.06	0.02
Uruguay[b]	2005	11.8	0	17	1	0.32	0.06	0.00
Switzerland[a]	2003	7.2	51	35	1	0.67	0.01	0.00
Spain	2007	10.2	21	28	2	0.60	0.00	0.00
Italy	2007	7.9	11	329	5	0.36	0.00	0.00
Poland	2009	6.1	64	10	1	0.61	0.00	0.00
Czech Republic	2005	4.2	109	31	1	0.68	0.00	0.00
Greece	2006	8.9	41	20	1	0.55	0.00	0.00
Syria	2006	14.3	0	53	2	1.00	0.00	0.00
South-Africa[b]	2005	12.1	0	13	1	0.72	0.00	0.00
Morocco	2010	13.1	0	30	2	0.45	0.00	0.00
Algeria	2010	0.1	28	28	1	0.86	0.00	0.00
Chile[a]	2005	n.a.	n.a.	28	1	0.70	0.00	0.00

N and n respectively indicate the number of individuals and the number of orchards analysed per country. 77 and 77/77 respectively correspond to kdr allele [16] and to homozygous kdr genotype [18]. No meteorological data were available for the Armenian and Chilean locations (n.a.).
[a]These samples correspond to the population samples A, C and 11 that were analysed in [16].
[b]These samples correspond to the population samples Ar2, NZ1, Ur2, SA1 that were analysed in [17].

orchards encompassed mainly class II pyrethroids (esfenvalerate, fluvalinate, deltamethrine).

Detection of the Kdr Mutation

Total DNA was extracted from the head of each individual following Wash *et al.* [23] with 200 μl of 10% Chelex 100 (Biorad) solution and 6 μl (10 mg/ml) of proteinase K (Eurobio). Tissues were digested over night at 56°C. After boiling for 30 minutes, supernatant was used as DNA template for PCR reaction. A PCR-RFLP test slightly modified from Franck *et al.* [16] was used to detect the *kdr* mutation. It was developed based on sequence variations in the *para* sodium channel gene of susceptible and deltamethrin resistant strains [15]. PCR amplifications were carried out with a Mastercycler thermocycler (Eppendorf) in a 25 μl reaction volume containing 1X reaction buffer (10 mM Tris-HCl, pH = 9, 50 mM KCl, 1.5 mM MgCl$_2$, and 0.1 mg/ml Bovine Serum Albumin), 200 μM of each dNTPs, 0.4 μM of each *CpNa-F* and *CpNa-R* primers (Table 3), 1 unit of Go Taq DNA Polymerase (Promega) and 2 μl of DNA template. The PCR conditions were: 3 minutes at 94°C followed by 35 cycles at 94°C for 30s, 55°C for 1 min, and 72°C of elongation for 1 min with a final extension step at 72°C for 2 min. PCR products were digested at 65°C for 16 hours with 2 unit of Tsp509I endonuclease and 1X of NEB1 Buffer (New England Biolabs) in 30 μl of reaction volume. Digested products were separated by electrophoresis on 6.5% polyacrylamide denaturing gel in a Li-Cor 4200 automatic DNA sequencer. The longest digested DNA fragments in acrylamide gels were visualised using 700 IRDye labelled *CpNa-F* primer with the SAGA software (Li-Cor Biosciences). Each

revealed fragment length was defined as an allele. The test was performed on 1784 individuals: 165 individuals were re-analysed according to this modified protocol [16,17] and 1619 individuals were newly investigated (Table 1).

Detoxification Activity by the P450 Cytochrome Oxidases

Enhanced activity of the P450 cytochrome oxidases confers heritable metabolic resistance to pyrethroid insecticides [18]. We assessed the activity of the P450 cytochrome oxidases measuring 7-ethoxycoumarin-O-deethylation (ECOD) activity on 557 moths (out of 771) collected in 19 French orchards (out of 21) to shed light on putative interaction between this resistance mechanism and pyrethroid target mutations in the sodium channel gene. ECOD activity was individually measured on abdomen samples using 0.4 mM ethoxycoumarin in 100 μL Hepes buffer [16,24]. After four hours of incubation at 30°C, the enzymatic reaction was stopped with 100 μL of glycine buffer (10^{-4} M), pH 10.4/ethanol (v/v) and fluorescence was measured with 380 nm excitation and 465 nm emission filters on a microplate reader (HTS 7000, Perkin Elmer). ECOD activity was estimated for each moth based on the amount of 7-hydroxycoumarine formed (pg/min).

DNA Sequencing in the Para Gene

DNA sequencing in the *para* gene (corresponding to trans-membrane segments 4 to 6 of the domain II region of the canal sodium protein, Figure 1) was performed on 50 codling moth individuals from various geographic origins and two *Grapholita*

Table 2. Proportions of alleles (*112* and *77*) and of homozygous genotypes (*77/77*), and expected and observed heterozygosities (H_E/H_O) in codling moth samples from 21 commercial apple orchards in South-eastern France.

Orchard	N	Protection	Pyrethroid treatment	Cytochrom P450 activity	*112*	*77*	*77/77*	H_E/H_O
154	36	Conventional	5	77	0.26	0.64	0.42	0.52/0.53
149	20	Conventional	4	687	0.43	0.38	0.10	0.66/0.60
122	37	Conventional	4	513	0.18	0.23	0.05	0.57/0.57
75	43	Conventional	4	328	0.48	0.17	0.02	0.63/0.67
65	40	Conventional	4	652	0.48	0.14	0.02	0.61/0.58
55	39	Conventional	3	635	0.45	0.24	0.10	0.65/0.56
68	21	Conventional	3	507	0.60	0.10	0.00	0.55/0.57
132	49	Conventional	3	862	0.46	0.06	0.00	0.56/0.67
35	33	Conventional	2	n.a.	0.41	0.38	0.15	0.65/0.70
84	23	Conventional	2	141	0.30	0.37	0.04	0.68/0.87
17	27	Conventional	2	542	0.30	0.35	0.11	0.56/068
140	36	Conventional	2	208	0.32	0.31	0.11	0.67/0.61
134	16	Conventional	2	340	0.31	0.16	0.00	0.62/0.81
10	29	Conventional	1	406	0.57	0.17	0.03	0.59/0.66
42	43	Conventional	1	n.a	0.52	0.14	0.02	0.60/0.56
145	41	Organic	0	188	0.28	0.45	0.22	0.65/0.63
51	59	Organic	0	32	0.38	0.25	0.05	0.66/0.66
125	52	Organic	0	70	0.41	0.23	0.04	0.66/0.71
124	45	Organic	0	90	0.39	0.17	0.02	0.63/0.76
119	58	Organic	0	34	0.45	0.16	0.03	0.62/0.57
126	24	Organic	0	265	0.40	0.13	0.04	0.61/0.54

References to crop protection practices, numbers of annual pyrethroid treatments, and cytochrom P450 oxidase activity were reported for each orchard and linked population sample. N was the number of individuals analysed per orchard. H_E and H_O were calculated with all the three alleles (*77*, *101* and *112*) detected with the PCR-RFLP. Values reported for cytochrom P450 oxidase activities were estimated as the average ECOD activity (pmol/min/individual) among the individuals collected at each orchard location. No ECOD measure was done on the individuals collected in orchards 35 and 134 (n.a.).

molesta (Lepidoptera: Tortricidae), collected in France and Brazil, to be used as an outgroup.

The whole region was sequenced for 38 codling moths displaying different homozygous genotypes according to the PCR-RFLP test (Table 4). The whole region was amplified in two independents PCRs with respectively the *SKdr-F/SKdr-R3* and the *Kdr-F/Kdr-R* primer pairs (Table 3). These PCR were

Table 3. Primers used to amplify and to sequence domain II S4-S6 region of the codling moth *para* sodium channel gene.

Primer	Sequence (5′-3′)
SKdr-F	GGCCGACACTTAATTTACTCATC
SKdr-R1	TTCCCGAAAAGTTGCATACC
SKdr-R2	GGGTTAACGAGCTAAACGTCCAA
SKdr-R3	GCAATCCCACATGCTCTCTA
CpNa-F	TAGAGAGCATGTGGGATTGC
CpNa-R	AATTTCGTAGCCCTTGATCG
Kdr-F	GGTGGAACTTCACCGACTTC
Kdr-R	GCAAGGCTAAGAAAAGGTTAAG

Primer positions are indicated in Figure1. *Kdr-F* and *Kdr-R* are slightly modified from *CgD1* and *CgD2*, respectively [15]. *SKdr-R3* is the reverse of *CpNa-F*.

performed in the same conditions as above but at annealing temperatures of 56°C and 54°C respectively. The PCR products were purified from agarose gel [25] then sequenced (GATC Biotech) using, as sequencing primers, *SKdr-R1*, *SKdr-R2* and the four PCR primers described above (Table 3). For twelve additional codling moths collected in orchard treated with pyrethroid (Table 2), the first exon coding for the transmenbrane segments 4 and 5 was partially sequenced using the *SKdrF* and *SKdr-R1* primers in order to check for the presence of the *super-kdr* mutation.

Data Analysis

The genetic variation at the sodium channel detected with the PCR-RFLP test was first analysed. In each orchard population sample, observed (H_O) and expected (H_E) heterozygosities were calculated and departure of genotype frequencies from Hardy-Weinberg proportions tested using the Genepop software [26] considering either all the detected alleles or only the *kdr* and the susceptible allele groups. Generalised linear models were used (*genmod* procedure, SAS version 9.1) to explain the proportions of *kdr* allele, and of homozygous *kdr* genotypes in the orchard population samples. These proportions were modelled as binomial variables with a logit link function considering the orchard as a random variable. First, the proportions of *kdr* allele or of homozygous *kdr* genotypes were modelled for the French population samples as functions of two factors (ECOD activity

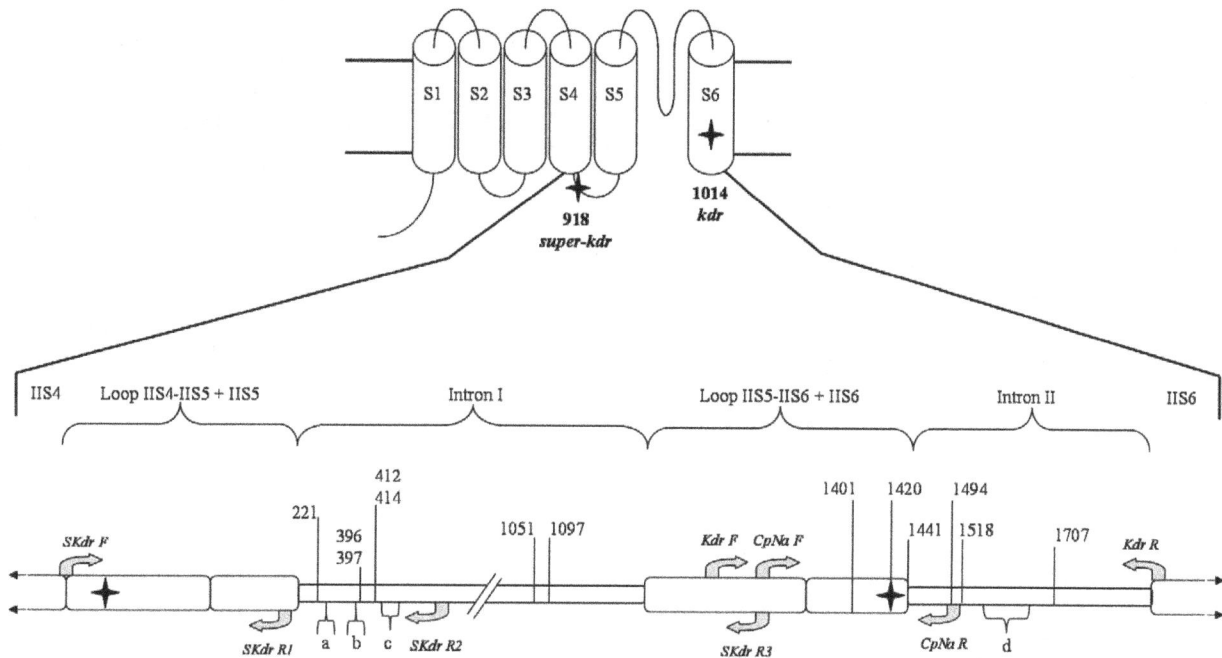

Figure 1. Trans-membrane segments in the domain II voltage-gated sodium channel with reference to codon position of the most frequent non-synonymous mutations and nucleotide variation in the codling moth sequences. Polymorphic sites in the nucleotide sequences are indicated as numbers (13 substitutions) or letter (4 indels) using *Syria1* sequence (1859 bp) as reference (Genbank accession number GU082334, Table 5). Arrows indicate the positions of the primers used for PCR-RFLP and sequencing analyses (Table 3).

in moths and number of pyrethroid treatments in orchard) and their interactions (Table 2). Second, the proportions of *kdr* allele or homozygous *kdr* genotypes were modelled considering all the sampled populations that displayed *kdr* polymorphism as functions of either the annual mean of the daily minimal temperatures or the annual number of freezing days at each orchard. Meteorological data were obtained from the National Climatic Data Center website (http://www.ncdc.noaa.gov/oa/ncdc.html) for each orchard location and sampling date.

Furthermore, sequence variation at the sodium channel gene was investigated. DNA sequences were manually aligned with the Bioedit software [27]. Recombination between DNA sequences was tested using six different methods implemented in the RDP3 software [28]: RDP [29], GENECONV [30], Chimaera [31], MaxChi [31,32], BootScan [33], and SiScan [34]. Minimum evolution trees were computed with the MEGA software [35] using a p-distance between the sequences that takes into account both substitution and indel polymorphisms. To assess the reliability of the tree, standard error tests were performed for every interior branch by resampling variable sites (1,000 bootstraps).

Table 4. Statistical results about six generalized linear models of the proportions of *kdr* allele and of *kdr* homozygote genotypes in the orchard population samples.

Statistical models		Observed data			Statistical test		
Dependent variables	Covariables	Orchard origins	N	n	chi2	(df)	P-value
Proportion of *kdr* allele	Pyrethroid treatments	France	1114	19	0.16	(1)	0.687
	ECOD activity				1.00	(1)	0.317
	Pyrethroid × ECOD				0.92	(1)	0.338
Proportion of *kdr* allele	Minimal temperature	World	1830	30	5.05	(1)	0.025
Proportion of *kdr* allele	Freezing days	World	1830	30	4.57	(1)	0.033
Proportion of *kdr* homozygote	Pyrethroid treatments	France	557	19	0.09	(1)	0.768
	ECOD activity				9.16	(1)	0.003
	Pyrethroid × ECOD				2.63	(1)	0.105
Proportion of *kdr* homozygote	Minimal temperature	World	915	30	3.13	(1)	0.077
Proportion of *kdr* homozygote	Freezing days	World	915	30	3.54	(1)	0.060

N and n respectively indicate the number of individuals and the number of orchards observed for each statistical model.

Results

Detection and Distribution of the *Kdr* Mutation in Populations

To detect the *kdr* mutation (L1014F) in codling moth populations we amplified a 170 bp region with the *CpNa-F* and *CpNa-R* primers (Figure 1 and Table 3), then digested it with the Tsp509I endonuclease that specifically cut ↓ AATT sites. Two out of five restriction sites in the 170 bp region were polymorphic (Figure 1, Table 4). This polymorphism was summed up by DNA fragments of three different detectable lengths (77, 101 and 112 bp), hereafter designed as three different alleles (Figure 2). Restriction at position 1417 generated the 77 bp fragment, which was interpreted as corresponding to the *kdr* allele [16]. Restrictions at positions 1441 and 1452 in intron II respectively generated 101 and 112 bp fragments, which were interpreted as corresponding to two different susceptible alleles.

A total of 1784 codling moths were genotyped using this PCR-RFLP test (Table 1). The *77* allele was observed in population samples from all the continents except Africa and from 9 out of the 19 countries analysed. It was observed in all the samples from South-eastern France in variable proportions (Table 2). It was the only observed allele in the Armenian sample. The *101* and *112* alleles were both observed at relatively high proportions in all the other samples, except the one from Syria that was monomorphic for the *112* allele (Tables 1 and 2). No departure from Hardy-Weinberg equilibrium was detected in any sample from the 52 orchards analysed when the *101* and *112* susceptible alleles were grouped (Fisher's exact test, $P > 0.28$). Significant heterozygote excesses were detected in two French samples (Fisher's exact test, $P = 0.03$ and $P = 0.04$ in the orchard sample 84 and 132, respectively) when all three alleles were considered (Table 2). The proportions of *77* allele and *77/77* genotype in the population samples from South-eastern France were slightly higher in orchards sprayed with pyrethroid insecticides (Tables 2 and 4), but these proportions did not depend on the number of pyrethroid treatments according to the generalized linear models (chi2 < 0.16, df = 1, $P > 0.68$). A significant effect of ECOD activities was detected on the proportion of the homozygous *77/77* genotype (chi2 = 9.16, df = 1, $P = 0.0025$), but not on the proportion of *77* allele (chi2 = 1.00, df = 1, $P = 0.317$). Consistently, ECOD activities were lower in homozygous *77/77* genotypes (284 pg/min in average) than in other genotypes (337 pg/min in average). The proportions of the *77* allele and the *77/77* genotype were modelled according to climatic covariables using the 30 population samples over the World that displayed *kdr* polymorphism. The annual mean of the daily minimal temperatures and the annual number of freezing days in the sampled regions were highly correlated (*number of freezing days* = −16.6 *minimal temperature* +184, $R^2 = 0.897$). Consequently, generalized linear models were performed independently with these two climatic variables (Table 4). The proportions of *77* allele (chi2 = 5.05, df = 1, $P = 0.025$) and, to a lesser extent, the proportions of *77/77* genotype (chi2 = 3.13, df = 1, $P = 0.077$) in the codling moth samples were positively correlated with temperature in the sampled regions. Consistently, very similar negative correlations were observed with the annual number of freezing days (Table 4).

DNA Sequencing in the Para Gene and Haplotype Divergences

Partial DNA sequences of the *para* gene (domain II, transmembrane segments 5 and 6, Figure 1) were obtained for two *G. molesta* (accession number GU082359 and GU082360) and 38 *C. pomonella* individuals (GU082334-GU082358 and JQ946336–

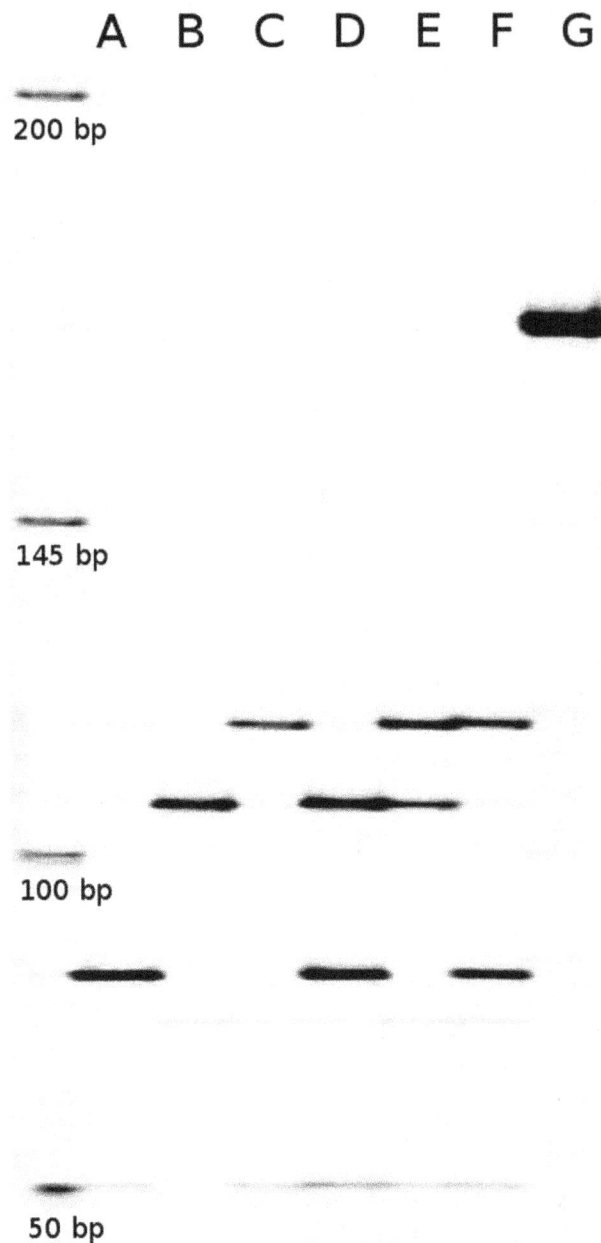

Figure 2. Detection of the *kdr* mutation by PCR-RFLP. PCR were conducted with the *CpNa-F* and *CpNa-R* primers (Table 3), and then digested with Tsp509I (700 IRDye labelled *CpNa-F* primer). Lengths of the restricted fragments were determined by electrophoresis in a Li-Cor 4200 automatic DNA sequencer (6.5% polyacrylamide denaturing gel) and visualised using the SAGA software (Li-Cor Biosciences). The 77 bp fragment corresponds to the *kdr* allele. The 101 and 112 bp fragments correspond to two different susceptible alleles. A–F corresponds to the different PCR-RFLP genotypes: *77/77, 101/101, 112/112, 77/101, 101/112* and *77/112*. G is the length of the non-restricted PCR product (170 bp). 50–350 pb sizing standard (Biosciences) is in the left well.

JQ946348): These 38 sequenced codling moths displayed homozygous genotypes according to the PCR-RFLP test; eleven were *77/77*, ten *101/101* and seventeen *112/112*. Introns I and II were both shorter in *G. molesta* than in *C. pomonella* (Figure 3) and we were not able to align and compare *G. molesta* and *C. pomonella* intron sequences. Twelve different haplotypes were identified among the 38 codling moth sequences. The sequences differed by their lengths (1649 to 1756 bp) because of the presence of three indels in intron I and one in intron II (Figure 3, Table 5). In addition, 13 substitutions were observed. Only one substitution at position 1420 was non-synonymous (L1014F). All the eleven *77/77* genotypes sequenced displayed at this *kdr* locus the phenylalanine amino acid. No variation was detected in the first exon that encompasses the *super-kdr* locus (50 moths sequenced including 23 displaying the *77/77* genotype). No evidence of recombination was detected among the 12 observed haplotypes with any of the six recombination tests performed with the RDP3 package. Consequently, we did not take into account recombination in the phylogenetic analyses. Thirteen out of the 17 polymorphic sites in *C. pomonella* were parsimonious informative sites. Minimum evolution trees were computed with the p-distance between the *C. pomonella* sequences using polymorphism at all the variable sites or all the variable sites except the *kdr* mutation (Figure 4). Both data sets highly supported the presence of two main clades, which differed at 11 sites in average (approximately 0.7% divergence in their nucleotide sequences). The first clade encompassed three different *112* sequences from the Bulgarian and Syrian samples. The second clade encompassed nine different *77, 101* and *112* sequences from samples distributed all over the World. Sequences within this second clade were lastly structured according to their PCR-RFLP profiles and their geographic origin. The four different *77* sequences differed from each other at three informative sites (Table 4, positions 380, 1051 and 1441) and were distributed in a least two non related subclades in addition with *101* and *112* sequences respectively (Figure 4).

Discussion

DNA sequences and SNPs analyses in a gene involved in insecticide resistance are complementary tools to shed light on recent evolutionary changes [13,36]. Proximal evolutionary processes can be assessed by analysing the DNA sequence that contains the mutations involved in resistance. SNPs analyses are convenient molecular tools that allow following the dynamic of insecticide resistance in natural populations and understanding

selective processes that may enhance or delay insecticide resistance evolution [21].

Although useful, SNP detection methods may fail to detect some genetic variations involved in insecticide resistance at a selected gene [37,38]. The L1014F replacement in the voltage-gated sodium channel was primarily observed in a deltamethrin resistant codling moth strain [15]. In the present study, a rapid PCR-RFLP test was developed to monitor this *kdr* mutation in codling moth populations. However, at least two additional mutations were reported in insect pest populations at this locus: L1014S in *Culex pipiens* [39] and *Anopheles gambiae* [40] and L1014H in *Heliothis virescens* [41] and *Musca domestica* [42]. As for the F1014 variant, the S1014 but not the H1014 variants would have produced a 77 bp restriction fragment in *C. pomonella* with the developed PCR-RFLP test. However, it is unlikely that these two additional mutations are present in the codling moth: L1014F was the only non-synonymous variation in the *para* gene observed along transmembrane segments 5 and 6 in domain II in individuals from various origins in the World. In absence of other proof, we assumed that L1014F was the only *kdr* mutation in this species.

A total of 1784 individuals collected all over the World were analysed using this PCR-RFLP test to estimate the distribution of *kdr* within and among populations. The *77* allele was observed almost worldwide and it is heterogeneously distributed among the codling moth populations. High proportions of *77/77* homozygous genotypes that are physiologically resistant to pyrethroid were only observed in Armenia, Argentina, Turkey and South-eastern France. These results confirm and extend previous observations [8,16,17,43].

At the orchard level, neither *kdr* selection by current pyrethroid treatments nor *kdr* counter-selection in absence of pyrethroid treatment was evident. The proportions of *kdr* allele were not significantly correlated with the number of pyrethroid treatments in the French apple orchards. Distributions of *kdr* genotypes did not significantly depart from Hardy-Weinberg proportions in any orchard population samples. The low impact of pyrethroid treatments observed on *kdr* proportions at the orchard level seems to be a general feature whatever the within-year generation of the codling moth [44]. These results contrast with observations in *Haematobia irritans* or in *Musca domestica* populations that clearly showed seasonal variations in the proportions of *kdr* allele as a function of pyrethroid treatments [45,46]. Three non exclusive hypotheses may explain such lack of structure of *kdr* in codling moth populations according to current insecticide applications. First, resistance management guidelines recommend alternation of pyrethroids with other insecticides among codling moth generations. Non-continuous use of pyrethroids largely limits the selection of sodium channel target mutations. This could explain why the *super-kdr* mutation is apparently absent in codling moth populations, a result also found in wild populations of horn flies [47]. Second, the usage of a large spectrum of insecticides selected various resistance mechanisms in the codling moth populations. Metabolic resistance associated with enhanced activity of the P450 cytochrome oxidases is largely spread over the World [16,17] and confers cross-resistance to numerous insecticides including pyrethroids [8,20]. In the present study, the activity of the P450 cytochrome oxidases was negatively correlated with the proportion of homozygous *kdr* genotypes. Metabolic resistance should be sufficient for the codling moth to resist pyrethroid treatments and it could limit selection of sodium channel target mutations in absence of strong pyrethroid selection [48]. Third, insignificant fitness cost associated with *kdr* was measured in laboratory codling moth strains [49], which could explain the maintenance of high *kdr* proportions in absence of pyrethroid selection as observed in

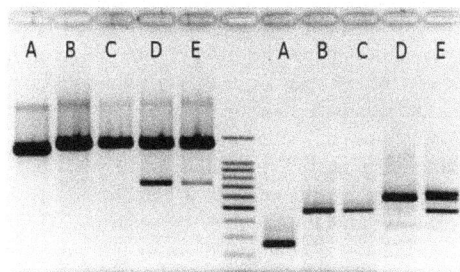

Figure 3. PCR product lengths obtained with the *SKdr-F/SKdr-R3* (left) and the *Kdr-F/Kdr-R* (right) primer pairs, respectively. In *Grapholita molesta*, lengths were respectively 1245 and 283 bp (A). In *Cydia pomonella*, lengths were respectively 1363 and 460 pb in specimens from France (B), 1359 and 460 pb in specimens from USA (C), 1359 and 460 or 576 pb in specimens from Syria (D and E). The electrophoresis was run on a 2% agarose gel at 100 volts for 30 minutes using the RunOne™ Electrophoresis System (Embi Technology). 100 pb DNA ladder (Promega) was in the central well.

Table 5. Sequence variability in the codling moth voltage gated sodium channel gene with variable positions numbered according to Figure 1.

Haplotype	221	265	a 380	b 396	397	412	414	447	c 1051	1097	1401	1420 kdr	1441	1494	1518	1528	d 1707	PCR RFLP	Intron I	Intron II
Syria 1	A	0	0	T	T	A	T	0	C	G	G	C	G	C	C	0	T	112	1106	411
Syria 2	A	0	0	T	T	A	T	−34	C	G	G	C	G	C	C	−116	T	112	1072	305
Bulgaria 1	A	0	0	T	T	A	T	0	C	G	G	C	G	C	C	−116	T	112	1106	305
Armenia 1	G	0	0	C	C	T	A	0	T	A	C	**T**	G	A	T	−116	C	77	1106	305
Turkey 1	G	0	0	C	C	T	A	0	T	A	C	C	**A**	A	T	−116	C	101	1106	305
Turkey 2	G	0	0	C	C	T	A	0	T	A	C	**T**	A	A	T	−116	C	77	1106	305
Czech 1	G	0	0	C	C	T	A	0	T	A	C	C	G	A	T	−116	C	112	1106	305
Czech 2	G	+7	0	C	C	T	A	0	T	A	C	C	G	A	T	−116	C	112	1113	305
Argentina 1	G	0	0	C	C	T	A	0	C	A	C	C	G	A	T	−116	C	112	1106	305
Argentina 2	G	0	+4	C	C	T	A	0	C	A	C	**T**	G	A	T	−116	C	77	1110	305
Argentina 3	G	0	0	C	C	T	A	0	C	A	C	**T**	**A**	A	T	−116	C	77	1106	305
Chile 1	G	0	0	C	C	T	A	0	C	A	C	**A**	A	A	T	−116	C	101	1106	305

Substitutions are indicated by the observed nucleotides and indels by the number of inserted or deleted base pairs using the *Syria1* sequence as reference. Nucleotides in bold characters refer to Tsp509I restriction sites. The three last columns respectively refer to lengths in base pairs of the largest Tsp509I digested fragment, and of introns I and II. Twelve different haplotypes were recognized among 38 sequences. The distribution of these twelve haplotypes among countries was reported in figure 4.

populations from organic orchards in South-eastern France (the granulosis virus was the only insecticide used to control codling moth in these orchards). However, fitness costs are difficult to predict when several resistance mechanisms interact [50] and may depend on the environmental conditions [51].

The lack of population structure of the *kdr* genotypes detected at the orchard level contrasts with the high structure observed among populations from different orchards at larger geographic scale. First, the selection of different resistance mechanisms among countries can partially explain regional *kdr* structure. Interestingly, the Armenian populations which were the only populations that fixed the *kdr* allele in the present study were also those in which metabolic resistance conferred by the cytochrome P450 oxidases were insignificant [8,16]. This is an additional argument supporting that *kdr* resistance evolved in interaction with other resistance mechanisms. Second, *kdr* proportions were negatively correlated with temperature in codling moth populations displaying *kdr* polymorphism in agreement with the hypothesis that fitness cost associated with sodium channel target mutations depends on temperature [51]. Consequently, fitness cost associated with *kdr* could be not equally distributed geographically and differently expressed along seasons in *C. pomonella* as previously noted in house fly populations [42,52]. It is to note that such temperature-dependent cost could also explain the latitudinal variation in the proportion of *kdr* allele previously observed in codling moth populations from France [16]. Finally, in the absence of strong resistance cost, ancient insecticide treatments may explain current *kdr* distribution [53]. Resistance to DTT was reported in some codling moth populations in the early 1950s [54]. The first cases of resistance to pyrethroid in the 1990s in some codling moth populations from France were observed less than five years after the beginning of treatments with these insecticides [7]. DTT treatments may have initiated the selection of the *kdr* mutation in the codling moth, which was secondarily selected by pyrethroid as suggested by observations in *Plutella xylostella* [55]. Consequently, any differences among countries in DTT and pyrethroid usage during the last sixty years may also explain current differences in

the proportions of *kdr* allele among codling moth populations at regional scales.

Variation in the *para* introns and exons was investigated in few insect pests – *Bemisia tabaci* [56], *Musca domestica* [42], *Mysus persicae* [57] and *Anopheles gambiiae* [40] – to shed light on the history of mutation events associated to insecticide resistance at this gene. In each species, several mutations, which likely arose independently in different populations, were linked with pyrethroid resistance. In contrast with these studies, only one substitution at the only *kdr* locus (L1014F) was observed in the codling moth. However, variations in *para* introns differentiated several haplotypes for both 1014L and 1014F variants. Phylogenetic relationships among these haplotypes confirmed the existence of two highly differentiated clades in *C. pomonella*, which is in agreement with previous analyses based on mitochondrial DNA sequences [58]. The four *kdr* haplotypes were all observed in only one clade (the most cosmopolitan one) but in several slightly differentiated sub-clades. In absence of recombination evidence, this result supports hypothesis of independent and convergent L1014F mutation events. A subclade encompassed *kdr* haplotypes only observed in Armenia and Bulgaria, which suggests that one mutation event may have occurred somewhere in Eastern Europe. The origins of the *kdr* mutations observed in America and Western Europe remain more dubious. Among the two different *kdr* haplotypes observed in Argentina, one was identical to a French haplotype. This suggests that the *kdr* mutation in the Argentine population may have two independent origins and that one origin would be shared with the French population. This similarity supports the idea that in some situations local selection of *kdr* resistance may have followed dispersal events in *C. pomonella* as postulated to explain the unique origin of A2 esterase over-production in *Culex pipiens* [59]. Previous observations also pointed at the importance of commercial exchanges on the dispersal of codling moth larvae among continents [16]. To prevent the spread of insecticide resistance, pest control programs should think management both at local and global levels [21,60].

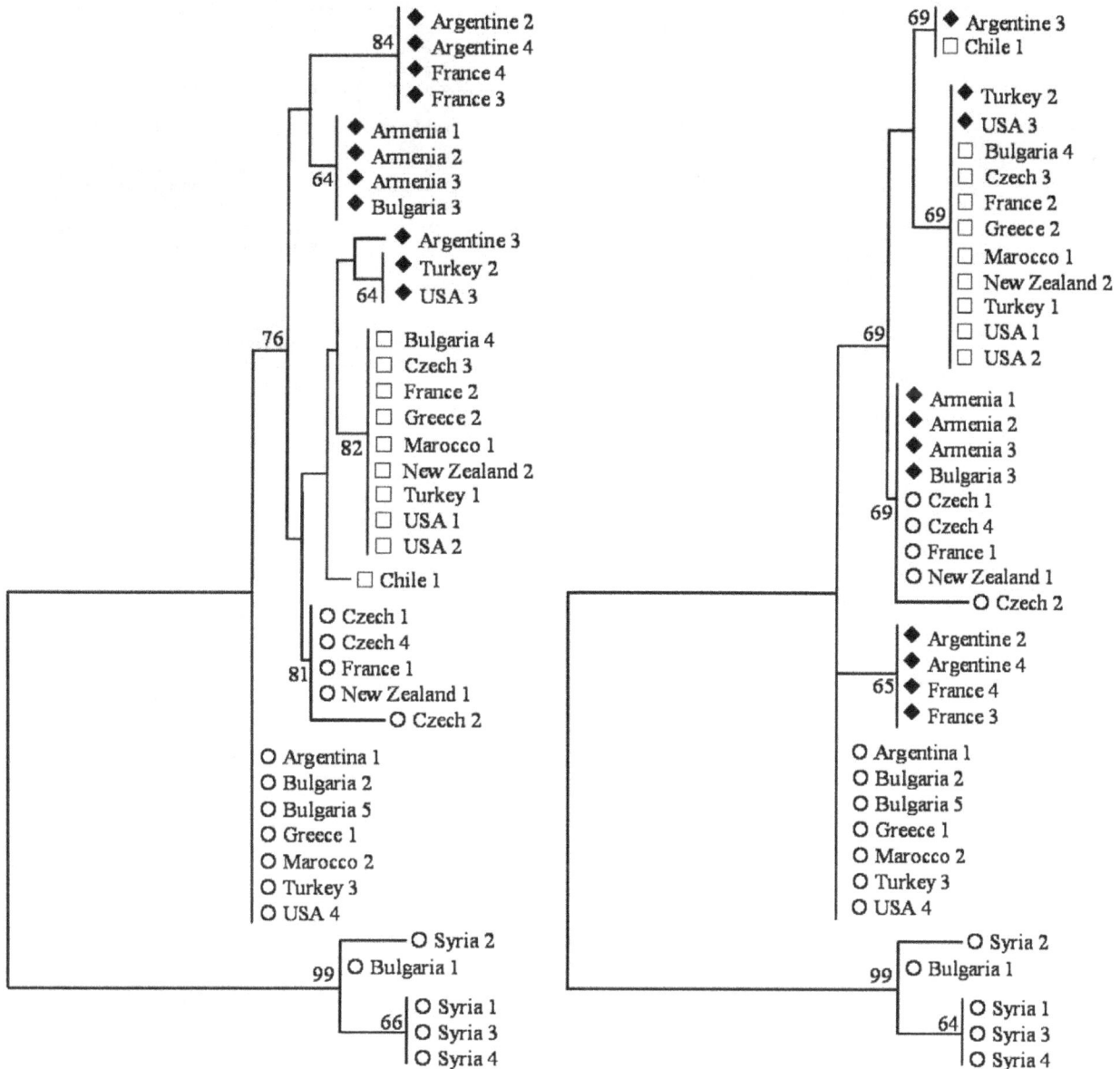

Figure 4. Minimum evolution trees established with the p-distance among 38 partial DNA sequences of the codling moth *para* gene. The trees were computed either with all the variable sites (left: 4 indels and 13 substitutions) or all the variable sites except the *kdr* locus (right: 4 indels and 12 substitutions). The *112* and *101* sequences (L1014) were represented by unfilled circles and squares, respectively. The *77* sequences (F1014) were represented by filled diamonds. Bootstraps values above 60% were reported.

Acknowledgments

We thank the EPI team for its help collecting codling moth larvae in the Basse-Durance Valley, France. We thank Maritza Reyes, Wilson Barros, Jim Walker, Claudio Ioriatti, Damian Gorzka, Costas Voudouris, John Margaritopoulos, Frantisek Marec, Oldrich Pultar, Alan Knight, Azya Ter-Hovhannesyan, Nadia Lombarkia, Salma Iraqui, Jesus Avilla, Marcella Rodriguez, Mohanad Ismail, and Recep Ay who kindly provided the codling moth samples from the other countries. We also thank Juliette Goussopoulos and Mathilde Willerval for their help extracting DNA and performing biochemistry analyses. Finally, we thank Benoît Moury for his suggestion about testing recombination and anonymous reviewers for their relevant comments and their corrections.

Author Contributions

Conceived and designed the experiments: PF MS CL. Performed the experiments: JO JFT. Analyzed the data: PF MS JO. Wrote the paper: PF MS.

References

1. Shel'Deshova GG (1967) Ecological factors determining distribution of the codling moth *Lapspeyresia pomonella* L. in the northern and southern hemispheres. Entomological Review 46: 349–361.

2. Thwaite WG, Williams DG, Hately AM (1993) Extent and significance of Azinphos-Methyl resistance in codling moth in Australia. Pest Control and Sustainable Agriculture: 166–168.

3. Fuentes-Contreras E, Reyes M, Barros W, Sauphanor B (2007) Evaluation of azinphos-methyl resistance and activity of detoxifying enzymes in codling moth (Lepidoptera : Tortricidae) from central Chile. Journal of Economic Entomology 100: 551–556.

4. Soleño J, Anguiano L, de D'Angelo AP, Cichon L, Fernandez D, et al. (2008) Toxicological and biochemical response to azinphos-methyl in *Cydia pomonella* L. (Lepidoptera : Tortricidae) among orchards from the Argentinian Patagonia. Pest Management Science 64: 964–970.

5. Varela LG, Welter SC, Jones VP, Brunner JF, Riedl H (1993) Monitoring and characterization of insecticide resistance in codling moth (Lepidoptera, Tortricidae) in 4 Western States. Journal of Economic Entomology 86: 1–10.

6. Reuveny H, Cohen E (2004) Evaluation of mechanisms of azinphos-methyl resistance in the codling moth *Cydia pomonella* (L.). Archives Of Insect Biochemistry And Physiology 57: 92–100.

7. Sauphanor B, Bovier JC, Brosse V (1998) Spectrum of insecticide resistance in *Cydia pomonella* (Lepidoptera : Tortricidae) in southeastern France. Journal of Economic Entomology 91: 1225–1231.

8. Reyes M, Franck P, Charlemillot P-J, Ioriatti C, Olivares J, et al. (2007) Diversity of insecticide resistance mechanisms and spectrum in European populations of the codling moth, *Cydia pomonella*. Pest Management Science 63: 890–902.

9. Sauphanor B, Brosse V, Bouvier JC, Speich P, Micoud A, et al. (2000) Monitoring resistance to diflubenzuron and deltamethrin in French codling moth populations (*Cydia pomonella*). Pest Management Science 56: 74–82.

10. Zlotkin E (1999) The insect voltage-gated sodium channel as target of insecticides. Annual Review of Entomology 44: 429–455.

11. O'Reilly AO, Khambay BPS, Williamson MS, Field LM, Wallace BA (2006) Modelling insecticide-binding sites in the voltage-gated sodium channel. Biochemical Journal 396: 255–263.

12. Williamson MS, Martinez-Torres D, Hick CA, Castells N, Devonshire AL (1996) Analysis of sodium channel gene sequences in pyrethroid-resistant houseflies. In: Brown TM, editor. Molecular Genetics and Evolution of Pesticide Resistance. Washington: American Chemical Society. 53–61.

13. Soderlund DM, Knipple DC (2003) The molecular biology of knockdown resistance to pyrethroid insecticides. Insect Biochemistry and Molecular Biology 33: 563–577.

14. Vais H, Williamson MS, Devonshire AL, Usherwood PNR (2001) The molecular interactions of pyrethroid insecticides with insect and mammalian sodium channels. Pest Management Science 57: 877–888.

15. Brun-Barale A, Bouvier J-C, Pauron D, Bergé J-B, Sauphanor B (2005) Involvement of a sodium channel mutation in pyrethroid resistance in *Cydia pomonella* L., and development of a diagnostic test. Pest Management Science 61: 549–554.

16. Franck P, Reyes M, Olivares J, Sauphanor B (2007) Genetic architecture in codling moth populations: comparison between microsatellite and insecticide resistance markers. Molecular Ecology 16: 3554–3564.

17. Reyes M, Franck P, Olivares J, Margaritopoulos J, Knight A, et al. (2009) Worldwide variability of insecticide resistance mechanisms in the codling moth, *Cydia pomonella* L. (Lepidoptera: Tortricidae). Bulletin of Entomological Research 99: 359–369.

18. Bouvier JC, Bues R, Boivin T, Boudinhon L, Beslay D, et al. (2001) Deltamethrin resistance in the codling moth (Lepidoptera : Tortricidae): inheritance and number of genes involved. Heredity 87: 456–462.

19. Sauphanor B, Brosse V, Monier C, Bouvier JC (1998) Differential ovicidal and larvicidal resistance to benzoylureas in the codling moth, *Cydia pomonella*. Entomologia Experimentalis et Applicata 88: 247–253.

20. Bouvier JC, Cuany A, Monier C, Brosse V, Sauphanor B (1998) Enzymatic diagnosis of resistance to deltamethrin in diapausing larvae of the codling moth, *Cydia pomonella* (L.). Archives of Insect Biochemistry and Physiology 39: 55–64.

21. Labbe P, Lenormand T, Raymond M (2005) On the worldwide spread of an insecticide resistance gene: a role for local selection. Journal of Evolutionary Biology 18: 1471–1484.

22. Ricci B, Franck P, Toubon J-F, Bouvier J-C, Sauphanor B, et al. (2009) The influence of landscape on insect pest dynamics: a case study in southeastern France Landscape Ecology 24: 337–349.

23. Walsh PS, Metzger DA, Higuchi R (1991) Chelex (R)100 as a medium for simple extraction of DNA for PCR-based typing from forensic material. Biotechniques 10: 507.

24. De Sousa G, Cuany A, Brun A, Amichot M, Rahmani R, et al. (1995) A microfluorometric method for measuring Ethoxycoumarin-o-deethylase activity on individual *Drosophila melanogaster* abdomens: interest for screening resistance in insect populations. Analytical Biochemistry 229: 86–91.

25. Joly E (1996) Purification of DNA fragments from agarose gels using glass beads. In: Harwood AJ, editor. Basic DNA and RNA Protocols. Berlin: Springer Verlag. 237–240.

26. Rousset F (2008) Genepop'007: a complete re-implementation of the genepop software for Windows and Linux. Molecular Ecology Resources 8: 103–106.

27. Hall TA (1999) Bioedit: a user-friendly biological sequence alignment editor and analysis program for Windows 95/98/NT. Nucleic Acids Symposium Series 41: 95–98.

28. Heath L, van der Walt E, Varsani A, Martin DP (2006) Recombination patterns in aphthoviruses mirror those found in other picornaviruses. Journal of Virology 80: 11827–11832.

29. Martin D, Rybicki E (2000) RDP: detection of recombination amongst aligned sequences. Bioinformatics 16: 562–563.

30. Padidam M, Sawyer S, Fauquet CM (1999) Possible emergence of new geminiviruses by frequent recombination. Virology 265: 218–225.

31. Posada D, Crandall KA (2001) Evaluation of methods for detecting recombination from DNA sequences: computer simulations. Proceedings of the National Academy of Sciences of the United States of America 98: 13757–13762.

32. Maynard Smith J (1992) Analysing the mosaic structure of genes. Journal of Molecular Evolution 34: 126–129.

33. Martin DP, Posada D, Crandall KA, C W (2005) A modified BOOTSCAN algorithm for automated identification of recombinant sequences and re-combination breakpoints. AIDS Research and Human Retroviruses 21: 98–102.

34. Gibbs MJ, Armstrong JS, Gibbs AJ (2000) Sister-Scanning: a Monte Carlo procedure for assessing signals in recombinant sequences. Bioinformatics 16: 573–582.

35. Tamura K, Dudley J, Nei M, Kumar S (2007) MEGA4: Molecular Evolutionary Genetics Analysis (MEGA) software version 4.0. Molecular Biology and Evolution 24: 1596–1599.

36. Lynd A, Weetman D, Barbosa S, Egyir Yawson A, Mitchell S, et al. (2010) Field, genetic, and modeling approaches show strong positive selection acting upon an insecticide resistance mutation in *Anopheles gambiae* s.s. Molecular Biology and Evolution 27: 1117–1125.

37. Martinez-Torres D, Chandre F, Williamson MS, Darriet F, Bergé JB, et al. (1998) Molecular characterization of pyrethroid knockdown resistance (*kdr*) in the major malaria vector *Anopheles gambiae s.s*. Insect Molecular Biology 7: 179–184.

38. Ranson H, Jensen B, Vulule JM, Wang X, Hemingway J, et al. (2000) Identification of a point mutation in the voltage-gated sodium channel gene of Kenyan *Anopheles gambiae* associated with resistance to DDT and pyrethroids. Insect Molecular Biology 9: 491–497.

39. Martinez-Torres D, Chevillon C, Brun-Barale A, Berge JB, Pasteur N, et al. (1999) Voltage-dependent Na+ channels in pyrethroid-resistant *Culex pipiens* L mosquitoes. Pesticide Science 55: 1012–1020.

40. Pinto J, Lynd A, Vicente JL, Santolamazza F, Randle NP, et al. (2007) Multiple origins of knockdown resistance mutations in the Afrotropical mosquito vector *Anopheles gambiae*. Plos One 2.

41. Park Y, Taylor MFJ (1997) A novel mutation L1029H in sodium channel gene hscp associated with pyrethroid resistance for *Heliothis virescens* (Lepidoptera: Noctuidae). Insect Biochemistry and Molecular Biology 27: 9–13.

42. Rinkevich FD, Zhang L, Hamm RL, Brady SG, Lazzaro BP, et al. (2006) Frequencies of the pyrethroid resistance alleles of *Vssc1* and *CYP6D1* in house flies from the eastern United States. Insect Molecular Biology 15: 157–167.

43. Voudouris CC, Sauphanor B, Franck P, Reyes M, Mamuris Z, et al. (2011) Insecticide resistance status of the codling moth *Cydia pomonella* (Lepidoptera: Tortricidae) from Greece. Pesticide Biochemistry and Physiology 100: 229–238.

44. Franck P, Timm AE (2010) Population genetic structure of *Cydia pomonella*: a review and case study comparing spatiotemporal variation. Journal of Applied Entomology 134: 191–200.

45. Cao XM, Song FL, Zhao TY, Dong YD, Sun CX, et al. (2006) Survey of deltamethrin resistance in house flies (*Musca domestica*) from urban garbage dumps in Northern China. Environmental Entomology 35: 1–9.

46. Guerrero FD, Alison MW, Kammlah DM, Foil LD (2002) Use of the polymerase chain reaction to investigate the dynamics of pyrethroid resistance in *Haematobia irritans* irritans (Diptera : Muscidae). Journal of Medical entomology 39: 747–754.

47. Jamroz RC, Guerrero FD, Kammlah DM, Kunz SE (1998) Role of the kdr and super-kdr sodium channel mutations in pyrethroid resistance: correlation of allelic frequency to resistance level in wild and laboratory populations of horn flies (*Haematobia irritans*). Insect Biochemistry and Molecular Biology 28: 1031–1037.

48. Brooke BD (2008) kdr: can a single mutation produce an entire insecticide resistance phenotype? Transactions of the Royal Society of Tropical Medicine and Hygiene 102: 524.

49. Boivin T, Chabert d'Hieres C, Bouvier JC, Beslay D, Sauphanor B (2001) Pleiotropy of insecticide resistance in the codling moth, *Cydia pomonella*. Entomologia Experimentalis et Applicata 99: 381–386.

50. Berticat C, Bonnet J, Duchon S, Agnew P, Weill M, et al. (2008) Costs and benefits of multiple resistance to insecticides for *Culex quinquefasciatus* mosquitoes. BMC Evolutionary Biology 8: 104–114.

51. Foster SP, Young S, Williamson MS, Duce I, Denholm I, et al. (2003) Analogous pleiotropic effects of insecticide resistance genotypes in peach-potato aphids and houseflies. Heredity 91: 98–106.

52. Rinkevich FD, Hamm RL, Geden CJ, Scott JG (2007) Dynamics of insecticide resistance alleles in house fly populations from New York and Florida. Insect Biochemistry and Molecular Biology 37: 550–558.

53. Davies ET, O'Reilly AO, Field LM, Wallace B, Williamson MS (2008) Perspective knockdown resistance to DDT and pyrethroids: from target-site mutations to molecular modelling. Pest Management Science 64: 1126–1130.

54. Smith LC (1955) DDT-resistant codling moth. A report on the 1954/55 control trials. Journal of the Department of Agriculture, Victoria 59: 12–15.

55. Kwon DH, Choi BR, Park HM, Lee SH, Miyata T, et al. (2004) Knockdown resistance allele frequency in field populations of *Plutella xylostella* in Korea. Pesticide Biochemistry and Physiology 80: 21–30.

56. Alon M, Benting J, Lueke B, Ponge T, Alon F, et al. (2006) Multiple origins of pyrethroid resistance in sympatric biotypes of *Bemisia tabaci* (Hemiptera: Aleyrodidae). Insect Biochemistry and Molecular Biology 36: 71–79.

57. Anstead JA, Williamson MS, Denholm I (2005) Evidence for multiple origins of identical insecticide resistance mutations in the aphid *Myzus persicae*. Insect Biochemistry and Molecular Biology 35: 249–256.

58. Meraner A, Brandstätter A, Thaler R, Aray B, Unterlechner M, et al. (2008) Molecular phylogeny and population structure of the codling moth (*Cydia pomonella*) in Central Europe: I. Ancient clade splitting revealed by mitochondrial haplotype markers. Molecular Phylogenetics and Evolution 48: 825.

59. Guillemaud T, Rooker S, Pasteur N, Raymond M (1996) Testing the unique amplification event and the wordwide migration hypothesis of insecticide resistance genes with sequence data. Heredity 77: 535–543.

60. Denholm I, Devine GJ, Williamson MS (2002) Insecticide resistance on the move. Science 297: 2222–2223.

3

Carnivore Use of Avocado Orchards across an Agricultural-Wildland Gradient

Theresa M. Nogeire[1]*, Frank W. Davis[2], Jennifer M. Duggan[1], Kevin R. Crooks[3], Erin E. Boydston[4]

1 School of Environmental and Forest Sciences, University of Washington, Seattle, Washington United States of America, **2** Bren School of Environmental Science and Management, University of California Santa Barbara, Santa Barbara, California, United States of America, **3** Department of Fish, Wildlife, and Conservation Biology, Colorado State University, Fort Collins, Colorado, United States of America, **4** U.S. Geological Survey, Western Ecological Research Center, Thousand Oaks, California, United States of America

Abstract

Wide-ranging species cannot persist in reserves alone. Consequently, there is growing interest in the conservation value of agricultural lands that separate or buffer natural areas. The value of agricultural lands for wildlife habitat and connectivity varies as a function of the crop type and landscape context, and quantifying these differences will improve our ability to manage these lands more effectively for animals. In southern California, many species are present in avocado orchards, including mammalian carnivores. We examined occupancy of avocado orchards by mammalian carnivores across agricultural-wildland gradients in southern California with motion-activated cameras. More carnivore species were detected with cameras in orchards than in wildland sites, and for bobcats and gray foxes, orchards were associated with higher occupancy rates. Our results demonstrate that agricultural lands have potential to contribute to conservation by providing habitat or facilitating landscape connectivity.

Editor: Mark S. Boyce, University of Alberta, Canada

Funding: Partial funding for this study was provided by The Bren School at UCSB, Orange County Great Park Corporation, and the U.S.G.S. The funders had no role in study design, data collection and analysis, decision to publish, or preparation of the manuscript.

Competing Interests: The authors have declared that no competing interests exist.

* E-mail: tnogeire@gmail.com

Introduction

Land-use change is a leading driver of loss of biological diversity globally [1]. As pressures from habitat loss increase, there is growing interest in agricultural landscapes as potential habitat or movement areas for wildlife. Agricultural landscapes are potentially rich in structure, food, and cover, and many native species forage and reproduce in these landscapes [2]. These lands can support moderate diversity of birds, mammals, arthropods, and plants, depending on the intensity of agriculture [3,4] and on configuration of natural land cover [5].

Mammalian carnivores are frequent targets of conservation efforts [6], and they play a key role in food webs, for example via mesopredator release [7] or trophic downgrading [8]. Because carnivores are typically wide-ranging, it is especially important to consider agricultural landscapes as well as protected areas when forming conservation plans for these species. Wildlife managers and conservation planners currently have little knowledge of carnivore use of agricultural landscapes, but this subject will become increasingly important as agricultural systems continue to expand and protected areas become more isolated. Connectivity between habitat patches is especially critical in human-dominated landscapes [9], but most connectivity models focus on natural vegetation types, not on differences between human-dominated land cover types within such landscapes [10]. When evaluating landscape connectivity for large carnivores, conservation planners have often relied on expert opinion and considered all agriculture as having uniformly low connectivity value (for example [11–13]).

Many members of the order Carnivora are omnivorous and feed, in part, on anthropogenic food sources [14–16]. Scat analysis has identified cultivated fruit in the diets of carnivores, particularly foxes and stone martens [15–17], and Borchert et al. [18] found that at least one orchard type – avocado – was regularly used by carnivores in California. California is a major producer of avocados, with 23,500 hectares of orchards [19] spread across five southern counties. Because avocados grow well on steep slopes, they are planted in a variety of landscape contexts, including hillslopes adjacent to native vegetation as well as valley bottoms adjacent to other types of crops.

We examined the use of avocado orchards by mammalian carnivores across agricultural-wildland gradients in southern California. We assessed whether occupancy of carnivores at motion-activated camera stations was a function of surrounding land cover, and in particular, whether area of orchards influenced carnivore occupancy. If orchards constitute poor quality carnivore habitat relative to natural areas, we would expect to observe carnivores less frequently in orchards than in nearby wildlands.

Methods

Study Area

Coastal southern California is highly urbanized and contains about two thirds of California's 38 million residents; it also has relatively little remaining undeveloped land [20], yet is experiencing rapid population growth [21]. This region has a Mediterranean-type climate, and the dominant natural vegetation types are oak woodland, riparian woodland, sage scrub, and annual grassland. Eleven native members of the order Carnivora occur in this region: American badger (*Taxidea taxus*), American black bear (*Ursus americanus*), bobcat (*Lynx rufus*), coyote (*Canis latrans*), gray fox (*Urocyon cinereoargenteus*), long-tailed weasel (*Mustela frenata*), mountain lion (*Puma concolor*), raccoon (*Procyon lotor*), ringtail (*Bassariscus astutus*), striped skunk (*Mephitis mephitis*), and Western spotted skunk (*Spilogale gracilis*).

Our study area included avocado orchards and wildlands in Santa Barbara and Ventura counties selected for their position in the landscape and for landowner cooperation (Figure 1). Avocado orchards (hereafter "orchards") grew on diverse topographies, from steep mountains to flat floodplains, and were surrounded by natural vegetation including sage scrub (Figure 2), oak woodland, grassland vegetation, other agriculture, or low-density development. Wildland sites with only natural vegetation were located at the University of California's Sedgwick Reserve and Gaviota State Park.

Land-cover Classification

No single existing land cover map met our habitat mapping needs in terms of scale, accuracy, and legend. We therefore created land cover maps from the National Land Cover Database (NLCD) [22], 2005 Southern California Association of Governments (SCAG), California Department of Water Resources, and the California Avocado Commission. Avocado orchards were identified by the California Avocado Association data, along with the SCAG and Department of Water Resources data. We used NLCD land cover to classify natural habitat types, which were not identified in the SCAG layer. Lands in classes which were essentially open space with substantial human activity (e.g., school yards, golf courses, dirt roads, urban parks, low-density development or developed open space) and which had less than 10% impervious surface were classified as 'disturbed'. When land-cover layers from the different sources were inconsistent, we verified classifications with ground visits or visual inspection of air photos (National Agriculture Imagery Program 2005, 1 m natural color).

Camera Stations

We used motion-activated digital cameras (Stealthcam, LLC, Grand Prairie, TX) at 38 sites to detect carnivore species from April 2007 to June 2008, resulting in 1,130 trap nights. Cameras

Figure 2. Orchard on hillslope. Typical landscape pattern of steep hills with orchards surrounded by wildlands.

were placed in and around 6 orchards (22 sites in orchards and 6 sites in natural vegetation adjacent to orchards) and 2 continuous wildlands (10 sites), with distance to nearest camera between 30–900 meters (mean = 193 meters).

At all sites, cameras were placed along similar-sized dirt roads, near signs of carnivore activity (e.g., scat) or at trail junctions when possible. We placed scent lure (Pred-a-Getter, Murray's Lures and Trapping Supplies, Walker, West Virginia) in front of the camera to encourage animals to approach the camera and to stop long enough to be photographed. For each carnivore species at each camera site, we tallied the number of nights in which the species was detected at least once. We considered each 24-hour trap night to begin at 6:30 am, and cameras were active continuously between 12 and 76 nights (average = 33 nights) at a particular site.

Detections and Occupancy

To determine the difference in carnivore species richness between land cover types, we examined whether the number of native carnivore species differed among cameras situated in orchards, natural vegetation adjacent to orchards, and continuous wildlands using a likelihood-ratio chi-squared test. We also tested for differences between pairs of land-cover categories with a post-hoc Tukey's Honest Significant Differences test.

We assessed the influence of landscape variables on carnivore presence at camera stations using a model selection framework to compare occupancy models. We used program PRESENCE v4.6 (PRESENCE, accessed 8/2/12, http://www.mbr-pwrc.usgs.gov/software/presence.html) to estimate occupancy (ψ) and detection rate (p, the probability of detecting a species if it is present) at each camera site [23] for bobcats, coyotes, and gray foxes (species with sufficient detections to permit analyses). This program uses a likelihood approach and has been used with camera-trap data [24,25], incorporating the effect of both site covariates and sample design. We considered each trap night as a survey. We used Akaike's Information Criterion corrected for small sample size (AICc) to choose the best-performing models [26].

We began the modeling process by selecting the best detection model for each species while holding occupancy rate constant, as in Negrões et al. [25] and Duggan et al. [27]. We expected that season (wet, November – March [28], versus dry) and land cover at the camera site (orchard, natural vegetation adjacent to orchards, or continuous wildland) could affect detection rate (in addition to occupancy) so we included both as covariates in detection models. Detection covariates, as well as predictors of

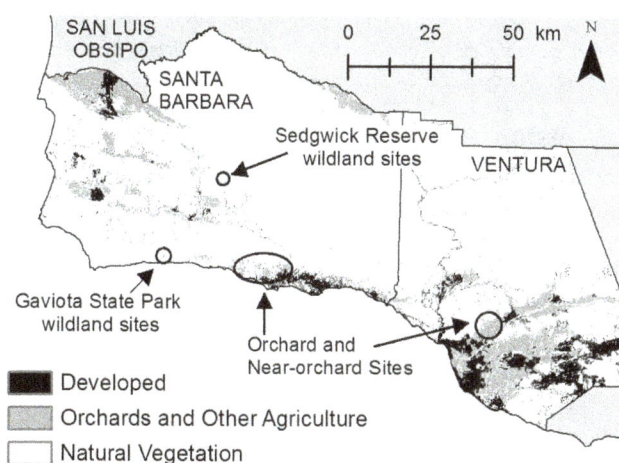

Figure 1. Study area in southern California. Study sites within Santa Barbara and Ventura counties.

occupancy described below, were standardized by z-score as described in Donovan and Hines [29].

Next, to determine if spatial clustering affected occupancy, and at what scale, we compared models including site (individual orchards at least 1 km from the perimeter of the nearest neighbor), meta-site (2–3 orchards 3–4 km from one another), or county (Ventura versus Santa Barbara) as predictors of occupancy while including any detection variables selected in the previous step. We then included the covariate from the top-ranked model of spatial scale as a predictor in the candidate model set for occupancy of that species.

Finally, for each species we modeled occupancy while including detection covariates from the top-ranked detection model for that species. Potential predictors of occupancy included land cover at the camera site, distance from each camera to the perimeter of continuous wildland (natural areas contiguous with Gaviota State Park, Los Padres National Forest or adjacent wildlands, ranging from 0–3.4 km), and season (wet versus dry). We also evaluated the degree to which area of orchards and other landscape variables in the neighborhood of a camera influenced carnivore occupancy. To do so, we used the land-cover map to quantify the extent (km^2) of orchards and covariates (disturbed, shrub/scrub, grassland/herbaceous, and woodland) within a 1,935 m-diameter circle centered on each camera, approximately the average size of a bobcat home range in this region and intermediate between range sizes of foxes and coyotes [30].

We had 38 sites, and therefore examined only single- and double-factor occupancy models to avoid overparameterization. We included all single- and double-factor models in our candidate model set and then conducted model averaging. We report results for the average model, but also include summaries of the top-ranked model and all models within 2 AICc points of this model, indicating substantial support [26]. To compare the selection support for each predictor variable, we also calculated variable importance weights, which are the sum of the model weights of all models that contain a given variable [26]. Averaged models include only models within 2 AICc points of the best model. Variable importance rates are assessed across all models and therefore each variable has equal representation.

Results

Camera Stations

Cameras were active for a total of 667 trap nights in orchards, 201 in natural vegetation near orchards, and 262 in wildlands. We detected 8 of the 11 native carnivore species in the study region. Seven native species were detected in orchards: coyote (38 detections), striped skunk (28), bobcat (25), gray fox (20), mountain lion (3), black bear (2), and raccoon (2). Eight native species were detected in natural vegetation: coyote (25), bobcat (21), mountain lion (4), gray fox (7), raccoon (2), badger (1), black bear (1), and striped skunk (1). The 3 native carnivore species not detected included ringtail, spotted skunk, and long-tailed weasel.

Detections and Occupancy

On average, the number of native species detected per site differed among land-cover classes ($\chi^2 = 6.69$, df = 2, p = 0.035), but differences between individual classes were not statistically significant (Tukey's HSD, all p>0.12). The number of native species detected was greatest in orchards (mean = 2.1, SE = 0.36), intermediate in sites with natural vegetation adjacent to orchards (mean = 1.8, SE = 0.48), and lowest in wildland sites (mean = 0.8, SE = 0.40).

The top-ranked detection rate models for coyote and gray fox included land cover at the camera station location (Tables 1, 2), with higher detections in avocado orchards (top model: $\beta = 2.62$, SE = 1.09 for coyote and $\beta = 2.21$, SE = 1.10 for fox) or near avocado orchard ($\beta = 3.19$, SE = 1.10 for coyote and $\beta = 1.47$, SE = 1.16 for fox) relative to wildlands. Season was also included in the top-ranked models; for fox, the direction of the effect could not be distinguished from 0 ($\beta = 1.02$; SE = 1.16), while for coyote, detection rate was lower in the dry season than in the wet season ($\beta = -0.78$, SE = 0.29). These detection covariates were included in all subsequent models. For bobcat, the intercept-only model was the top model, so subsequent bobcat models did not include detection covariates. For all three species, the intercept-only occupancy model was the top-ranked model for spatial variation, thus we did not include spatial variables as predictors of occupancy in our final set of candidate models.

Avocado orchards, either at the camera site or in the neighborhood of the camera, were included in at least one competitive occupancy model for all three carnivore species (Tables 1, 2, 3). The area of avocado orchard in the neighborhood of a camera was the most important predictor of bobcat occupancy (model average: $\beta = 0.56$, SE = 0.32; Table 4) and was included in all top four models for bobcat occupancy (Table 3). The area of avocado orchards in the neighborhood of a camera was the third most important predictor for gray fox occupancy ($\beta = 0.17$, SE = 0.14). For coyote, the area of orchard in the neighborhood had a weak negative effect ($\beta = -0.25$, SE = 0.20), but both avocado orchard and 'near orchard' at the camera site had a positive effect (avocado orchard: $\beta = 13.41$, SE = 6.73; near orchard: $\beta = 13.23$, SE = 6.67). Land cover at the camera site was not included in any competitive bobcat or gray fox occupancy models. Distance to continuous wildland was the most important variable for predicting gray fox occupancy (model average: $\beta = -1.16$, SE = 0.93) and third most important variable for bobcats ($\beta = -0.16$, SE = 0.14; Table 4), with occupancy increasing closer to or within wildland habitat. Distance to continuous wildland was not, however, included in any competitive coyote models. The area of disturbed land in the neighborhood of the camera was included in competitive models for both coyote ($\beta = 4.81$, SE = 2.20) and gray fox ($\beta = 0.31$, SE = 0.39) occupancy (Tables 1, 2), but large standard error values for fox occupancy suggested a weak influence. Disturbed land was not included in models for bobcat occupancy (Table 3). Woodland, shrub, and grassland/herbaceous vegetation in the neighborhood of a camera had a positive effect on occupancy in all models for all species, except that woodland had a negative effect on gray fox occupancy.

Discussion

Carnivores were detected with surprising frequency in avocado orchards. We detected most carnivore species native to coastal southern California in avocado orchards, and these orchards were used frequently by bobcats, coyotes and gray foxes. Further, we detected more carnivore species in orchards than in wildland sites. Although orchards are often adjacent to wildlands, the presence of carnivores in orchards does not appear to be simply an artifact of landscape context. If this were the case, we would expect to find more carnivores in wildlands than in orchards, which we did not. We would also expect to find that distance to continuous wildland was a more consistently important predictor in our models; although it was the strongest predictor of occupancy for gray fox, it was present in only one competitive model for bobcat occupancy and no competitive models for coyote.

Table 1. Top-ranked models of site occupancy (ψ) and detection rate (p) for gray fox.

Model I.D.	K	−2*log-likelihood	ΔAICc	Relative likelihood	ω	ψ̂ (SE)	p̂ (SE)
Detection (p) ~							
Land cover + season	5	230	0.00	1.00	0.45	0.47 (0.13)	0.018 (0.025)
Land cover	4	234	0.55	0.76	0.34	0.41 (0.11)	0.017 (0.033)
Season	3	238	2.89	0.24	0.11	0.45 (0.13)	0.015 (0.029)
Intercept only	2	241	2.98	0.23	0.10	0.38 (0.10)	0.059 (0.012)
Occupancy (ψ) ~							
Distwild	6	227	0	1.00	0.14	0.47 (0.14)	0.009 (0.010)
Distwild + woodland	7	224	0.40	0.82	0.11	0.61 (0.09)	0.066 (0.019)
Intercept only	5	230	0.67	0.72	0.10	0.47 (0.13)	0.067 (0.020)
Distwild + Avocado orchard	7	225	0.98	0.61	0.08	0.39 (0.17)	0.068 (0.021)
Shrub	6	228	1.33	0.51	0.07	0.56 (0.12)	0.066 (0.019)
Distwild + disturbed	7	225	1.51	0.47	0.06	0.55 (0.14)	0.065 (0.018)
Averaged model					*0.50*	*0.51 (0.13)*	*0.049 (0.016)*

Footnote: All models with ΔAICc <2.0, plus the intercept-only models, are reported. K is the number of parameters, ΔAICc is the difference between the AICc of the model and the lowest-AICc model, ω is the AICc model weight (summed for the averaged model), ψ is the predicted occupancy at a site and p is the probability of detecting the species at a given site. Covariate abbreviations: distwild is distance to continuous wildland, land cover is land cover (avocado orchard, near orchard, or wildland) at the camera site, and woodland, avocado orchard, shrub and disturbed refer to the area of that land cover in the neighborhood of the camera site.

The food subsidy value of avocados may explain why omnivores such as bears, coyotes, and raccoons were present in orchards. Indeed, remote cameras have recorded these species eating avocados in southern California (M. Borchert, U.S. Forest Service, personal communication), but why obligate carnivores like mountain lions and bobcats would be present in orchards is less clear. Orchards may provide good cover for carnivores; many of these species are habitat generalists, and orchards often replace oak woodlands with structurally similar vegetation. Irrigation lines in orchards act as a rare source of perennial water in arid landscapes. In our study, we did not find an effect of wet versus dry season on occupancy, as might be expected if carnivores were attracted by water sources. However, irrigation lines, combined with abundant avocados, might simulate year-round wet-season conditions for small mammals, perhaps leading to bottom-up effects in these agricultural systems. Future research could assess whether orchards are providing more food and water for small mammals than native vegetation, and whether a relative increase in prey might help explain the use of these lands by carnivores [31]. Finally, further study could evaluate whether the presence of infrequently-used dirt roads in orchards might appeal to animals moving across densely vegetated landscapes.

Table 2. Top-ranked models of site occupancy (ψ) and detection rate (p) for coyote.

Model I.D.	K	−2*log-likelihood	ΔAICc	Relative likelihood	ω	ψ̂ (SE)	p̂ (SE)
Detection (p) ~							
Land cover+season	5	440	0.00	1.00	0.86	0.68 (0.10)	0.079 (0.019)
Land cover	4	447	4.12	0.13	0.11	0.71 (0.10)	0.069 (0.014)
Season	3	452	6.93	0.03	0.03	0.55 (0.095)	0.10 (0.019)
Intercept only	2	486	38.70	0.00	0.00	0.44 (0.10)	0.056 (0.0068)
Occupancy (ψ) ~							
Disturbed	6	433	0.00	1.00	0.18	0.76 (0.08)	0.056 (0.012)
Grass/herbaceous	6	434	0.68	0.71	0.13	0.75 (0.10)	0.055 (0.012)
Avocado orchard + disturbed	7	432	1.24	0.54	0.10	0.77 (0.10)	0.055 (0.012)
Land Cover + grass/herbaceous	8	429	1.42	0.49	0.09	0.52 (0.10)	0.070 (0.027)
Intercept only	5	440	3.74	0.15	0.03	0.68 (0.10)	0.055 (0.013)
Averaged model					*0.53*	*0.72 (0.09)*	*0.06 (0.015)*

Footnote: All models with ΔAICc <2.0, plus the intercept-only models, are reported. K is the number of parameters, ΔAICc is the difference between the AICc of the model and the lowest-AICc model, ω is the AICc model weight (summed for the averaged model), ψ is the predicted occupancy at a site and p is the probability of detecting the species at a given site. Covariate abbreviations: distwild is distance to continuous wildland, land cover is land cover (avocado orchard, near orchard, or wildland) at the camera site, and grass/herbaceous, avocado orchard and disturbed refer to the area of that land cover in the neighborhood of the camera site.

Table 3. Top-ranked models of site occupancy (ψ) and detection rate (p) for bobcat.

Model I.D.	K	-2*log-likelihood	ΔAICc	Relative likelihood	ω	$\hat{\psi}$ (SE)	$\hat{p}$ (SE)
Detection (p) ~							
Intercept only	2	368	0	1.00	0.48	0.44 (0.10)	0.080 (0.012)
Season	3	367	1.26	0.53	0.26	0.52 (0.14)	0.066 (0.019)
Land cover	4	365	2.10	0.35	0.17	0.47 (0.11)	0.071 (0.019)
Land cover+season	5	364	3.27	0.20	0.09	0.55 (0.14)	0.060 (0.022)
Occupancy (ψ) ~							
Avocado orchard + grass/herbaceous	4	363	0	1.00	0.13	0.44 (0.15)	0.081 (0.012)
Avocado orchard + woodland	4	363	0.17	0.92	0.12	0.26 (0.17)	0.081 (0.012)
Avocado orchard	3	365	0.20	0.90	0.12	0.42 (0.13)	0.081 (0.012)
Avocado orchard + distwild	4	363	0.29	0.87	0.11	0.42 (0.15)	0.081 (0.012)
Intercept only	2	368	0.34	0.84	0.11	0.44 (0.10)	0.080 (0.012)
Woodland	3	366	1.11	0.57	0.07	0.28 (0.16)	0.080 (0.012)
Woodland + grass/herbaceous	4	365	1.88	0.39	0.05	0.35 (0.15)	0.079 (0.012)
Averaged model					0.71	0.38 (0.14)	0.080 (0.012)

Footnote: All models with ΔAICc <2.0, plus the intercept-only models, are reported. K is the number of parameters, ΔAICc is the difference between the AICc of the model and the lowest-AICc model, ω is the AICc model weight (summed for the averaged model), ψ is the predicted occupancy at a site and p is the probability of detecting the species at a given site. Covariate abbreviations: distwild is distance to continuous wildland, land cover is land cover (avocado orchard, near orchard, or wildland) at the camera site, and woodland, grass/herbaceous, shrub, avocado orchard and disturbed refer to the area of that land cover in the neighborhood of the camera site.

There is growing interest in managing for movement of wild animals through agricultural areas [10,32]. Knowing the value of different land-cover types for habitat or movement can inform conservation decisions regarding which lands should be purchased or put under easements, or which areas are most suitable for the placement of highway crossings [12]. Avocado orchards appear to serve as both foraging and movement habitat for most carnivore species in California, and conservation easements or other incentives to keep land in orchards could offer a cost-effective conservation strategy. Such alternative conservation strategies are particularly important when considering agriculture (including avocados) in Mediterranean-type ecosystems, which are highly threatened [33].

Acknowledgments

M. Borchert, B. Kendall, P. Jantz, S. Riley, J. Orrock, and S. Rothstein contributed ideas during early phases of the project. E. Fleishman, B. McRae, J. Yee, J. Hilty and M. Ricca provided helpful comments on the manuscript. The use of trade, product, or firm names in this publication is for descriptive purposes only and does not imply endorsement by the U.S. Government.

Author Contributions

Conceived and designed the experiments: FD TN. Performed the experiments: TN. Analyzed the data: TN JD FD KC EB. Contributed reagents/materials/analysis tools: FD TN. Wrote the paper: TN FD EB JD KC.

References

Table 4. Variable importance weights (ω) for predictors of occupancy for bobcats, coyotes, and gray foxes.

Bobcat		Coyote		Gray fox	
Covariate	Σω	Covariate	Σω	Covariate	Σω
Avocado orchard	0.58	Disturbed	0.50	Distance to wildland	0.51
Woodland	0.29	Grass/herbaceous	0.47	Shrub	0.21
Grass/herbaceous	0.28	Avocado orchard	0.15	Avocado orchard	0.18
Distance to wildland	0.13	Shrub	0.12	Woodland	0.18
Season	0.09	Land cover at site	0.10	Disturbed	0.13
Shrub	0.07	Woodland	0.10	Grass/herbaceous	0.11
Disturbed	0.06	Season	0.08	Season	0.09
Land cover at site	0.03	Distance to wildland	0.06	Land cover at site	0.05

1. Sala OE, Chapin FS, Armesto JJ, Berlow E, Bloomfield J, et al. (2000) Global Biodiversity Scenarios for the Year 2100. Science 287: 1770–1774.
2. Brosi BJ, Daily GC, Davis FW (2006) Agriculture and Urban Landscapes. In: Scott JM, Goble DG, Davis FW, editors. The Endangered Species Act at 30: Volume 2. Washington, D.C.: Island Press. 256–274.
3. Benton TG, Vickery JA, Wilson JD (2003) Farmland biodiversity: is habitat heterogeneity the key? Trends in Ecology & Evolution 18: 182–188.
4. Flynn DFB, Gogol-Prokurat M, Nogeire T, Molinari N, Richers BT, et al. (2009) Loss of functional diversity under land use intensification across multiple taxa. Ecology Letters 12: 22–33.
5. Daily GC, Ceballos G, Pacheco J, Suzán G, Sánchez-Azofeifa A (2003) Countryside Biogeography of Neotropical Mammals: Conservation Opportunities in Agricultural Landscapes of Costa Rica. Conservation Biology 17: 1814–1826.
6. Ray JC (2005) Large carnivores and the conservation of biodiversity. Washington, D.C.: Island Press.
7. Crooks KR, Soulé ME (1999) Mesopredator release and avifaunal extinctions in a fragmented system. Nature 400: 563–566.
8. Estes J, Terborgh J, Brashares J (2011) Trophic downgrading of planet earth. Science 333: 301–306.
9. Crooks KR, Sanjayan MA (2006) Connectivity conservation. Cambridge Univ Pr.
10. Cosentino BJ, Schooley RL, Phillips CA (2011) Connectivity of agroecosystems: dispersal costs can vary among crops. Landscape ecology 26: 1–9.

11. Beier P, Penrod K, Luke C, Spencer W, Cabañero C (2006) South Coast Missing Linkages: restoring connectivity to wildlands in the largest metropolitan area in the United States. In: Crooks KR, Sanjayan MA, editors. Connectivity Conservation. New York: Cambridge Univ Pr. 555–586.

12. Beier P, Garding E, Majka D (2008) Arizona Missing Linkages: Linkage Designs for 16 Landscapes. Phoenix, AZ: Arizona Game and Fish Department.

13. Singleton P, Gaines W, Lehmkuhl J (2002) Landscape permeability for large carnivores in Washington: a geographic information system weighted-distance and least-cost corridor assessment. Portland, Oregon: US Department of Agriculture Forest Service.

14. Fedriani J, Fuller T, Sauvajot R (2001) Does availability of anthropogenic food enhance densities of omnivorous mammals? An example with coyotes in southern California. Ecography 24: 325–331.

15. Padial J, Avila E, Sanchez J (2002) Feeding habits and overlap among red fox (*Vulpes vulpes*) and stone marten (*Martes foina*) in two Mediterranean mountain habitats. Mammalian Biology 67: 137–146.

16. Shapira I, Sultan H, Shanas U (2007) Agricultural farming alters predator–prey interactions in nearby natural habitats. Animal Conservation 11: 1–8.

17. López-Bao J V, González-Varo JP (2011) Frugivory and spatial patterns of seed deposition by carnivorous mammals in anthropogenic landscapes: a multi-scale approach. PloS one 6: e14569.

18. Borchert M, Davis F, Kreitler J (2008) Carnivore use of an avocado orchard in southern California. California Fish and Game 94: 61–74.

19. California Avocado Commission (2010). Available: www.avocado.org.

20. Landis JD, Reilly M (2003) How We Will Grow: Baseline Projections of the Growth of California's Urban Footprint through the Year 2100.

21. Conservation International (2010) Biodiversity Hotspots. Available: www.biodiversityhotspots.org.

22. Homer C, Huang C, Yang L (2004) Development of a 2001 national landcover database for the United States. Photogrammetric Engineering and Remote Sensing 70: 829–840.

23. MacKenzie DI, Nichols JD, Lachman GB, Droege S, Andrew Royle J, et al. (2002) Estimating site occupancy rates when detection probabilities are less than one. Ecology 83: 2248–2255.

24. Linkie M, Dinata Y, Nugroho A, Haidir IA (2007) Estimating occupancy of a data deficient mammalian species living in tropical rainforests: Sun bears in the Kerinci Seblat region, Sumatra. Biological Conservation 137: 20–27.

25. Negrões N, Sarmento P, Cruz J, Eira C, Revilla E, et al. (2010) Use of camera-trapping to estimate puma density and influencing factors in central Brazil. Journal of Wildlife Management 74: 1195–1203.

26. Burnham KP, Anderson DR (2002) Model selection and multimodel inference: a practical information-theoretic approach. New York: Springer Verlag.

27. Duggan J, Schooley R, Heske E (2011) Modeling occupancy dynamics of a rare species, Franklin's ground squirrel, with limited data: are simple connectivity metrics adequate? Landscape ecology 26: 1477–1490.

28. Keeley JE (2000) Chaparral. In: Barbour M, Billings W, editors. North American terrestrial vegetation. 203–254.

29. Donovan TM, Hines J (2007) Exercises in Occupancy Estimation and Modeling. Available: http://www.uvm.edu/envnr/vtcfwru/spreadsheets/occupancy/occupancy.htm.

30. Crooks K (2002) Relative sensitivities of mammalian carnivores to habitat fragmentation. Conservation Biology 16: 488–502.

31. Thibault K, Ernest S, White E, Brown J, Goheen J (2010) Long-term insights into the influence of precipitation on community dynamics in desert rodents. Journal of Mammalogy 91: 787–797.

32. Muntifering JR, Dickman AJ, Perlow LM, Hruska T, Ryan PG, et al. (2006) Managing the matrix for large carnivores: a novel approach and perspective from cheetah (*Acinonyx jubatus*) habitat suitability modelling. Animal Conservation 9: 103–112.

33. Cox RL, Underwood EC (2011) The importance of conserving biodiversity outside of protected areas in mediterranean ecosystems. PloS ONE 6: e14508.

4

The Effect of Urbanization on Ant Abundance and Diversity: A Temporal Examination of Factors Affecting Biodiversity

Grzegorz Buczkowski*, Douglas S. Richmond

Department of Entomology, Purdue University, West Lafayette, Indiana, United States of America

Abstract

Numerous studies have examined the effect of urbanization on species richness and most studies implicate urbanization as the major cause of biodiversity loss. However, no study has identified an explicit connection between urbanization and biodiversity loss as the impact of urbanization is typically inferred indirectly by comparing species diversity along urban-rural gradients at a single time point. A different approach is to focus on the temporal rather than the spatial aspect and perform "before and after" studies where species diversity is cataloged over time in the same sites. The current study examined changes in ant abundance and diversity associated with the conversion of natural habitats into urban habitats. Ant abundance and diversity were tracked in forested sites that became urbanized through construction and were examined at 3 time points - before, during, and after construction. On average, 4.3 ± 1.2 unique species were detected in undisturbed plots prior to construction. Ant diversity decreased to 0.7 ± 0.8 species in plots undergoing construction and 1.5 ± 1.1 species in plots 1 year after construction was completed. With regard to species richness, urbanization resulted in the permanent loss of 17 of the 20 species initially present in the study plots. Recovery was slow and only 3 species were present right after construction was completed and 4 species were present 1 year after construction was completed. The second objective examined ant fauna recovery in developed residential lots based on time since construction, neighboring habitat quality, pesticide inputs, and the presence of invasive ants. Ant diversity was positively correlated with factors that promoted ecological recovery and negatively correlated with factors that promoted ecological degradation. Taken together, these results address a critical gap in our knowledge by characterizing the short- and long-term the effects of urbanization on the loss of ant biodiversity.

Editor: Deborah M. Gordon, Stanford University, United States of America

Funding: Grzegorz Buczkowski was supported by the Industrial Affiliates Program at Purdue University. The funders had no role in study design, data collection and analysis, decision to publish, or preparation of the manuscript.

Competing Interests: The authors have declared that no competing interests exist.

* E-mail: gbuczkow@purdue.edu

Introduction

Urbanization is a major threat to biodiversity [1–5] and is responsible for species extinctions and biotic homogenization. The disturbance created by urbanization destroys the habitat of a wide array of unique endemic species and often creates an attractive habitat for relatively few species able to adapt to urban conditions [6]. This may lead to biotic homogenization whereby the genetic, taxonomic, or functional similarity of regional biota increases over time [7–8]. Emerging evidence suggests that biotic homogenization is occurring in a variety of ecosystems [4,9–10] with important ecological and evolutionary consequences [11]. As urbanization spreads rapidly across the globe, a key question for urban ecology and a basic challenge for conservation is to understand how it affects biodiversity [3].

Although urbanization provides excellent opportunities to test the effects of habitat alteration, degradation, and fragmentation on ecological communities, urbanizing landscapes have received relatively little attention with most research efforts being focused on more natural processes [12–14]. Studying ecological processes in urban environments is a relatively new direction in ecology [15–16]. To date, most studies have focused on birds [17–19] and we know much less about other vertebrates and very little about arthropod communities [20]. A recent review [19] of invertebrates from a variety of urbanized habitats reports that diversity decreased in 64% of studies, increased in 30% of studies, and remained unchanged in 6% of studies with the losses driven mostly by native species extinction and the gains by non-native species additions. Such variability in findings likely reflects the wide range of taxa and functional groups represented by the invertebrates.

Arthropods are excellent candidates for studying the effects of urbanization because they perform a wide range of ecosystem services and serve as important bioindicators of ecological change [19,21–22]. Ants in particular are important because they represent a variety of trophic levels, have relatively short generation times and therefore respond quickly to environmental change, and they are important economic components of human-altered habitats [19,20,22–25]. Ants are a remarkable example of animals adapting to urban habitats [26–27] and the ecological and economic impacts of ants, especially invasive species, are well documented [28–29]. Ants are also abundant, highly diverse, and easy to collect and identify [30–32].

Numerous studies have examined the effect of urbanization on ant species richness [24,33–39] and most studies implicate urbanization as the major cause of extinctions [28,36,40]. However, very few studies have identified an explicit connection between urbanization and biodiversity loss. Doing so requires long-term observations to document temporal changes in species inventories over time, and such data is logistically difficult to obtain and typically unavailable. As a result, the impact of urbanization is typically inferred indirectly by comparing species diversity along spatial gradients, typically by examining diversity along urban-rural gradients at a single time point [24,27,35,41]. Urban-rural gradient studies are clearly a simplification of the complex patterns produced by urbanization [42]. A typical approach is to compare species abundance and diversity along gradients of urbanization (e.g. urban vs. rural areas, urban edge vs. inner city, urban green spaces vs. residential areas) and subsequently correlate species diversity and composition with habitat characteristics. Clearly, there are problems with this approach because the effect of urbanization is confounded by numerous extraneous factors and the evidence for the role of urbanization is only correlational rather than direct. For example, subtle differences in microhabitat characteristics between intact and urbanized sites might contribute to the observed differences in diversity making it difficult to isolate the role of urbanization. A different approach is to focus on the temporal rather than the spatial aspect and perform "before and after" studies where species diversity is carefully cataloged over time in the same sites and the role of habitat disturbance is examined directly over time.

The current study is a large-scale, long-term, survey-based examination of changes in ant abundance and diversity associated with the conversion of natural habitats into urban habitats. It represents a novel approach to studies on the effect of urbanization on native communities because it emphasizes the temporal component (i.e. comparing biodiversity in the same site before and after disturbance) rather than the spatial component (i.e. comparing biodiversity across disturbance gradients at a single time point). Therefore, it allows the opportunity to isolate the effect of urbanization on biodiversity without the confounding effects of other environmental factors. The study had two main objectives. The first objective was to track changes in ant abundance and diversity in forested sites that became urbanized through residential construction. Ant diversity was examined at 3 distinct time points - before, during, and after construction - with the prediction that biodiversity would decline as a result of disturbance. The second objective was to examine the recovery of ant fauna in developed residential lots based on several factors such as time since construction, neighboring habitat quality, pesticide inputs, and the presence of dominant, invasive ant species. The prediction was that ant diversity would be positively correlated with factors that promote ecological recovery (e.g. proximity to undisturbed sites) and negatively correlated with factors that promote ecological degradation (presence of invasive ant species, pesticide inputs). Taken together, the results of these two objectives address a critical gap in our knowledge by investigating how urbanization affects the richness and abundance of ants.

Methods

Study Sites and Research Plots

The study was carried out in The Orchard, a 94 acre (38 ha) residential development site centered at 40.44°N and 86.95°W in West Lafayette, Indiana, U.S.A. The Orchard is an abandoned apple orchard where commercial apple production ceased approximately 20 years ago and residential development begun approximately 10 years ago. The Orchard is successionally advanced with dense shrub understory (dominant species include bush honeysuckle, *Lonicera spp.* and autumn olive, *Elaeagnus umbellata*), thick herbaceous ground cover, and numerous hardwood trees overtopping the naturalized apple trees. Old, abandoned apple orchards provide extremely important habitat to a myriad of species that require early successional habitat, especially insects, birds, reptiles, and small mammals [43]. The site is comprised of approximately 145 lots that are available for individual purchase prior to construction. The majority of lots are approximately 1,000 square meters (0.25 acres) in size. Developed lots consist of housing of various age constructed within the last 10 years. These houses are interspersed among undeveloped orchard lots. Prior to construction, the lots are cleared of all vegetation (with the exception of a few desirable hardwood saplings) and the topsoil is removed. After construction, the landscaping is installed consisting mainly of sodded lawn, mulched flower beds, and additional landscape trees.

Effect of House Construction on Ant Abundance and Diversity

The impact of house construction activities on ant abundance and diversity was examined in 15 lots throughout The Orchard. For each lot, ant abundance and diversity were estimated at 3 time points: before, during, and after construction (Fig. 1). The houses were sampled approximately 6–12 months prior to construction, during construction (typically within 1–2 weeks after construction begun), and 1 year after construction was completed. A combination of baiting and visual searching was used at all lots to estimate ant abundance and diversity. Ten note cards baited with a blend of peanut butter and corn syrup (50:50, v:v) [44] were placed on the ground in each lot. Prior to construction, the cards were placed uniformly throughout the plot. After construction, the cards were placed around the foundation of the house and throughout the yard. The bait cards were collected 2 hours after placement, placed in individual plastic bags, and the ants were later identified to species in the lab. In addition to baiting, visual searches were conducted throughout the sites. This involved turning over rocks and logs, inspecting debris on the ground, and looking for signs of ant activity on the ground. The searching effort was standardized across plots by having 2 people sample each plot for approximately 15 minutes. Ant abundance (the percentage of bait stations that had at least 1 ant on it) and ant diversity (the total number of ant species present) were then calculated for each site.

Ant Fauna Recovery in Residential Lots

The long-term recovery of ant fauna in developed lots was examined by sampling ant abundance and diversity at houses of various age. The objective was to examine the relationship between house age (years since construction) and ant abundance and diversity. The sampling procedure was as above, with 10 bait cards per lot. In total, 51 houses were sampled. All houses were single family dwellings with traditional landscaping that included ornamental plants around the foundation, mulched patches of shrubs or trees not adjacent to the house, and a mowed lawn covering majority of the yard. In addition, the importance of various factors that could potentially affect ant communities was investigated. The homeowners were surveyed regarding pesticide use around homes to determine the potential effect of chemical insect control on ant presence. They were asked whether any ant control products had been used on the property in the last 3 years and responses were recorded as either yes (1) or no (0). In addition, the type of property bordering the sampled houses on either side

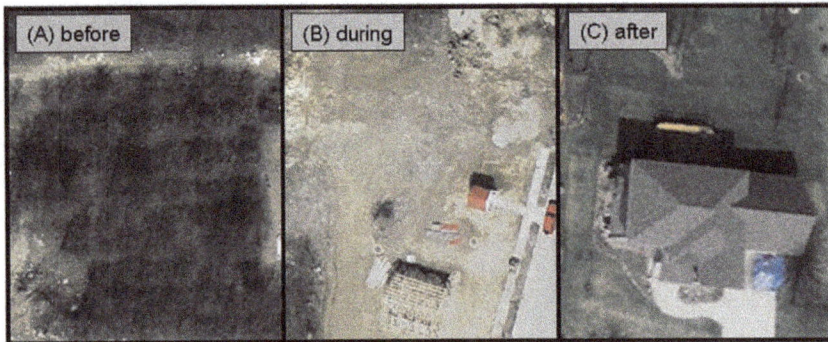

Figure 1. Aerial photos of research plots representative of each stage of habitat disturbance. (A) before construction: naturalized apple orchard, (B) during construction: trees and top soil are removed to prepare the ground for construction, (C) after construction: sodded lawn and landscape trees are installed, the house covers majority of the plot.

was recorded. The surrounding lots were categorized either as developed (another house) or undeveloped (orchard). Undeveloped orchard lots serve as refugia for a variety of ant species and could potentially serve as a source of ants for nearby developed lots. For data analysis, houses surrounded by two undeveloped lots received a value of 0, surrounded by 1 developed lot and 1 undeveloped lot a value of 1, and surrounded by 2 developed lots a value of 2.

Statistical Analysis

The effect of house construction on ant abundance and diversity was estimated by using an ANOVA test (PROC GLM procedure) in SAS 9.2 [45] with time (before, during, after) as an independent variable and abundance or diversity as dependent variables. The ANOVA analyses were followed by post-hoc Tukey's HSD tests to separate the means. The relationship between house age and the percentage of developed adjacent lots was examined using simple linear regression. A homogeneity of slopes ANCOVA model was used to determine if the relationship between house age and ant abundance and diversity parameters varied depending on the history of insecticide use around the structure and a separate slopes ANCOVA model was then used to describe the relationship between house age and ant abundance and diversity parameters separately according to insecticide history.

Results

Effect of House Construction on Ant Abundance and Diversity

The degree of environmental disturbance (before, during, or after construction) had a significant effect on ant abundance (ANOVA, $F = 48.35$, df = 2, $P < 0.0001$). In undisturbed orchard plots, the ants were present on 6.3 ± 2.0 bait stations. In contrast, the ants were present on only 0.9 ± 1.1 bait stations in plots undergoing construction (85% decline; t test, $t = 9.81$, df = 28, $P < 0.0001$) and 2.1 ± 1.4 bait stations in plots 1 year after the completion of construction (64% decline; t test, $t = 9.81$, df = 28, $P < 0.0001$). The degree of environmental disturbance also had a significant effect on ant diversity (ANOVA, $F = 50.86$, df = 2, $P < 0.0001$). On average, 4.3 ± 1.2 unique species were detected in undisturbed orchard plots prior to construction (Table 1). Ant diversity decreased to only 0.7 ± 0.8 species in plots undergoing construction (84% decline; t test, $t = 9.81$, df = 28, $P < 0.0001$) and 1.5 ± 1.1 species in plots 1 year after the completion of construction (63% decline; t test, $t = 9.81$, df = 28, $P < 0.0001$). With regard to species richness (S), a total of 20 ant species were detected in

undisturbed orchard plots (Table 1, Figure 2). Construction activities resulted in the permanent loss of 17 species (85% decline) and only 3 species were present right after construction was completed and 4 species were present 1 year after construction was completed. No statistical difference was detected in ant abundance or diversity between experimental plots during and after construction (Table 1). Species identity for the ants discovered in the experimental plots is shown in Figure 2. Of the 20 species detected in undisturbed plots, 3 were relatively abundant prior to construction, persisted during construction, and experienced a relatively fast recovery: *Lasius neoniger* (LNE), *Tetramorium caespitum* (TCA), and *Tapinoma sessile* (TSE). Other species, such as *Crematogaster cerasi* (CCE) or *Prenolepis imparis* (PIM) were relatively common prior to disturbance, but were unable to recover once construction was completed and were absent from the plots. Still, other species such as *Solenopsis molesta* (SMO) were relatively rare, but were able to persist and recover.

Ant Fauna Recovery in Residential Lots

Ant communities were sampled around 51 houses of varying age and a total of 22,560 ants belonging to 7 species were detected (Figure 3). The ants were present on 116 out of 510 (23%) bait stations placed around the houses. Ant activity, as indicated by the number of bait stations with ants present, ranged between 0 and 10 (out of 10 bait stations) and averaged 5.0 ± 2.8 baits per house.

Table 1. Ant faunal diversity in plots before, during, and after construction.

sampling time	ant abundance[1]	ant diversity[2]	S[3]	D[4]	H[5]	J[6]
before construction	6.3±2.0 a	4.3±1.2 a	20 a	0.11	1.05	0.81
during construction	0.9±1.1 b	0.7±0.8 b	3 b	0.26	0.47	0.99
after construction	2.1±1.4 b	1.5±1.1 b	4 b	0.27	0.55	0.91

[1]Mean (± SD) number of bait stations with ants present (out of 10 stations, averaged over 15 plots).
[2]Mean (± SD) number of ant species discovered (averaged over 15 plots).
[3]Ant species richness (the total number of ant species discovered in experimental plots).
[4]Simpson index.
[5]Ant species diversity (Shannon index).
[6]Ant species equitability.
Numbers within columns followed by the same letter are not different based on a Tukey test ($P = 0.05$).

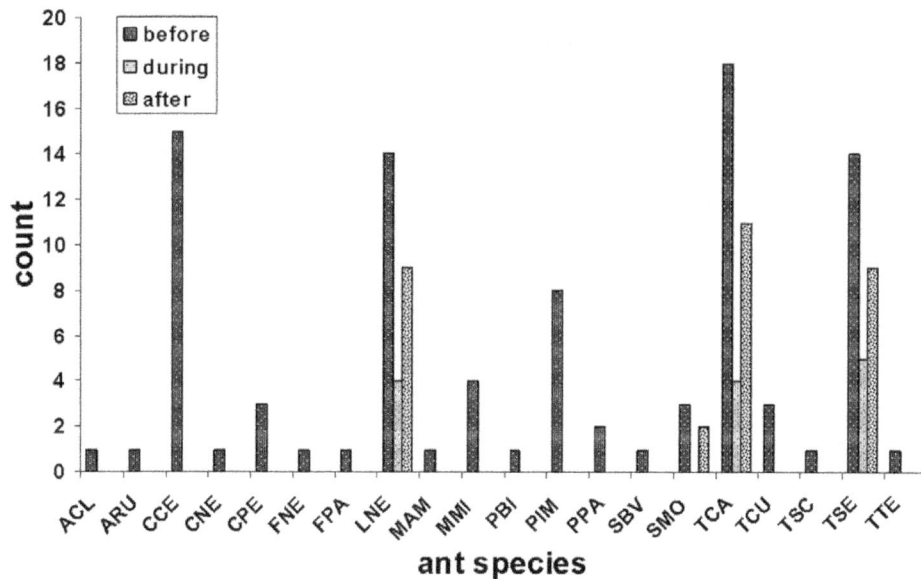

Figure 2. Ant species abundance before, during, and after construction. For each time category, count is the total number of baits stations where a given species was detected. In total, 20 species were detected in the study: ACL (*Acanthomyops claviger*), ARU (*Aphaenogaster rudis*), CCE (*Crematogaster cerasi*), CNE (*Camponotus nearcticus*), CPE (*Camponotus pennsylvanicus*), FNE (*Formica neogagates*), FPA (*Formica palleidefulva*), LNE (*Lasius neoniger*), MAM (*Myrmica americana*), MMI (*Monomorium minimum*), PBI (*Pheidole bicarinata*), PIM (*Prenolepis imparis*), PPA (*Paratrechina parvula*), SBV (*Stenamma brevicorne*), SMO (*Solenopsis molesta*), TCA (*Tetramorium caespitum*), TCU (*Temnothorax curvispinosus*), TSC (*Temnothorax schaumii*), TSE (*Tapinoma sessile*), TTE (*Temnothorax texanus*).

The number of species per house ranged between 0 and 4 and averaged 2.3±1.0 species. Pavement ants, *Tetramorium caespitum* (TCA) dominated the counts (Figure 3). They comprised 75% of all ants encountered at the bait stations and were present at 44/51 (85%) of the houses. Odorous house ants, *Tapinoma sessile* (TSE) were the second most frequently encountered ant. They comprised 18% of all ants encountered at the bait stations and were present at 33/51 (65%) of the houses. The remaining 5 species accounted for the remaining 7% of the ants. A significant correlation was detected between house age and ant abundance (Figure 4A, Pearson's correlation, $r = 0.79$, $P < 0.0001$) suggesting that ant counts increase around older houses. However, this increased abundance is mainly due to high numerical presence of a few species able to persist in urban environments, not high species diversity. Likewise, a significant correlation was detected between house age and ant diversity (Fig. 4B, Pearson's correlation, $r = 0.51$, $P = 0.0001$) and between house age and the number of baits with ants present (Fig. 4C, Pearson's correlation, $r = 0.82$, $P < 0.0001$). No significant relationship was detected between insecticide use and the total ant count around homes (ANCOVA, $F = 0.02$, df = 1, $P = 0.894$). Of the 51 houses that participated in the study, 21 (41%) had used some form of outdoor pest control in the last 3 years, and 30 (59%) did not (t test, $t = -1.85$, df = 49, $P = 0.071$). Interestingly, the average number of ants found around homes that used pesticides, 520±364, was higher than the number of ants found around homes that did not, 331±354, although not significantly (t test, $t = 2.01$, df = 44, $P = 0.070$). A significant negative correlation was detected between the total number of *T. caespitum* and the total number of all other ant species present around the houses (Pearson's correlation, $r = 0.79$, $P < 0.0001$) suggesting that *T. caespitum* may negatively affect native ant diversity in urban environments. Proximity of developed lots to undisturbed lots did not. Of the 51 houses included in the study, 10 (20%) were surrounded by 2 undeveloped lots (307±343 ants present), 16 (31%) were surrounded by 1 developed lot and 1 undeveloped lot (271±314 ants present), and 25 (49%) were surrounded by 2 developed lots (606±350 ants present).

Effect of House Age on Ant Abundance and Diversity

As house age increased, so did the probability that adjacent lots were developed ($F = 15.1$, df = 1, 49, $P = 0.0003$). The influence of house age on ant abundance and diversity varied according to the history of insecticide use around the structure ($F = 1.8$, df = 1, 41, $P = 0.11$) with significant interactions between house age and insecticide use observed for the total number of ant species recorded, H, and J (Table 2). On properties with no recent history of insecticide use, the total number of ant species, H, and J all increased as house age increased (Table 3, Fig. 4). On properties with a recent history of insecticide use, there was no significant relationship between house age and any of these parameters. Total ant numbers, total invasive ant numbers, total ants at bait stations, and D all increased significantly with house age regardless of insecticide use.

Discussion

This study represents a novel approach to studies on the effect of urbanization on native communities because it emphasizes the temporal component (i.e. comparing biodiversity in the same site before and after disturbance) rather than the spatial component (i.e. comparing biodiversity across disturbance gradients at a single time point). Previous studies were largely a simplification of the complex patterns produced by urbanization [19,42,46] because they largely failed to account for any climatic, geographic, historical, or spatial scale factors that were unique to each site. Habitat and landscape factors are known to be important determinants of ant communities [23,38–39,47–49] and comparisons of different sites along urban gradients carry a significant bias. The current study allowed a unique opportunity to document the process of urbanization through time at a single location,

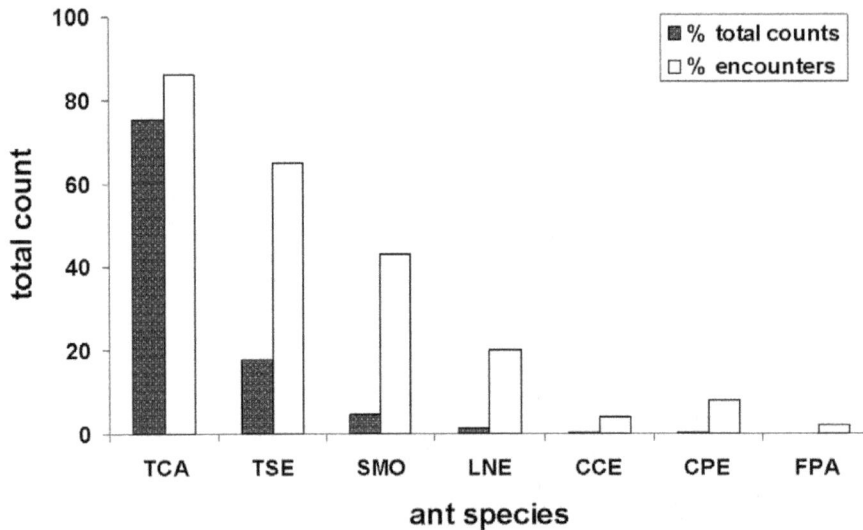

Figure 3. The relative abundance of the seven ant species found in post-construction plots. (A) relative abundance expressed as the percentage of the total number of ants collected at the bait stations, (B) relative abundance expressed as the percentage of the homes where each species was encountered. In both (A) and (B), $n = 510$ bait stations; 51 houses with 10 bait stations per house. Species names as in Figure 2.

avoiding the potentially confounding effects of other location-related factors associated with many similar experiments. Biodiversity was tracked in natural sites that subsequently experienced urban disturbance and a profound effect of urbanization was discovered. Urbanization resulted in the permanent loss of 17 of the 20 species initially present in the study plots and recovery was slow as indicated by the lack of significant improvement in species richness 1 year after construction was completed. Environmental disturbance had a severely negative effect on ant abundance and diversity which declined by 85% and 84%, respectively. Species richness also experienced a significant decline. This suggests that previous studies, which focused mainly on the spatial component and discovered relatively minor diversity losses, may have underestimated the impact of urbanization.

Urbanization creates intensively managed, homogenous landscapes and forces native species that adapt to a relatively uniform environment that is often radically different from the undeveloped habitat. Under such scenario, many ecological specialists become locally extinct and are replaced by a few ecological generalists that are broadly adapted and able to tolerate or even benefit from human activity [6]. This may lead to biotic homogenization, where the rapid and drastic environmental change promotes the geographic reduction of some species ('losers') and the geographic expansion of others ('winners') [6]. In the current study the 'losers' were species with relatively sensitive nesting and/or feeding requirements that were unable to tolerate disturbance. The 'winners' were typically disturbance specialists that were able to tolerate disturbance and recover fairly quickly. House construction created a highly uniform disturbance where all lots were cleared of trees and topsoil. Previous results show that this type of disturbance has the greatest effect on epigeic ant species which utilize above-ground organic debris as nesting and feeding sites, and the lowest effect on hypogaeic ant species which have subterranean nests [38]. Arboreal species such as *Crematogaster cerasi* (CCE) and *Camponotus pennsylvanicus* (CPE) and cavity-nesting species such as *Temnothorax curvispinosus* (TCU) and *Monomorium minimum* (MMI) were fairly common prior to disturbance and completely absent following disturbance. In contrast, subterranean

species such as *Lasius neoniger* (LNE), *Solenopsis molesta* (SMO), and *Tetramorium caespitum* (TCA) appeared largely unaffected by the disturbance. In fact, *L. neoniger* and *T. caespitum* were frequently observed rebuilding their nests in heavily compacted, clayey subsoil soon after the lots were cleared.

Of the 20 species detected in undisturbed plots, 3 were relatively abundant before, during, and after development: cornfield ants (*Lasius neoniger*), pavement ants (*Tetramorium caespitum*), and odorous house ants (*Tapinoma sessile*). *Lasius neoniger* is the dominant open habitat species in the northeastern United States [50–51] and is common in urban areas [37,44]. It is probably best classified as an urban adapter – a species that can adapt to urban habitats, but also utilizes more natural environments. The majority of the colonies nested in turf and did not seem to be closely associated with the structures themselves.

Tetramorium caespitum is an introduced species that has spread widely across the United States and is almost invariably associated with human disturbed sites [40,52,53,54]. In a study by [37], *T. caespitum* comprised 53% of all ants collected in a highly urbanized habitat and they were the most abundant species in an ant survey conducted in West Lafayette, Indiana, approximately 2 km from the present study site [54]. In addition, [35] reported that ant richness in urban sites negatively correlated with the abundance of *T. caespitum* and [37] reported that *T. caespitum* abundance correlated negatively with tree density, indicating this species' preference for open, disturbed sites.

Tapinoma sessile is widespread throughout North America and has the widest geographic range and greatest ecological tolerance of any ant in North America [55]. It is very opportunistic and inhabits a variety of nesting sites, both natural and man-made and in urban areas it is classified a pest species [56]. Recent work demonstrated that *T. sessile* is a highly plastic species with a flexible social structure [38,57]. In natural habitats, *T. sessile* is a subdominant species comprised of small, single-queen colonies. In urban areas, *T. sessile* exhibits the characteristics common to most invasive ant species such as extreme polygyny (thousands of queens), extensive polydomy (multiple nests), and ecological dominance over native ant species [38,57–59]. Furthermore,

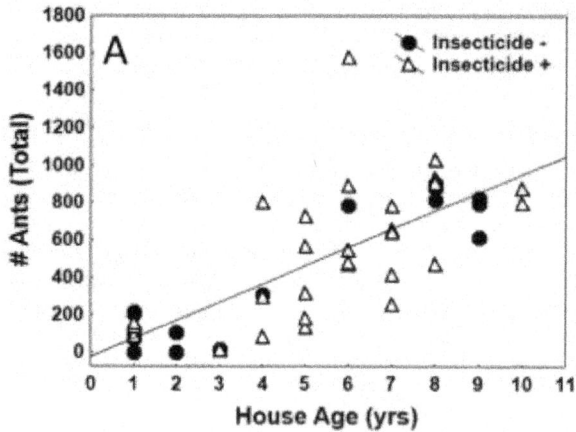

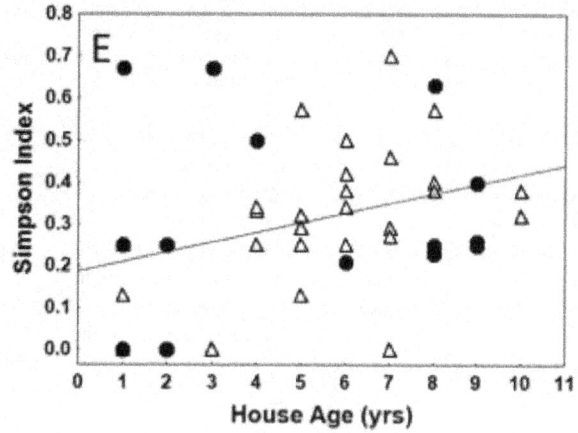

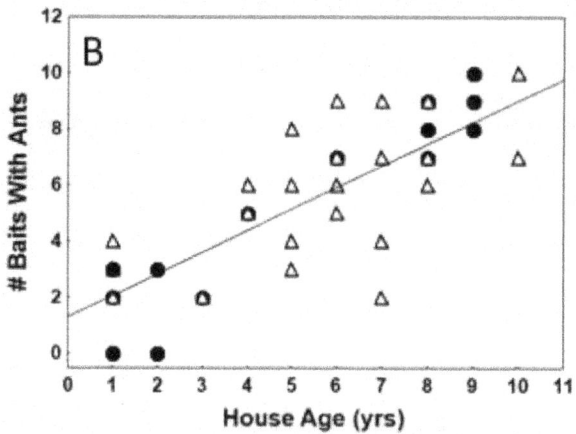

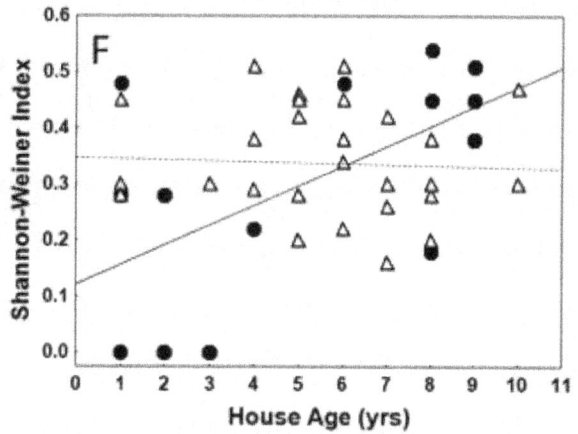

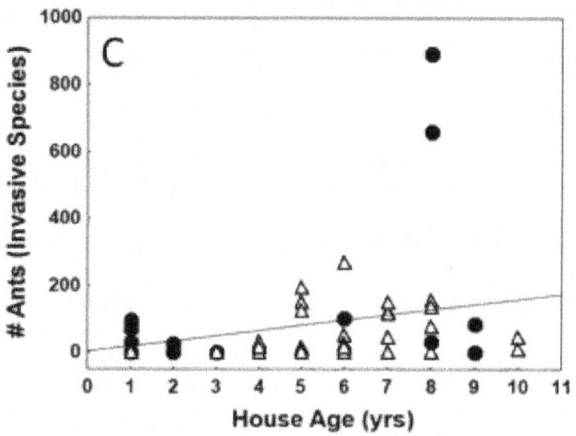

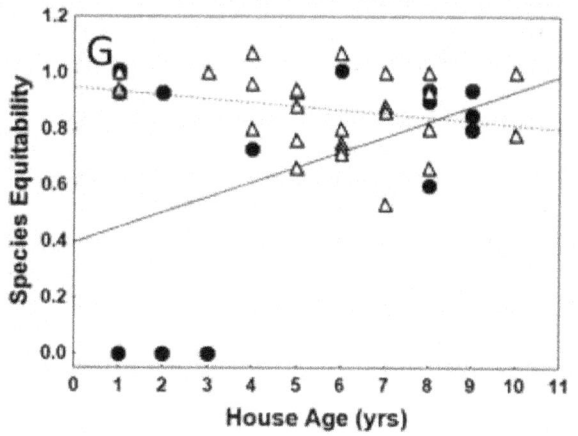

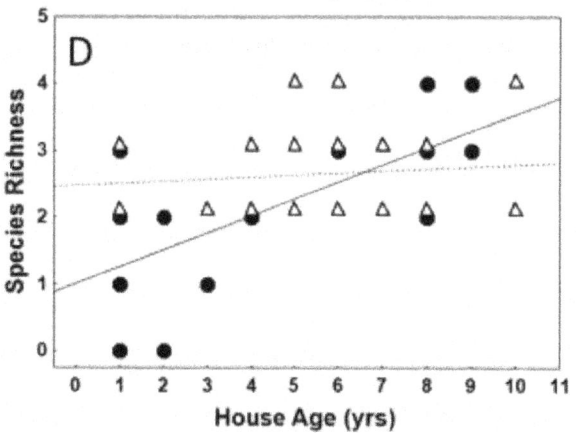

Figure 4. The relationship between house age and various metrics of ant diversity and abundance. (A) the total number of ants discovered, (B) the number of bait stations with ants present (C) the total number of invasive ants discovered, (D) ant species richness, (E) ant species diversity (Simpson), (F) ant species diversity (Shannon-Weiner) and (G) ant species equitability at each site. When two regression lines are present, separate models were necessary based on history of insecticide use.

T. sessile has been recently reported as an invasive species in human-altered habitats in Hawaii [44]. Both *T. caespitum* and *T. sessile* are best categorized as urban exploiters or species that become dependent on humans for food and shelter [59–60]. The great majority of the colonies nested in mulch beds around the foundation of the house and under concrete pathways associated with the house. The ability to exploit abundant resource subsidies offered by humans was likely the primary reason for *T. caespitum* and *T. sessile* attaining such high population densities on converted sites.

Urbanization affected species richness in a variety of ways. The majority of ant species, especially those that nested in above-ground material, were physically removed from the site when the tree cover and the topsoil were removed. Urbanization also affected species richness through the species-area effect: the negative relationship between the area of a habitat and the number of species found within that area. In all urban habitats, large expanses of impervious concrete and asphalt pavement reduce and fragment the area available for life to survive. The typical size of a residential plot in this study was approximately 1,000 square meters. A typical footprint for the house, including concrete driveway and sidewalks, was approximately 350 square meters. Therefore, the area available for nesting and foraging was reduced by approximately 35%.

Another negative impact on biodiversity is related to the severe structural simplification of vegetation in urbanized areas. Trees serve as important nesting sites for many ant species and trees colonized by honeydew-producing hemipterans provide important feeding sites for many species. During urban development, mature trees were removed and replaced with various landscaping plants once construction was completed. The remaining area was covered by a monoculture of grass to create lawns. Previous studies show that the percentage canopy cover is an important factor influencing ant species richness [37–38,41,61]. A study by [24] demonstrated that land development can significantly affect ant diversity, even in areas that retain a substantial component of native vegetation. Land development and disturbance of 30–40% appeared to be the level above which ant diversity began to decline.

The second objective examined the recovery of ant fauna in developed residential lots based on several factors such as time since construction, neighboring habitat quality, and pesticide inputs, and the presence of dominant, invasive ant species (*Tetramorium caespitum*). The prediction was that ant diversity would be positively correlated with factors that promote ecological recovery (e.g. longer post-recovery time and proximity to undisturbed sites) and negatively correlated with factors that promote ecological degradation (e.g. pesticide use and the presence of invasives).

A significant relationship was discovered between house age (time since construction) and various metrics of ant presence (the total number of ants discovered on the bait stations, number of bait stations occupied by ants). House age is an important factor driving ant abundance as older houses have had more opportunities for colonization events to occur, either via the influx of ants from neighboring plots or the arrival of alate queens from more distant areas. Older houses typically have more mature landscaping which may provide the ants with various ecological niches not available around newly constructed homes. However, older houses were also more likely to be surrounded by developed (built-on) lots, making the possibility of emigration from adjacent undisturbed patches less likely. Not surprisingly, ants largely experienced a numerical recovery over time (as indicated by the total number of ants present on the bait stations), rather than a rebound in species richness. Older houses had significantly more ants relative to newer houses, but the increase was driven mostly by the presence of the invasive *T. caespitum*. In contrast, species richness never recovered and only 7 species were found around older construction (average 2.3 ± 1.0 species per house; range 0–4), substantially lower than the 20 species present in undisturbed plots.

Houses categorized as affected by insecticide use were those that received some type of outdoor insecticide application in the last 3 years. Insecticides commonly used in such cases may range from neonicotinyls and phenyl pyrazoles with relatively long-term residual activity, to pyrethroids and organophosphates with much shorter persistence. As a result, the long-term effects of such treatments can vary widely [62]. Even targeted ant control treatments around structures typically do not result in complete elimination of all colonies, but rather offer temporary suppression

Table 2. F-statistics and degrees of freedom (df) for a homogeneity of slopes model used to examine the relationship between house age and ant abundance and diversity parameters as a function of the history of insecticide use around the structure.

factor	df	total[1]	baits[2]	invasive[3]	S	D	H'	J'
insecticide	1, 47	0.1	3.1	0.1	8.4**	0.6	9.1**	12.1**
house age	1, 47	66.1****	83.6****	5.9*	11.7**	6.5*	6.1*	2.0
insecticide×house age	1, 47	0.0	4.0	2.1	7.1*	0.7	7.5**	5.6*

*P≤0.05,
**≤P 0.01,
***≤P 0.001,
****P≤0.0001.
[1] the total number of ants discovered at 10 bait stations placed around the house.
[2] the total number of bait stations with ants present (x/10).
[3] the total number of pavement ants, *Tetramorium caespitum*, discovered at 10 bait stations placed around the house.
ant diversity parameters as in Table 1.

Table 3. Mean responses, parameter estimates and performance statistics for separate slopes models used to describe the influence of structure age and insecticide use around the structure on the density and composition of the ant community in a rural housing development established in a former apple orchard.

response variable[1]	insecticide usage	mean (±SE)	parameter estimate (±SE)	t	p	R^2
total	NA	442.16±51.61	97.68±11.01	8.87	<0.000001	0.62
baits	NA	5.00±0.39	0.77±0.08	9.79	<0.000001	0.69
invasive	NA	78.14±21.99	15.49±7.25	2.13	0.037523	0.18
S	No	1.95±0.28	0.25±0.06	4.60	0.000032	0.37
S	Yes	2.53±0.12	0.03±0.06	0.50	0.616778	–
D	NA	0.30±0.03	0.02±0.01	2.62	0.011791	0.14
H'	No	0.25±0.04	0.04±0.01	3.94	0.000270	0.31
H'	Yes	0.34±0.02	−0.01±0.01	−0.18	0.855200	–
J'	No	0.60±0.10	0.05±0.02	2.85	0.006394	0.31
J'	Yes	0.87±0.02	−0.01±0.02	−0.63	0.529624	–

*NA = not applicable as determined by homogeneity of slopes ANCOVA.
[1]response variables as in tables 1 and 2.

[54]. In some cases, insecticide treatments may actually create empty ecological niches that are then filled by invading ant species previously absent from the sites [54]. Not surprisingly, insecticide use had mixed effects on ant abundance and diversity following the establishment of a structure.

While ant abundance generally increased with house age, ant diversity (Shannon), species richness, and equitability increased with house age only at locations without a history of insecticide use. These parameters remained flat over time at locations where insecticides were recently used. However, the likelihood of future increases in ant diversity, species richness and species equitability at insecticide free sites remains questionable due to the relatively high impact of just a few data points on the regression line for this class. The positive relationship between house age, and these indices appeared to be driven largely by three or four younger structures (<3 yrs old) registering zero or very low values compared to their insecticide-treated counterparts. Although any number of factors including insecticide usage on adjacent lots, misapplication of insecticides by homeowners, or the failure of ants to recolonize these particular sites for an extended period of time following disturbance could potentially explain this pattern, such

speculation is beyond the scope of the current study. Further data collection aimed at clarifying a potential mechanism underlying these observations will be required. As such, the current study provides strong evidence that although ant abundance may recover within a few years after urbanization, gains are likely to be driven by a relatively few pest/invasive species. Recovery of ant species richness and diversity to pre-disturbance levels appears to be highly unlikely under current post-development landscape management regimes.

Acknowledgments

We thank numerous homeowners for permission to work at their properties, A. Salyer for assistance in the field, J. Datta for advice on data analysis, and T. Carroll for identifying ants.

Author Contributions

Conceived and designed the experiments: GB. Performed the experiments: GB. Analyzed the data: GB DR. Contributed reagents/materials/analysis tools: GB. Wrote the paper: GB DR.

References

1. Czech B, Krausman PR, Devers PK (2002) Economic associations among causes of species endangerment in the United States. BioScience 50: 593–601.
2. Sala OE, Chapin FS III, Armesto JJ, Berlow E, Bloomfield J, et al. (2000) Global Biodiversity Scenarios for the year 2100. Science 287: 1770–1774.
3. McKinney ML (2002) Urbanization, biodiversity, and conservation, BioScience 52: 883–890.
4. McKinney ML (2004) Measuring floristic homogenization by non-native plants in North America. Global Ecology and Biogeography 13: 47–53.
5. Vitousek P, Mooney H, Lubchenco J, Melillo J (2007) Human domination of earth's ecosystems. Science 277: 494–499.
6. McKinney ML, Lockwood JL (1999) Biotic homogenization: a few inners replacing many losers in the next mass extinction. Trends in Ecology and Evolution 14: 450–453.
7. Lockwood JL, McKinney ML (2001) Biotic homogenization: a sequential and selective process. In: Lockwood JL, McKinney M, eds. Biotic homogenization. Kluwer Academic, New York. 1–17.
8. Olden JD, Poff NL (2003) Toward a mechanistic understanding and prediction of biotic homogenization. American Naturalist 162: 442–460.
9. Rahel FJ (2000) Homogenization of fish faunas across the United States. Science 288: 854–856.
10. Rooney TP, Wiegmann SM, Rogers DA, Walker DM (2004) Biotic impoverishment and homogenization in unfragmented forest understory communities. Conservation Biology 18: 787–798.
11. Olden JD, Poff NL, Douglas MR, Douglas ME, Fausch KD (2004) Ecological and evolutionary consequences of biotic homogenization. Trends in Ecology and Evolution 19: 18–24.
12. McDonnell MJ, Pickett STA (1990) Ecosystem structure and function along urban-rural gradients: an unexploited opportunity for ecology. Ecology 71: 1232–1237.
13. Adams LW (1994) Urban Wildlife Habitats: A Landscape Perspective. University of Minnesota Press, Minneapolis, MN, USA.
14. Pickett STA, Cadenasso ML, Grove JM, Nilon CH, Pouyat RV, et al. (2001) Urban ecological systems: linking terrestrial ecological, physical, and socioeconomic components of metropolitan areas. Annual Review of Ecology and Systematics 32: 122–157.
15. Grimm NB, Grove M, Pickett STA, Redman C (2000) Integrated approaches to long-term studies of urban ecological systems. BioScience 50: 571–584.
16. Zipperer WC, Wu J, Pouyat RV, Pickett STA (2002) The application of ecological principles to urban and urbanizing landscapes. Ecological Applications 10: 685–688.

17. Marzluff MJ (2001) Worldwide urbanization and its effects on birds. In: Marzluff JM, Bowman R, Donnelly R, eds. Avian ecology in an urbanizing world. Kluwer, Norwell, Massachusetts. 19–47.

18. Chace JF, Walsh JJ (2006) Urban effects on native avifauna: a review. Landscape and Urban Planning 74: 46–69.

19. McKinney ML (2008) Effects of urbanization on species richness: a review of plants and animals. Urban Ecosystems 11: 161–176.

20. McIntyre NE (2000) Ecology of urban arthropods: a review and a call to action. Annals of the Entomological Society of America 93: 825–835.

21. Kremen C, Colwell RK, Erwin TL, Murphy DD, Noss RF, et al. (1993) Terrestrial arthropod assemblages – their use in conservation planning. Ecography 27: 57–64.

22. McIntyre NE, Rango J, Fagan WF, Faeth SH (2001) Ground arthropod community structure in a heterogenous urban environment. Landscape and Urban Planning 52: 257–274.

23. Gibb H, Hochuli DF (2002) Habitat fragmentation in an urban environment: large and small fragments support different arthropod assemblages. Biological Conservation 106: 91–100.

24. Sanford MP, Manley PN, Murphy DD (2008) Effects of urban development on ant communities: implications for ecosystem services and management. Conservation Biology 23: 131–141.

25. Stringer L, Stephens A, Suckling D, Charles J (2009) Ant dominance in urban areas. Urban Ecosystems 12: 503–514.

26. Angilletta MJ Jr, Wilson RS, Niehaus AC, Sears MW, Navas CA, et al. (2007) Urban physiology: city ants possess high heat tolerance. PLoS ONE 2: e258.

27. Menke SB, Guenard B, Sexton JO, Weiser MD, Dunn RR, Silverman J (2011) Urban areas may serve as habitat and corridors for dry-adapted, heat tolerant species; an example from ants. Urban Ecosystems 14: 135–163.

28. Holway DA, Suarez AV (2006) Homogenization ant communities in the mediterranean California: the effects of urbanization and invasion. Biological Conservation 127: 319–326.

29. Passera L (1994) Characteristics of tramp species. In: Williams DF, ed. Exotic ants: impact and control of introduced species. Westview Press, Boulder, Colorado. 23–43.

30. Alonso LE, Agosti D (2000) Biodiversity studies, monitoring and ants: an overview. In: Agosti D, Majer JD, Alonson E, Shultz TR, eds. Ants: standard methods for measuring and monitoring biodiversity. Smithsonian Institution, Washington. 1–8.

31. Andersen AN, Hoffman BD, Muller WJ, Griffiths AD (2002) Using ants as bioindicators in land management: simplifying assessment of ant community responses. Journal of Applied Ecology 39: 8–17.

32. Underwood EC, Fisher BL (2006) The role of ants in conservation monitoring: if, when, and how. Biological Conservation 132: 166–182.

33. Dauber J, Bengtsson J, Lenoir L (2006) Evaluating effects of habitat loss and land-use continuity on ant species richness in seminatural grassland remnants. Conservation Biology 20: 1150–1160.

34. Clarke KM, Fisher BL, LeBuhn G (2008) The influence of urban park characteristics on ant (Hymenoptera, Formicidae) characteristics. Urban Ecosystems 11: 317–334.

35. Uno S, Cotton J, Philpott SM (2010) Diversity, abundance, and species composition of ants in urban green spaces. Urban Ecosystems 13: 425–441.

36. Thompson B, McLachlan S (2007) The effects of urbanization on ant communities and myrmecochory in Manitoba, Canada. Urban Ecosystems 10: 43–52.

37. Pecarevic M, Danoff-Burg J, Dunn RR (2010) Biodiversity on Broadway – enigmatic diversity of the societies of ants (Formicidae) on the streets of New York City. PLoS ONE, 5(10): e13222. doi:10.1371/journal.pone.0013222.

38. Buczkowski G (2010) Extreme life history plasticity and the evolution of invasive characteristics in a native ant. Biological Invasions, 12: 3343–3349.

39. Buczkowski G (2011) Suburban sprawl: environmental features affect colony social and spatial structure in the black carpenter ant, *Camponotus pennsylvanicus*. Ecological Entomology 36: 62–71.

40. Lessard JP, Buddle CM (2005) The effects of urbanization on ant assemblages (Hymenoptera: Formicidae) associated with the Molson Nature Reserve, Quebec. Canadian Entomologist 137: 215–225.

41. Toennisson TA, Sanders NJ, Klingeman WE, Vail KM (2011) Influences on the structure of suburban ant (hymenoptera: Formicidae) communities and the abundance of *Tapinoma sessile*. Environmental Entomology 40: 1397–1404.

42. Alberti M, Botsford E, Cohen A (2001) Quantifying the urban gradient: linking urban planning and ecology. In: Marzluff JM, Bowman R, Donnelly R, eds. Avian ecology in an urbanizing world. Kluwer, Norwell, Massachusetts. 68–85.

43. Wilson JM (2006) Managing abandoned orchards and apple trees for wildlife. In: Oehler JD, Covell DF, Capel S, Long B, eds. Managing grasslands, shrublands, and young forests for wildlife. The Northeast Upland Habitat Technical Committee, Massachusetts Division of Fisheries and Wildlife. 51–57.

44. Buczkowski G, Kruskelnycky P (2012) The odorous house ant, *Tapinoma sessile* (Hymenoptera: Formicidae), as a new temperate-origin invader. Myrmecological News 16: 61–66.

45. SAS Institute (2008) SAS/STAT guide for personal computers, version 9.2. SAS Institute, Cary, NC.

46. Hahs AK, McDonnell MJ (2006) Selecting independent measures to quantify Melbourne's urban-rural gradient. Landscape and Urban Planning 78: 435–448.

47. Wang C, Strazanac JS, Butler L (2001) Association between ants (Hymenoptera: Formicidae) and habitat characteristics in oak-dominated mixed forests. Environmental Entomology 30: 842–847.

48. Lassau SA, Hochuli DF (2004) Effects of habitat complexity on ant assemblages. Biological Conservation 106: 91–100.

49. Philpott SM, Perfecto I, Vandermeer J (2006) Effects of management intensity and season on arboreal ant diversity and abundance in coffee agrosystems. Biodiversity Conservation 15: 139–155.

50. Wilson EO (1955) A monographic revision of the genus *Lasius*. Bulletin Museum of Comparative Zoology Harvard 113: 1–201.

51. Buczkowski G (2012) Colony spatial structure in polydomous ants: complimentary approaches reveal different patterns. Insectes Sociaux 59: 241–250.

52. Coovert GA (2005) The ants of Ohio (Hymenoptera: Formicidae). Ohio Biological Survey, Inc. Columbus, OH. 196 p.

53. Steiner FM, Schlick-Steiner BC, VanDerWal J, Reuther KD, Christian D, et al. (2009) Combined modeling of distribution and niche in invasion biology: a case study of two invasive *Tetramorium* ant species. Diversity and Distributions 14: 538–545.

54. Scharf ME, Ratliff CR, Bennett GW (2004) Impacts of residual insecticide barriers on perimeter-invading ants, with particular reference to the odorous house ant, *Tapinoma sessile*. Journal of Economic Entomology 97: 601–605.

55. Fisher BL, Cover SP (2007) Taxonomic descriptions. In: Fisher BL, Cover SP, eds. Ants of North America: a guide to the genera. University of California Press. 145–146.

56. Thompson RC (1990) Ants that have pest status in the Unites States. In: Vander Meer RK, Jaffe K, Cedeno A, eds. Applied Myrmecology: A World Perspective. Westview Press, Boulder, Colorado. 51–67.

57. Menke SB, Booth W, Dunn RR, Schal C, Vargo EL, et al. (2010) Is it easy to be urban? Convergent success in urban habitats among lineages of a widespread native ant. PLoS ONE, 5: e9194. doi:10.1371/journal.pone.0009194.

58. Buczkowski G, Bennett GW (2006) Dispersed central-place foraging in the polydomous odorous house ant, *Tapinoma sessile* as revealed by a protein marker. Insectes Sociaux 53: 282–290.

59. Buczkowski G, Bennett GW (2008) Seasonal polydomy in a polygynous supercolony of the odorous house ant, *Tapinoma sessile*. Ecological Entomology 33: 780–788.

60. Meissner HE, Silverman J (2001) Effects of aromatic cedar mulch on the Argentine ant and the odorous house ant (Hymenoptera: Formicidae). Journal of Economic Entomology 94: 1526–1531.

61. Yasuda M, Fomito K (2009) The contribution of the bark of isolated trees as habitat for ants in an urban landscape. Landscape and Urban Planning 92: 276–281.

62. Peck DC (2009) Comparative impacts of white grub (Coleoptera: Scarabaeidae) control products on the abundance of non-target soil-active arthropods in turfgrass. Pedobiologia 52: 287–299.

5

Quasi-Double-Blind Screening of Semiochemicals for Reducing Navel Orangeworm Oviposition on Almonds

Kevin Cloonan[1], Robert H. Bedoukian[2], Walter Leal[1]*

1 Honorary Maeda-Duffey Laboratory, University of California Davis, Davis, California, United States of America,, 2 Bedoukian Research Inc., Danbury, Connecticut, United States of America

Abstract

A three-step, quasi-double-bind approach was used as a proof-of-concept study to screen twenty compounds for their ability to reduce oviposition of gravid female navel orangeworm(NOW), *Ameylois transitella* (Lepidoptera: Pyralidae). First, the panel of compounds, whose identity was unknown to the experimenters, was tested by electroantennogram (EAG) using antennae of two-day old gravid females as the sensing element. Of the twenty compounds tested three showed significant EAG responses. These three EAG-active compounds and a negative control were then analyzed for their ability to reduce oviposition via small-cage, two-choice laboratory assays. Two of the three compounds significantly reduced oviposition under laboratory conditions. Lastly, these two compounds were deployed in a field setting in an organic almond orchard in Arbuckle, CA using black egg traps to monitor NOW oviposition. One of these two compounds significantly reduced oviposition on black egg traps under these field conditions. Compound 9 (later identified as isophorone) showed a significant reduction in oviposition in field assays and thus has a potential as a tool to control the navel orangeworm as a pest of almonds.

Editor: Raul Narciso Carvalho Guedes, Federal University of Viçosa, Brazil

Funding: This work was supported by Bedoukian Research Inc. (BRI) under a research agreement with the University of California-Davis (number 201118886). Under the terms of the UC Davis-BRI agreement, the findings were disclosed to the sponsor prior to publication. The funder (BRI) had no role in study design, data collection and analysis, decision to publish, or preparation of the manuscript

Competing Interests: One of the authors (RHB) is affiliated with Bedoukian Research Inc. - a funder of this project. RHB is the President of Bedoukian Research Inc. RHB selected the compounds to be tested, KC and WSL (UC Davis) designed the indoor and field experiments, and the three investigators discussed the results.

* E-mail: wsleal@ucdavis.edu

Introduction

Allelochemicals eliciting oviposition repellence and deterrence have been extensively studied for agriculturally important Lepidopteran pest species [1–13]. Several of these studies focused primarily on other Pyralid moth species [14–17].

The Navel Orangeworm (NOW), *Ameylois transitella*, (Walker) is the most serious insect pest [18–20] of the $3.6 billion dollar almond industry in California [21]. First instar larvae tunnel through the almond hull and into the nutmeat leaving behind larval frass and webbing as they develop into subsequent larval instars [22–25]. In addition to this direct feeding damage NOW infestation may also lead to infection of *Aspergillus* spp., which in turn produces aflatoxins [26–28].

Current management for NOW focuses primarily on winter sanitation, early harvest, and in-season insecticide applications [29,30]. Other integrated pest management (IPM) practices include the use of biological control agents and mating disruption [27,31–34]. Currently the combined effects of NOW infestation may greatly exceed 1% damage to the almond industry [27]. In an industry valued around $3.6 billion dollars, a 1% reduction results in millions of dollars of loss. Since 2002 the industry standard for NOW damage was less than 2% [19]. New approaches for control may be considered for a more robust IPM program of this most serious insect pest of almonds. Oviposition repellents and

deterrents may provide, in conjunction with current IPM strategies, further control of the NOW in almonds.

In this proof of concept study, we screened a panel of 20 compounds, including known insect repellents, to determine if any of them would cause reduction of NOW oviposition under field conditions. This was a blind approach as we did not know the identity of the test compounds until the end of the experiments. Given that evaluation of 20 compounds under field conditions is tedious, time-consuming, and costly, we devised a three-step approach. We reasoned that active compounds would be detected by antennae of NOW gravid female moths [35]. Thus, non-electroangennogram (EAG) active compounds could be ruled out from an intermediate step: indoor two-choice behavioral bioassay. The field evaluation would then be performed with only a handful of promising compounds. As described here, we were able to reduce the panel to 3 compounds by EAG, then to 2 compounds by indoor behavioral bioassay, and finally tested 2 compounds in the field, one of which has practical potential applications in IPM strategies to control NOW populations.

Materials and Methods

Insects

The *A. transitella* moths used in this study were from a 2-year-old laboratory colony maintained at UC Davis. The UC Davis colony was initiated with moths kindly provided by Dr. Charles Burks, USDA-ARS (United States Department of Agriculture - Agricultural Research Service, Parlier, CA, USA) from his colony, which in turn was founded in 2005 [36]. At UC Davis larvae were reared in 1.5-L glass jars on a wheat bran, brewer's yeast, honey, and Vanderzant vitamins (Sigma-Aldrich, St. Louis, MO, USA) diet [37]. Jars were filled with 300 ml of diet to which ca. 300 eggs were added. Cultures were maintained in growth chambers (Percival Scientific, Perry, IA, USA) at 27°C, 70% RH, and a 16:8 h (light:dark) photo regime. For colony maintenance newly emerged male and female moths were separated (ca. 50 males and ca. 50 females) into 12×12×5 cm plastic containers (Rubbermaid, Fairlawn, OH, USA) and lined at the bottom with one layer of moist paper towels and lined at the top with one layer of dry paper towels (Thirsty Ultra Absorbent, 27.9×27.9 cm; Safeway, Phoenix, AZ, USA) and left in rearing conditions for 72 h. After 72 h the top sheet containing red fertilized eggs was washed in a 10% formaldehyde solution for 15 min and rinsed with double-distilled water and allowed to air dry overnight. These eggs were then used for mass colony rearing.

Electroantennogram (EAG) Recordings

For EAG assays last instar male and female larvae were separated into 1.5-L glass jars filled with 50 ml of artificial diet and allowed to pupate. Eclosed males and females of the same age were separated into 12×12×5 cm plastic containers (Rubbermaid, Fairlawn, OH, USA) and allowed to copulate. Copulation usually occurred the night following eclosion [20]. Mated pairs who mated on the first night after eclosion were separated into individually capped culture tubes (17 mm×10 cm; Fisher Scientific) and allowed to uncouple. Males were discarded. Female NOW moths are gravid and able to oviposit fertile eggs 24 h after mating [38]. For this reason two-day old mated females were used for all EAG and laboratory assays. We chose to use gravid females in all assays since this is the physiological state of females we would target in a field setting. The antennae of these females were excised and positioned on a fork electrode using an electrolytic gel and connected to an EAG Probe with an internal gain of 10×(Syntech, Kirchzarten, Germany). The EAG signals were recorded and analyzed with EAG 2000 software (Syntech). Antennal preparations were held in a constant stream of humidified air and stimuli delivery procedures were as previously described [39]. Briefly, stimulus pulses were delivered at a rate of 4 ml/s for 500 ms from a 5 ml polypropylene syringe containing a 2 mm strip of filter paper (70 mm diameter, Whatmanone Qualitative, GE Healthcare, Piscataway, NJ) impregnated with 10 µl of a test or control compound and recorded for 10 s. Each antennal preparation was stimulated with a battery of twenty test compounds at the same concentration: 10 µg/µl, 1 µg/µl, 0.1 µg/µl) and hexane as a control. The order of stimulation was randomized and a gap of 1 min was allowed between stimulations. Those antennal preparations that showed no response to our hexane controls were discarded and not included in our analysis. For each of the above concentrations three antennal preparations from three different females from three different cohorts were used.

Small-Cage, Two-Choice Laboratory Oviposition Assays

Two-choice oviposition assays were performed under laboratory conditions and carried out in 30 cm×30 cm×30 cm green mesh cages (Bioquip, CA, USA). Each cage contained two-day old mated females ($N = 20$), isolated as above, and partitioned with a cotton ball soaked in a 10% sucrose solution on the floor of the cage. To monitor oviposition two black egg traps (Pherocon IV NOW, Black, Trece Inc., Adair, OK) were hung in opposite

corners of each cage ca. 30 cm apart [36,40]. These egg traps attract gravid female moths utilizing almond meal and almond oil as an attractant. The traps are lined with vertical grooves acting as substrate for the gravid moths to oviposit on. All black egg traps contained 1 g of larval diet to focus female oviposition [40]. Though no-choice assays have shown that in the absence of a preferred oviposition substrate female moths will oviposit on black egg traps alone [36], we wanted to focus our study on the effects these compounds have on the olfactory system of gravid female moths. Each black egg trap contained a 2 mm strip of filter paper (70 mm diameter, Whatmanone Qualitative, GE Healthcare, Piscataway, NJ) impregnated with either a control or a test compound. For all replicates control strips were impregnated with 10 µl of hexane. Test compound strips were impregnated 1 mg of test compound. We chose to use 1 mg of test compounds after several preliminary laboratory oviposition assays. These assays revealed the minimum dose eliciting a reduction in oviposition to be 1 mg. For each test compound 4 cages, each with 20 mated females, from three different cohorts of adult moths were tested for a total of 12 cages and 240 gravid females analyzed per test compound. To prevent any positional effects egg traps in all 4 cages for each trial were rotated clockwise to occupy a different location in each cage. All trials were setup 4 h before the scotophase and allowed to run for the length of the scotophase. All black egg traps were collected 2 h into the photophase and the number of eggs on each trap was counted. After each trial females were discarded and black egg traps, green mesh, and metal cage supports were then cleaned with hot water and Alconox, soaked in a 70% ethanol solution and allowed to air dry.

Field Oviposition Assay

The almond orchard for this experiment was located at the Nickels Soils Lab (Arbuckle, CA, USA, University of California Cooperative Extension). The ca. 6 acre organic orchard chosen contained Nonpareil and Fritz varieties at a 3:1 ratio with ca. 30 trees/row. This orchard was adjacent to a 23 years old ca. 20 acre conventional orchard containing: 33.3% Nonpareil, 33.3% Butte, 16.7% Carmel and 16.7% Monterey varieties. This field site was chosen for proximity to Davis, CA as well as a history of NOW infestation and little chemical control (personal communication, F. Niederholzer, UC Extension). The experimental plot consisted of two perpendicular border rows, one West facing and the other South facing. Nonpareil trees comprise the majority of the almond acreage planted in California [19,21]. For this reason only Nonpareil trees were used in the experiment. Every fifth or sixth Nonpareil tree was chosen for the experiment starting from the West facing row and continuing down the South facing row for a total of 18 trees. The trunk of each tree was vertically fixed with a 1.8 m PVC pipe (10 mm in diameter, Ace Hardware, Davis, CA, USA) and fastened with a 0.3 m long bungee cord. To this another 1.8 m PVC pipe was attached for a total height of ca. 3.6 m. These second 6 ft1.8 m PVC pipes had three holes (1 mm in diameter) drilled ca. 0.3 m apart. The wire end of the black egg traps were then secured into these holes to prevent them from falling out of the trees. Each of these 1.8 m PVC pipes was fixed with either two or three black egg traps depending on the experiment. All black egg traps were filled with 15 g of almond meal, (Pherocon® IV Bait,Trece Inc., Adair, OK, USA) commonly used to monitor ovipositing female NOW moths in the field [36,40–42]. Nestled in the center of the almond meal of each egg trap was a rubber septa impregnated with 10 mg of either a control or test compound (Precision Seal® rubber septa red, 8 mm O.D. glass tubing; Sigma-Aldrich, St. Louis, MO, USA). Each day between 9:00 AM and 12:00 PM Pacific Standard

Time (PST) NOW eggs on each black egg trap were counted and destroyed with a toothbrush. The positions of egg traps on the last segment of 1.8 m PVC pipe were then rotated. Black egg traps, almond meal, and rubber septa were changed every three days. Black egg traps were cleaned with hot water and Alconox, soaked in a 70% ethanol solution and allowed to air dry after use in the field.

To determine the biofix, or first of two days with 75% increase in detected oviposition, 18 black egg traps filled only with 15 g of almond meal and were hung on the 18 experimental trees in the organic block and checked every other day starting April 1st, 2012. Black egg traps and almond meal were replaced on a weekly basis. The biofix was recorded on May 9th, 2012 as the first of two dates with a 75% increase in eggs [30].

For the first trial all materials were placed in the field on May 12th, 2012 and pulled from the field on June 23rd, 2012 following two weeks of no egg detection. Each of the 18 trees was a single competitive assay between compounds 8, 9, and hexane, all vertically spaced 60 cm apart on the PVC pipe. Each of the 18 trees contained three black egg traps on the second 1.8 m PVC pipe. Each black egg trap contained a rubber septa impregnated with 10 mg of either compound 8 or compound 9; control had hexane only.

After this first trial each of the 18 trees was once again fixed with just black egg traps and 15 g of almond meal and treated as above to determine when oviposition could once again be detected. NOW eggs were detected again on September 10th, 2012.

For the second trial all materials were placed in the field on September 13th, 2012 and pulled from the field on October 18th, 2012 following two weeks of no egg detection. Each of the 18 trees was a two-choice assay between compound 9 and hexane, vertically spaced 120 cm apart on the PVC pipe. Each of the 18 trees contained two black egg traps on the second 1.8 m PVC pipe. Each black egg trap contained a rubber septa impregnated with 10 mg of either compound 9 or hexane as a control.

Chemical Preparation

The panel of compounds, which was decoded at the end of the field results, was prepared at Bedoukian Research Inc (BRI, Danbury, CT, USA). It includes 20 compounds covered by US Patent Application No. 61/687,920 for application on vegetation. Compounds were originally discovered when testing compounds closely related to naturally occurring ketones and lactones, finally focusing in on cyclic ketones and lactones and includes: **1**: Farnesyl cyclopentanone, **2**: (E,E)-farnesol, **3**: methyl dihydrojasmonate, **4**: methyl jasmonate, **5**: γ-decalactone, **6**: δ-tetralactone, **7**: ethyl palmitate, **8**: isophrol, **9**: isophorone, **10**: prenyl dihydrojasmonate, **11**: 2-pentadecanol, **12**: 3,5,5-trimethyl-cyclohexanol, **13**: methyl apritol, **14**: methyl dihydrojasmolate, **15**: dihydrojasmonic acid, **16**: methyl apritone (= miranone), **17**: dihydrojasminlactone, **18**: dihydrojasmindiol, **19**: ethyl dihydrojasmonate, and **20**: 2-pentadecanone. All test compounds were diluted in hexane to make stock solutions of 100 µg/µl. Decadic solutions were then made in hexane for desired concentrations of 10 µg/µl, and 1 µg/µl and 0.1 µg/µl. The 2 mm strips of filter paper used in both the EAG and laboratory oviposition assays were allowed to air dry under a fume hood for 2 min before use. For field oviposition assays 100 µl of the 100 µg/µl stock solution were used for application of 10 mg in rubber septa. This volume was allowed to soak into the rubber septa for 1 h before nestled inside the black egg traps.

Statistical Analysis

To analyze EAG recordings we did not normalize our data [43] because the variation between individual antennal preparations was low. Mean responses for all compounds were compared to the hexane control within all concentrations. All data were first tested for normality via the D'Agostino-Pearson omnibus test. Data were then analyzed with either ANOVA or the Kruskal-Wallis test using GraphPad Prism v5.0 software (GraphPad Software Inc., CA, USA). For both the laboratory and field oviposition assays mean eggs per trap were analyzed. Since each tree in the field assay contained all treatments each tree was considered an experimental unit. Those trees that had no black egg traps with eggs were excluded from each day analysis and therefore the final analysis.

Results

This proof of concept study was approached in a quasi-double-blind test. Although we knew the controls, the identity of the test compounds was unknown to the investigators at UC Davis until the end of the field tests. The panel of 20 compounds was prepared at BRI and the experimenters were aware that it included insect repellents and non-insect repellents (placebos). To make the preparer blind, we changed their code into compound numbers. Here, we will initially describe the results with compound numbers (as it happened during the tests) and decode them at the end of the section.

Initially we screened the 20 compounds via EAG using gravid, two-day old female NOW moth antennae as the sensing element. Of the three concentrations that were tested (10 µg/µl, 1 µg/µl, and 0.1 µg/µl) the greatest responses were recorded at 10 µg/µl. Of the twenty compounds tested compounds **8**, **9**, and **12** showed significantly higher EAG responses compared to the hexane

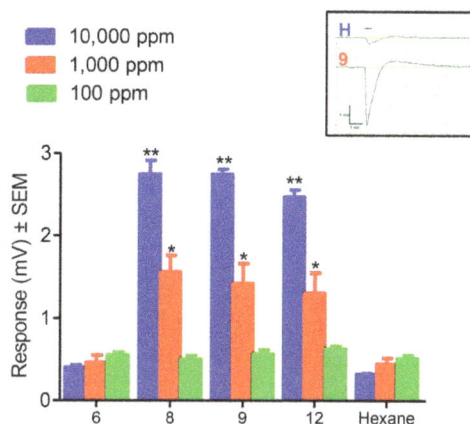

Figure 1. EAG responses from gravid female moths to test compounds. EAG responses recorded from two-day old mated female NOW moth antennae. Dose dependence responses ($N = 9$) for compounds **6**, **8**, **9**, and **12** at concentrations 10 µg/µl ($\approx$10,000 ppm), 1 µg/µl ($\approx$1,000 ppm), and 0.1 µg/µl ($\approx$100 ppm). There were significantly higher responses for compounds **8**, **9**, and **12** as compared to the hexane controls at concentrations 10 µg/µl ($P = 0.0001$; Tukey's Multiple Comparison Test) and 1 µg/µl ($P = 0.0001$). EAG responses at concentration of 0.1 µg/µl were not significantly different ($P > 0.05$) from hexane controls. Compound **6** did not show EAG activity ($P > 0.05$). *Inset* Representative EAG traces of hexane control (H) and compound **9** (9) at 10 µg/µl. The solid line above the traces represents stimulus duration (500 ms). The time and voltage scales are shown at the bottom left of the graph.

control ($P<0.05$; $P<0.05$; $P<0.05$; respectively) for 10 µg/µl and 1 µg/µl (Fig. 1).

Next we examined the effect of EAG-active compounds on the oviposition of female NOW moths in small-cage, two-choice laboratory assays. For these experiments we tested compounds **6**, **8**, **9**, and **12**. Compounds **8**, **9**, and **12** all showed significantly greater EAG responses compared to the hexane controls (Fig. 1). From those compounds that did not show a significant EAG response compared to the hexane control (Fig. 1) we chose compound **6** to serve as a negative control. There was a significant difference of mean eggs per trap between those black egg traps containing filter papers impregnated with hexane and those containing filter papers impregnated with compound **8** ($P<0.005$). Similar results were seen for compound **9** ($P<0.02$). There was no significant difference of mean eggs per trap between those black egg traps containing filter papers impregnated with hexane and those containing filter papers impregnated with compound **12** ($P = 0.80$). Similar results were observed for compound 6 ($P = 0.74$) (Fig. 2).

Lastly we tested the effect of those compounds that showed significantly reduced oviposition from the small-cage two-choice laboratory assay to female NOW moths in a field setting. The first trial ran for 27 days from May 13th, 2012 through June 9th, 2012. Each of the 18 trees in this first trial was a competitive assay between hexane, compound **8**, and compound **9**. Oviposition events were monitored 30 times on 15 of the 18 trees during these 27 days. Ovipositional events can be defined as newly laid eggs on an egg trap within a 24 h period. There was no pattern of ovipositional preference between the 18 trees during this 27 day period. There was no significant difference of mean eggs per trap between those black egg traps containing rubber septa impregnated with hexane and those containing rubber septa impregnated with compound **8** ($P = 0.16$). There was a significant difference of mean eggs per trap between those black egg traps containing rubber septa impregnated with hexane and those containing rubber septa impregnated with compound **9** ($P = 0.0002$) (Fig. 3).

The second trial of this field experiment ran for 34 days from September 14th, 2012 through October 18th, 2012. Each of the 18 trees in this second trial was a two-choice assay between hexane and compound **9**. We chose to exclude compound **8** from this second trial to minimize or avoid intertrap competition. Ovipo-

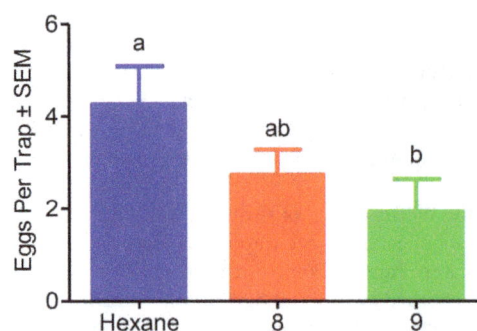

Figure 3. Multiple-choice field evaluations of compounds pre-screened by EAG and indoor bioassay. This first trial was conducted in Arbuckle, CA from May 13th, 2012 through June 9th, 2012. The number of eggs laid on black egg traps baited with compound **9** was significantly lower than those oviposited on control (hexane) traps ($P = 0.003$; Dunn's Multiple Comparisons Test). Oviposition in traps baited with compound **8** was not significantly different ($P>0.05$) from those on control traps. Lastly, there was no significant difference between the two treatments: compounds **8** and **9** ($P>0.05$; $N = 30$).

sition events were monitored 36 times on 14 of the 18 trees during these 34 days. There was no pattern of ovipositional preference between the 18 trees during this 34 day period. There was a significant difference of mean eggs per trap ($P<0.0001$) between those black egg traps containing rubber septa impregnated with hexane and those containing rubber septa impregnated with compound **9** (Fig. 4).

Once we discussed the field test results, the preparer (R.H.B.) disclosed the code names he used for the 20 test compounds, and the experimenters (K.R.C. and W.S.L.) matched these with their own compound numbers to identify the complete panel of test compounds. Thus, compound **6** is δ-tetradecalactone (CAS# 2721-22-4), **8**, isophorol (CAS# 470-99-5), **9**, isophorone (CAS# 78-59-1), and **12** is 3,5,5-trimethylcyclohexanol (CAS# 116-02-09); the complete list is provided in Methods and Materials section. Therefore, our study suggests that isophorone has potential practical applications in IPM strategies as it reduced oviposition by the navel orageworm under field conditions.

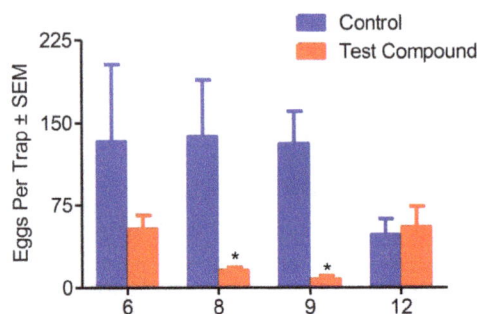

Figure 2. Small-cage two-choice laboratory oviposition assay. For each compound tested there were four cages each with twenty two-day old mated female NOW moths replicated three times from three different cohorts ($N = 12$). Gravid female moths laid significantly fewer eggs on those black egg traps treated with compound **8** ($P = 0.03$) as well as compound **9** ($P = 0.003$; analyzed via the Mann-Whitney Test) when compared to traps treated with hexane (controls). No significant differences were observed for oviposition on black egg traps spiked with compound **6** ($P = 0.95$) or compound **12** ($P = 1.0$) as compared to control traps.

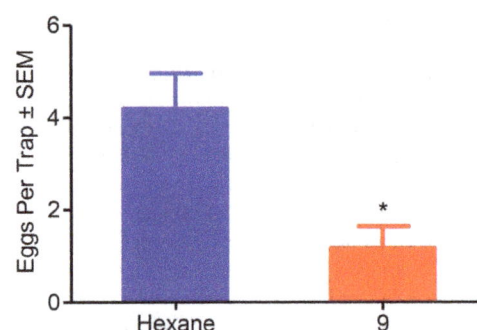

Figure 4. In-depth field evaluations of an active compound. This second field trial was performed in Arbuckle, CA from September 14th, 2012 through October 18th, 2012. In these direct comparison compound **9** was highly significant ($P = 0.0001$; Mann-Whitney Test, $N = 34$).

Discussion

Our ultimate goal was to identify semiochemicals that effect the oviposition of NOW in a field setting. Since deploying these twenty compounds in the field would prove cumbersome and inefficient, we developed a 3-step screening method to identify those behaviorally significant compounds. First, through EAG analysis, we identified three of the twenty compounds that gravid female moth antennae could detect using female moth antennae as the sensing element. Next, using small-cage, two-choice assays, we examined the effect of these three compounds on female NOW oviposition. Of these three compounds two significantly reduced oviposition under these laboratory conditions. As was our ultimate goal, we then deployed these two compounds in the field in Arbuckle, CA to examine their effects on female oviposition. After two field trials in 2012 we found that isophrone (= compound **9**) significantly reduced oviposition under field conditions, thus showing tremendous potential to control NOW populations in almonds.

The mean eggs per trap for both of the field experiments were lower compared to previous studies using black egg traps to monitor NOW oviposition [27,36,44]. This is likely due to the lower accumulation of degree days in more Northerly almond orchards as compared to more Southerly orchards [30,45,46] leading to a condition of reduced NOW development and abundance in our Northern field site. Though our mean eggs per trap counts were low, previous work has shown that at low mean eggs per trap the presence or absence of eggs on black egg traps is more important for monitoring the behavior of female NOW oviposition than the actual number of eggs per trap [41].

Previous research has shown allelochemicals that reduce oviposition in other agriculturally important pyralid moth species [14–17]. There is an ever-increasing demand for these reduced-risk insecticides in agriculture [47–49]. Due to their low mammalian toxicity, non-lethal effects, and high selectivity to insects, repellent and deterrent allelochemicals may present a viable reduced-risk addition to current IPM practices for NOW as a pest of almonds [49,50].

For future work we would like to examine the longevity of isophorone (= compound 9) under field settings and higher NOW densities, potentially in more Southern almond orchards of California. The hope is that behaviorally significant compounds, identified through these described screening methods, will eventually be used in large-scale field experiments. Phytotoxity experiments should be coupled with behavioral assays in the field as previously described [51]. As a generalist scavenger of stonefruit and nut crops [25], it may be the case that the insect is able to identify a suite of plant produced semiochemicals as host cues, and can discriminate different blends of these semiochemicals as host attractants. If this is the case, it may also be likely that the insect can perceive and respond to a blend of plant produced semiochemicals to a greater degree than single plant compounds. There then lies potential to combine compounds showing repellency in a mixture that may show an even greater reduction in oviposition than compound 9 alone. For later field assays in more southern orchards we would also like to examine the ability of compound 9 to reduce oviposition in no-choice assays against hexane controls in egg traps. In these future trials, one tree should be fixed with a PVC pipe as described above and only one egg trap, impregnated with hexane or compound 9, and compared with an adjacent tree fixed with the alternative treatment. We would also like to examine the intriguing question of what behavioral modality these compounds elicit in the navel orangeworm. Are these compounds acting as oviposition repellents or oviposition deterrents [52]? Though very interesting and important scientific questions, this study sought to develop a series of assays one could use to viably and efficiently screen many compounds to identify those that can reduce oviposition of the navel orangeworm.

Acknowledgments

We are grateful to Dr. Franz Niederholzer for his help with the field portion of this research, Dr. Frank Zalom for his advice, Dr. Charles Burks for supplying us with the initial navel orangeworm stock to start our colony, and all members of the Leal lab for their continued support and advice throughout the research.

Author Contributions

Conceived and designed the experiments: KC WL. Performed the experiments: KC. Analyzed the data: KC. Contributed reagents/materials/analysis tools: RHB. Wrote the paper: KC WL.

References

1. Anderson P, Hansson BS, Lofqvist J (1995) Plant-odour-specific receptor neurones on the antennae of female and male *Spodoptera littoralis*. Physiological Entomology 20: 189–198.
2. Gajmer T, Singh R, Saini RK, Kalidhar SB (2002) Effect of methanolic extracts of neem (*Azadirachta indica* A. Juss) and bakain (*Melia azedarach* L) seeds on oviposition and egg hatching of *Earias vittella* (Fab.) (Lep., Noctuidae). Journal of Applied Entomology-Zeitschrift Fur Angewandte Entomologie 126: 238–243.
3. Ge XS, Weston PA (1995) Ovipositional and feeding deterrent from Chinese prickly ash against angoumois grain moth (Lepidoptera, Gelechiidae). Journal of Economic Entomology 88: 1771–1775.
4. Gokce A, Stelinski LL, Isaacs R, Whalon ME (2006) Behavioural and electrophysiological responses of grape berry moth (Lep., Tortricidae) to selected plant extracts. Journal of Applied Entomology 130: 509–514.
5. Guerra PC, Molina IY, Yabar E, Gianoli E (2007) Oviposition deterrence of shoots and essential oils of *Minthostachys* spp. (Lamiaceae) against the potato tuber moth. Journal of Applied Entomology 131: 134–138.
6. Honda K (1995) Chemical basis of differential oviposition by Lepidopterous insects. Archives of Insect Biochemistry and Physiology 30: 1–23.
7. Houghgoldstein J, Hahn SP (1992) Antifeedant and oviposition deterrent activity of an aqueous extract of *Tanacetum vulgare* L on 2 cabbage pests. Environmental Entomology 21: 837–844.
8. Liu ML, Yu HJ, Li GQ (2008) Oviposition deterrents from eggs of the cotton bollworm, *Helicoverpa armigera* (Lepidoptera: Noctuidae): Chemical identification and analysis by electroantennogram. Journal of Insect Physiology 54: 656–662.
9. Rojas JC, Virgen A, Cruz-Lopez L (2003) Chemical and tactile cues influencing oviposition of a generalist moth, *Spodoptera frugiperda* (Lepidoptera: Noctuidae). Environmental Entomology 32: 1386–1392.
10. Scott IM, Jensen H, Nicol R, Lesage L, Bradbury R, et al. (2004) Efficacy of piper (Piperaceae) extracts for control of common home and garden insect pests. Journal of Economic Entomology 97: 1390–1403.
11. Seljasen R, Meadow R (2006) Effects of neem on oviposition and egg and larval development of *Mamestra brassicae* L: Dose response, residual activity, repellent effect and systemic activity in cabbage plants. Crop Protection 25: 338–345.
12. Tabashnik BE (1985) Deterrence of diamondback moth (Lepidoptera: Plutellidae) oviposition by plant-compounds. Environmental Entomology 14: 575–578.
13. Tingle FC, Mitchell ER (1986) Behavior of *Heliothis virescens* (F) (Lepidoptera: Noctuidae) in presence of oviposition deterrents from elderberry. Journal of Chemical Ecology 12: 1523–1531.
14. Agboka K, Mawufe AK, Tamo' M, Vidal S (2009) Effects of plant extracts and oil emulsions on the maize cob borer *Mussidia nigrivenella* (Lepidoptera: Pyralidae) in laboratory and field experiments. International Journal of Tropical Insect Science 29: 185–194.
15. Binder BF, Robbins JC (1997) Effect of terpenoids and related compounds on the oviposition behavior of the European corn borer, *Ostrinia nubilalis* (Lepidoptera: Pyralidae). Journal of Agricultural and Food Chemistry 45: 980–984.
16. Singh R, Koul O, Rup PJ, Jindal J (2011) Oviposition and feeding behavior of the maize borer, *Chilo partellus*, in response to eight essential oil allelochemicals. Entomologia Experimentalis Et Applicata 138: 55–64.
17. Varshney AK, Babu BR, Singh AK, Agarwal HC, Jain SC (2003) Ovipositional responses of *Chilo partellus* (Swinhoe) (Lepidoptera: Pyralidae) to natural products from leaves of two maize (*Zea mays* L.) cultivars. Journal of Agricultural and Food Chemistry 51: 4008–4012.

18. Burks CS, Higbee BS, Brandl DG, Mackey BE (2008) Sampling and pheromone trapping for comparison of abundance of *Amyelois transitella* in almonds and pistachios. Entomologia Experimentalis Et Applicata 129: 66–76.

19. Higbee BS, Siegel JR (2009) New navel orangeworm sanitation standards could reduce almond damage. California Agriculture 63: 24–28.

20. Parra-Pedrazzoli AL, Leal WS (2006) Sexual behavior of the navel orangeworm, *Amyelois transitella* (Walker) (Lepidoptera: Pyralidae). Neotropical Entomology 35: 769–774.

21. (USDA-NASS) USDoA-NASS (2012) California Almond Acreage Report.

22. Beck JJ, Higbee BS, Merrill GB, Roitman JN (2008) Comparison of volatile emissions from undamaged and mechanically damaged almonds. Journal of the Science of Food and Agriculture 88: 1363–1368.

23. Eilers EJ, Klein AM (2009) Landscape context and management effects on an important insect pest and its natural enemies in almond. Biological Control 51: 388–394.

24. Johnson JA, Vail PV (1989) Damage to raisins, almonds, and walnuts by irradiated indianmeal moth and navel orangeworm larvae (Lepidoptera: Pyralidae). Journal of Economic Entomology 82: 1391–1394.

25. Wade WH (1961) Biology of the navel orangeworm, *Paramyelois transitella* (Walker), on almonds and walnuts in northern California. Hilgardia 31: 129–171.

26. Campbell BC, Molyneux RJ, Schatzki TF (2003) Current research on reducing pre- and post-harvest aflatoxin contamination of US almond, pistachio, and walnut. Journal of Toxicology-Toxin Reviews 22: 225–266.

27. Niu GD, Pollock HS, Lawrance A, Siegel JP, Berenbaum MR (2012) Effects of a naturally occurring and a synthetic synergist on toxicity of three insecticides and a phytochemical to navel orangeworm (Lepidoptera: Pyralidae). Journal of Economic Entomology 105: 410–417.

28. Niu GD, Siegel J, Schuler MA, Berenbaum MR (2009) Comparative toxicity of mycotoxins to navel orangeworm (*Amyelois transitella*) and corn earworm (*Helicoverpa zea*). Journal of Chemical Ecology 35: 951–957.

29. Engle CE, Barnes MM (1983) Cultural control of navel orangeworm in almond orchards. California Agriculture 37: 19–19.

30. Zalom FG, Pickel C, Bentley WJ, Coviello RL, Van Steenwyk RA (2009) Navel Orangeworm. UC IPM Pest Management Guidelines: Almond: 61–71.

31. Agudelosilva F, Zalom FG, Hom A, Hendricks L (1995) Dormant season application of *Steinernema carpocapsae* (Rhabditida, Steinernematidae) and *Heterorhabditis sp* (Rhabditida, Heterorhabditidae) on almond for control of overwintering *Amyelois transitella* and *Anarsia lineatella* (Lepidoptera, Gelechiidae). Florida Entomologist 78: 516–523.

32. Higbee BS, Burks CS (2008) Effects of mating disruption treatments on navel orangeworm (Lepidoptera: Pyralidae) sexual communication and damage in almonds and pistachios. Journal of Economic Entomology 101: 1633–1642.

33. Legner EF, Gordh G (1992) Lower navel orangeworm (Lepidoptera: Phycitidae) population-densities following establishment of *Goniozus legneri* (Hymenoptera: Bethylidae) in California. Journal of Economic Entomology 85: 2153–2160.

34. Siegel J, Lacey LA, Fritts R, Higbee BS, Noble P (2004) Use of *steinernematid* nematodes for post harvest control of navel orangeworm (Lepidoptera: Pyralidae, *Amyelois transitella*) in fallen pistachios. Biological Control 30: 410–417.

35. Syed Z, Leal WS (2011) Electrophysiological measurements from a moth olfactory system. J Vis Exp.

36. Burks CS, Higbee BS, Siegel JP, Brandl DG (2011) Comparison of trapping for eggs, females, and males of the naval orangeworm (Lepidoptera: Pyralidae) in almonds. Environmental Entomology 40: 706–713.

37. Tebbets JS, Curtis CE, Fries RD (1978) Mortality of immature stages of the navel orangeworm (Lepidoptera: Pyralidae) stored at 3.5-degrees-c. Journal of Economic Entomology 71: 875–876.

38. Landolt PJ, Curtis CE (1991) Mating frequency of female navel orangeworm moths (Lepidoptera: Pyralidae) and patterns of oviposition with and without mating. Journal of the Kansas Entomological Society 64: 414–420.

39. Liu ZO, Vidal DM, Syed Z, Ishida Y, Leal WS (2010) Pheromone binding to general odorant-binding proteins from the navel orangeworm. Journal of Chemical Ecology 36: 787–794.

40. Rice RE, Sadler LL, Hoffman RA (1976) Egg traps for the navel orangeworm, *Paramyelois transitella* (Walker). Environmental Entomology 5: 697–700.

41. Higbee BS, Burks CS (2011) Effect of bait formulation and number of traps on detection of navel orangeworm (Lepidoptera: Pyralidae) oviposition ssing egg traps. Journal of Economic Entomology 104: 211–219.

42. Rice RE, Johnson TW, Profita JC, Jones RA (1984) Improved attractant for navel orangeworm (Lepidoptera: pyralidae) egg traps in almonds. Journal of Economic Entomology 77: 1352–1353.

43. Park KC, Ochieng SA, Zhu JW, Baker TC (2002) Odor discrimination using insect electroantennogram responses from an insect antennal array. Chemical Senses 27: 343–352.

44. Kuenen LPSB, Bentley W, Rowe HC, Ribeiro B (2008) Bait formulations and longevity of navel orangeworm egg traps tested. California Agriculture 62: 36–39.

45. Sanderson JP, Barnes MM, Seaman WS (1989) Synthesis and validation of a degree-day model for navel orangeworm (Lepidoptera: Pyralidae) development in California almond orchards. Environmental Entomology 18: 612–617.

46. Seaman WS, Barnes MM (1984) Thermal summation for the development of the navel orangeworm in almond (Lepidoptera: Pyralidae). Environmental Entomology 13: 81–85.

47. Atanassov A, Shearer PW, Hamilton G, Polk D (2002) Development and implementation of a reduced risk peach arthropod management program in New Jersey. Journal of Economic Entomology 95: 803–812.

48. Jenkins PE, Isaacs R (2007) Reduced-risk insecticides for control of grape berry moth (Lepidoptera: Tortricidae) and conservation of natural enemies. Journal of Economic Entomology 100: 855–865.

49. Regnault-Roger C, Vincent C, Arnason JT (2012) Essential oils in insect control: Low-risk products in a high-stakes world. Annual Review of Entomology, Vol 57 57: 405–424.

50. Koul O, Walia S, Dhaliwal GS (2008) Essential oils as green pesticides: potential and constraints. Biopesticide International 4: 63–84.

51. Van Steenwyk RA, Barnett WR (1987) Disruption of navel orangeworm (Lepidoptera: pyralidae) oviposition by almond by-products. Journal of Economic Entomology 80: 1291–1296.

52. Miller JR, Siegert PY, Amimo FA, Walker ED (2009) Designation of chemicals in terms of the locomotor responses they elicit from insects: An update of Dethier et al. (1960). Journal of Economic Entomology 102: 2056–2060.

6

Comparative Toxicities and Synergism of Apple Orchard Pesticides to *Apis mellifera* (L.) and *Osmia cornifrons* (Radoszkowski)

David J. Biddinger[1,3], **Jacqueline L. Robertson**[2]*, **Chris Mullin**[3], **James Frazier**[3], **Sara A. Ashcraft**[3], **Edwin G. Rajotte**[3], **Neelendra K. Joshi**[1,3], **Mace Vaughn**[4]

1 Fruit Research and Extension Center, Pennsylvania State University, Biglerville, Pennsylvania, United States of America, 2 USDA Forest Service PSW Station, Albany, California, United States of America, 3 Department of Entomology, Pennsylvania State University, University Park, Pennsylvania, United States of America, 4 The Xerces Society, Portland, Oregon, United States of America

Abstract

The topical toxicities of five commercial grade pesticides commonly sprayed in apple orchards were estimated on adult worker honey bees, *Apis mellifera* (L.) (Hymenoptera: Apidae) and Japanese orchard bees, *Osmia cornifrons* (Radoszkowski) (Hymenoptera: Megachilidae). The pesticides were acetamiprid (Assail 30SG), λ-cyhalothrin (Warrior II), dimethoate (Dimethoate 4EC), phosmet (Imidan 70W), and imidacloprid (Provado 1.6F). At least 5 doses of each chemical, diluted in distilled water, were applied to freshly-eclosed adult bees. Mortality was assessed after 48 hr. Dose-mortality regressions were analyzed by probit analysis to test the hypotheses of parallelism and equality by likelihood ratio tests. For *A. mellifera*, the decreasing order of toxicity at LD_{50} was imidacloprid, λ-cyhalothrin, dimethoate, phosmet, and acetamiprid. For *O. cornifrons*, the decreasing order of toxicity at LD_{50} was dimethoate, λ-cyhalothrin, imidacloprid, acetamiprid, and phosmet. Interaction of imidacloprid or acetamiprid with the fungicide fenbuconazole (Indar 2F) was also tested in a 1:1 proportion for each species. Estimates of response parameters for each mixture component applied to each species were compared with dose-response data for each mixture in statistical tests of the hypothesis of independent joint action. For each mixture, the interaction of fenbuconazole (a material non-toxic to both species) was significant and positive along the entire line for the pesticide. Our results clearly show that responses of *A. mellifera* cannot be extrapolated to responses of *O.cornifrons*, and that synergism of neonicotinoid insecticides and fungicides occurs using formulated product in mixtures as they are commonly applied in apple orchards.

Editor: Nicolas Desneux, French National Institute for Agricultural Research (INRA), France

Funding: The authors thank the USDA NIFA for a SCRI grant (# PEN04398) on sustainable fruit pollination and the State Horticultural Association of Pennsylvania for their financial support. The funders had no role in study design, data collection and analysis, decision to publish, or preparation of the manuscript.

Competing Interests: The authors have declared that no competing interests exist.

* E-mail: thesmokesdude@aol.com

Introduction

Pollinator species such as honey bee [*Apis mellifera* (L.) (Hymenoptera: Apidae)] and the Japanese orchard bee [*Osmia cornifrons* (Radoszkowski) (Hymenoptera: Megachilidae)] provide important services in orchard and other agricultural ecosystems [1]–[6]. Members of the genus *Osmia* are important and efficient pollinators of tree fruit crops such as apples, plums and cherries, among other economically important fruit and nut crops [7]–[9]. The Japanese orchard bee (also known as the Japanese horn-faced bee) is commercially used for pollination of pears and apples in Japan [9]. This species was introduced into the mid-Atlantic region of the United States by USDA scientists in 1977 [10], [11], has since become commercially available, and is now an important (and efficient) wild pollinator in tree fruit orchards in Pennsylvania [12], [13]. One *O. cornifrons* can set up to 80% more apple flowers per day compared with flowers set by one honey bee, *A. mellifera* (L.), worker [14]. These two species are complementary pollinators of apple, peach and pear orchards in the Northeastern United States [12], [13], [15]–[19].

A recent significant decline in bee populations in the United States [20]–[23] and elsewhere [24] has led researchers to investigate the effects of possible factors such as pesticide residues [25]–[33], different types of pathogens [34]–[36], adjuvants [37], various parasites [38], floral resource availability and diversity [3], and pest management and agricultural practices [39] on the health, abundance and diversity of different species of bees and pollinators. Although most past research has emphasized the effects of these factors on either *A. mellifera* or on bumble bee species, the relative effects of pesticide mixtures (e.g., insecticides and fungicides) on two different species of pollinators (for instance, *O. cornifrons* versus *A. mellifera*) have not been investigated.

In agricultural production systems, various classes of chemicals are used for management of various pests and diseases. Commercial and farmer tank-mixes of insecticide and fungicide are used to reduce operational or production costs (or both). In tree fruit orchards in Pennsylvania, pesticides are applied for management of a complex of diverse pest species. These include fruit-feeding Lepidoptera, leaf rollers, mites, aphids, plum curculio and stink bugs [40]–[42]. The most critical period to reduce pollinators' exposure to pesticide application in apple orchards is bloom. Different pesticides are applied for control of pests such as rosy apple aphid, European apple sawfly, and plum curculio just before, during, and immediately after bloom. This timing is also critical for control of fungal diseases such as apple scab, *Venturia inequalis* (Cooke) and apple powdery mildew, *Podosphaera leucotricha* (Ell. & Evherh.) [40]. However, information about the toxicity of these toxicant mixtures to beneficial invertebrates including pollinators such as *O. cornifrons* and *A. mellifera* must be obtained if rational conservation plans for pollinators are to be implemented. Well-designed laboratory bioassays provide the scientific basis for such decisions.

Our studies were done to provide basic information about the comparative toxicities of commonly-used orchard pesticides and pesticide-fungicide mixtures to *O. cornifrons* and *A. mellifera*. Results of our experiments were used to test the general premise that one species such as *A. mellifera* can serve as a surrogate species for a larger taxonomic group such as several families of bees, or as a surrogate for an arbitrary group such as all terrestrial arthropods [43]. Our study expands a previous investigation of differential toxicity of imidacloprid to *A. mellifera* versus *Bombyx mori* [44] and suggests the direction in which the study of ecotoxiciology of pollinators can progress [45]. Of most importance, our study provides an example of the type of information necessary to improve the sensitivity of testing pesticides on diverse species in the superfamily Apoidea [46] in the future.

Materials and Methods

Insects

O.cornifrons were purchased from a single source in Wisconsin where they had been reared in an organic apple orchard. Because fruit from this orchard were used only for cider production, pesticide use was minimal. Larvae were reared in natural *Phragmites* reed bundles and in wooden blocks lined with paper straws. Cocoons containing the overwintering adults were removed and refrigerated at 3°C until 1 April to ensure that their chilling requirements had been met. Loose cocoons were then held inside an incubator (25°C, constant darkness) until adults emerged. Adults were held in darkness until treated 24–72 hr after emergence (24 hr for the males; 24–72 hr for the females). Emergence occurred from April through May. *A. mellifera* used in this study were purchased as new packages from Gardner Apiaries, Spell Bee (Baxley, GA). Each package included a queen and workers. The packages were exposed to the miticides fluvalinate and coumaphos for bee mite control. They were then put into hives pre-sterilized by irradiation. Colonies were established in the spring and kept in an isolated area at least 6 km from any pesticide applications.

Treatments

For treatment, cages were made of a Petri dish (100×20 mm) encasing a 100 mm-long wire mesh cylinder constructed of hardware cloth (with 3×3 mm openings). One side of the Petri dish had two holes made with a heated cork borer. The larger hole was used to put treated bees into the dish. Once the cage was full, the hole was sealed with tape. The other (smaller) hole held a glass vial for *ad lib* feeding of a 50% sucrose solution.

For each replication (per chemical or combination) with a species, 10 bees were placed in a 50 ml centrifuge tube with a pair of forceps. The tube was placed in an ice bath for 1–2 min. to immobilize the bees, which were then poured onto a paper towel. If a bee did not move after the cold anesthesia, it was discarded from the bioassay. Each bee was picked up at the base of the

wings, treated on the thorax, and placed in the cage. One µl/bee was applied with a Hamilton repeating dispenser (Hamilton Company, Reno, NV) that held a 50 µl syringe. After each group of 10 bees was treated, the cage was set aside and checked after ~3 min. to ensure that all bees appeared to be healthy. Six cages were placed on their sides inside a plastic container, which also contained a moist paper towel and a jar of a saturated NaCl solution to maintain ~75% RH. Six caps from 20 ml scintillation vials were used to separate and hold the cages in place.

Water was used as the solvent to mimic what growers use in the field to apply the pesticides. No problems with formulation solubility were observed. The commercial formulations (AI% ; manufacturer) were Assail 30SG (acetamiprid 30%; United Phosphorous Inc., King of Prussia, PA), Dimethoate 4EC (dimethoate 43.5%; Drexel Chemical Company, Memphis, TN), Imidan 70W (phosmet 70%; Gowan Company, Yuma, AZ), Provado 1.6F (imidacloprid 17.4%; Bayer CropScience, Research Triangle Park, NC) and Warrior II (λ - cyhalothrin 22.8%; Syngenta, Wilmington, DE). Interactions of the fungicide Indar 2F (fenbuconazole 22.86%; Dow AgroSciences LLD, Indianapolis, IN) in a 1:1 proportion with Assail 30SG and in a 1:2 proportion with Provado 1.6F were also tested.

Experimental Designs and Data Analyses

In all experiments, mortality was tallied after 48 hr. Control mortality was <5%. Average control mortality was 2.7%. At least 6 doses of each pesticide plus a control (water only) were tested in each replication per pesticide per bee species. Each replication included a total of 60–135 bees of each species depending on species' availability. Dose-mortality regressions were estimated assuming the normal distribution (i.e., probit model) with the computer program PoloPlus [47] as described by Robertson et al. [48]. We used a two-step procedure to analyze data for each chemical. In the first step, we examined plots of standardized residuals for outliers, which were then eliminated from the data sets. The second and final probit analysis was done to test hypotheses of parallelism (slopes not significantly different) and equality (slopes and intercepts not significantly different) with likelihood ratio tests [48]. PoloPlus also calculated lethal dose ratios (LDR's) of the most toxic chemical compared with all other chemicals for each species. An LDR provides a means to test whether two LD's are significantly different (i.e., when the 95% CI for the LDR did not include the value 1.0 [47], [48]).

For tests with a mixture, at least 5 doses of the mixture that bracketed 5–95% mortality were tested concurrently with experiments with at least 5 doses of individual mixture components. As before, 60–135 bees of each species were tested depending on species' availability. To test the hypothesis of independent joint action of fenbuconazole with acetamiprid or imidacloprid, we used the computer program PoloMix [49]. Assuming independent joint action of two mixture chemicals, test subjects can die of three possible causes. The first cause is natural mortality, with a probability p_o (a constant). The other two causes of mortality are the probabilities of mortalities for chemical 1 or chemical 2. For the first chemical, the probability of response ($p1$) is a function of dose $D1$. Usually, the probit or logit of dose X of chemical 1 is $\log(D1)$ (i.e., $X1 = \log[D1]$). For the second chemical, the probability of response ($p2$) is a function of dose $D2$. If these three causes of mortality are independent, the probability of death (p) is $p = p0+(1-p0)p1+(1-p0)(1-p1)$ $p2$. When each "+" sign means "or" and each product means "and," this equation means that the total probability of death equals death from natural causes ($p0$), or no death from natural causes ($1-p0$) and death from the first chemical [e.g., $(1-p0)p1$], or no death from natural causes or from the first chemical [i.e., $(1-p0)(1-p1)$], but death from the second chemical [i.e., $(1-p0)(1-p1)\,p2$]. The $\chi2$ statistic produced by PoloMix [49] was used to test the hypothesis of independent joint action. This test statistic is calculated by obtaining an estimate for the probability of mortality (p) for several dose levels of the two components and then comparing $\hat{p}$ (the estimate of p) with the observed proportion killed at the corresponding dose levels. The three contributions to p are estimated separately. First, $p0$ is calculated as the proportional mortality observed in the control group. Next, $p1$ and $p2$ are estimated from bioassays of chemical 1 and chemical 2, with test statistics estimated from PoloPlus [47].

Results and Discussion

Responses of O. cornifrons (Table S1)

The hypotheses of parallelism and equality of response of *O. cornifrons* to the five pesticides were both rejected ($P = 0.05$). At LD_{50} (the most reliable point of comparison for dose-response regressions [48]), the decreasing order of toxicity was dimethoate$>\lambda$-cyhalothrin$>$imidacloprid$>$acetamiprid$>$phosmet. LDR50's indicated three groups of significantly increasing toxicity. The least toxic group consisted of phosmet, acetamiprid and imidacloprid. λ-cyhalothrin was significantly more toxic than the first group, and the most toxic pesticide, dimethoate, was ~ 70 times more toxic than phosmet (the least toxic pesticide). At LD90, the increasing order of toxicity was dimethoate$>\lambda$-cyhalothrin$>'''$ phosmet$>$imidacloprid$>$acetamiprid. Responses to acetamiprid and imidacloprid were not significantly different at LD_{90}. Significantly increased susceptibilities to phosmet and λ-cyhalothrin were indicated by their LDR90's. Dimethoate was ~ 60 times more toxic than acetamiprid.

Responses of A.mellifera (Table S1)

The pattern of responses for *A.mellifera* differed considerably from that of *O. cornifrons*. For *A.mellifera*, responses were also neither parallel nor equal ($P = 0.05$). The decreasing order of toxicity at LD_{50} was imidacloprid$>\lambda$-cyhalothrin $=$ dimethoate$>$phosmet$>$acetamiprid. LDR50's indicated two levels of decreasing toxicity relative to imidacloprid: λ-cyhalothrin or dimethoate $<$phosmet. At LD_{90}, the increasing order of toxicity was dimethoate$>$imidacloprid$>\lambda$- cyhalothrin$>$phosmet$>$acetamiprid. Dimethoate was ~ 1200 times more toxic than acetamiprid as this response level. LDR90's suggested three groups of response: dimethoate$>\lambda$- cyhalothrin or imidacloprid$>$phosmet$>$acetamiprid.

Comparative Responses of the Pollinator Species (Table S1)

Neither species was consistently more susceptible than the other. Their responses to the two neonicotinoid pesticides were parallel, but not equal. *O. cornifrons* was significantly more susceptible than *A.mellifera* to acetamiprid at LD_{50} (i.e., the LDR_{50} did not bracket the value 1.0 [47,48]). At LD_{50}, *O. cornifrons* ~ 12 times more susceptible than *A.mellifera*. In contrast, *A.mellifera* was significantly more susceptible to imidacloprid at the 50% response level (LDR50 did not bracket the value 1.0). At LD_{50}, *A.mellifera* were ~ 26 times more susceptible than *O. cornifrons*. Responses to the two organophosporous pesticides — dimethoate and phosmet — were also not consistent by species. The hypotheses of parallelism and equality were each rejected in tests with these chemicals. For dimethoate, *O. cornifrons* was 3.7 times more susceptible than honey bees at LD_{50}, but the regression slope (7.62) was so steep for *A.mellifera* that relative toxicities were reversed at the LD_{90} (at 90% response, honey bees were 3.2 times more susceptible that *O.

cornifrons*). With phosmet, the slope of the regression line for *A.mellifera* was very shallow and its line crossed the dose-response line for *O. cornifrons* at the upper end of response. At LD_{50}, *A.mellifera* workers were ~ 3 times more susceptible than *O. cornifrons*, but at LD_{90}, *O. cornifrons* was ~ 2 times more susceptible than *A.mellifera*.

Responses to the only pyrethroid tested, λ- cyhalothrin, were neither parallel nor equal. At LD_{50}, *A. mellifera* was 3 times more susceptible than *O. cornifrons*. LD90's for the two species were not significantly different.

Responses to Mixtures (Table 1, Figure 1)

Fenbuconazole was minimally toxic to both species. In combination with acetamiprid, the 1:1 mixture was ~ 5 times more toxic than acetamiprid alone to *A. mellifera* at LD_{50} (Fig. 1a). The toxicity of the mixture was 2 times greater than acetamiprid to *O. cornifrons* at LD_{50} (Fig. 1b). Fenbuconazole enhanced responses of imidacloprid slightly, but significantly, for both species. Although responses at the LD_{50} were not significantly different, a greater effect was apparent at higher levels of response (Fig. 1c, d). $\chi2$ values of observed vs. expected values were small but still significant. Rejection of the null hypothesis of independent joint action of the fenbuconazole with either of the two pesticides indicated that significant synergism occurred in both bee species. In contrast, another neonicotinoid – thiacloprid – applied as technical material in 100% ethanol, was synergized 1141- and 559-fold by the addition, respectively, of the fungicides triflumizole and propiconazole, which are in the same class as fenbuconazole [49].

Conclusions

Because of their highly controlled conditions and rigorous experimental design, laboratory bioassays provide the ideal means to estimate comparative responses of *A. mellifera* and wild bees to pesticides. Our results clearly show that the response of *A. mellifera* cannot be extrapolated to the response of *O.cornifrons*. Such results might be expected given the extensive body of information from analogous (and as rigorously designed and statistically analyzed) experiments showing that responses of the Lepidopteran families Tortricidae and Lymantriidae vary significantly among genera, within a single genus, among populations or even among sibling groups of the same genetic strain [47], [50], [51] tested with the same pesticide. Equally rigorously designed bioassays among Apoidea populations of the same species, species of the same genus, and among genera done to evaluate variation at all levels of response are clearly needed. Natural variation in response for a populations of single species to single pesticides, whether applied as pure active ingredient or formulated material, also needs to be estimated [52]. This systematic approach should be part of the overhaul of the pesticide registration process as suggested by Decourtye et al. [46]. Finally, without direct access to raw data, we have no way to compare our results statistically with previously published information from previous experiments with *A. mellifera* tested with neonicotinoid insecticides [53], let alone explain why these differences occurred. Until such basic comparisons are made possible even for one species such as *A. mellifera*, bee toxicologists and ecologists will continue to debate possible explanations for different contact and oral LD_{50}'s. One solution to this situation would be establishment of a free-access network for raw data from all experiments that have been used in applications for pesticide registrations by the US EPA (US EPA) and the European Union (EU).

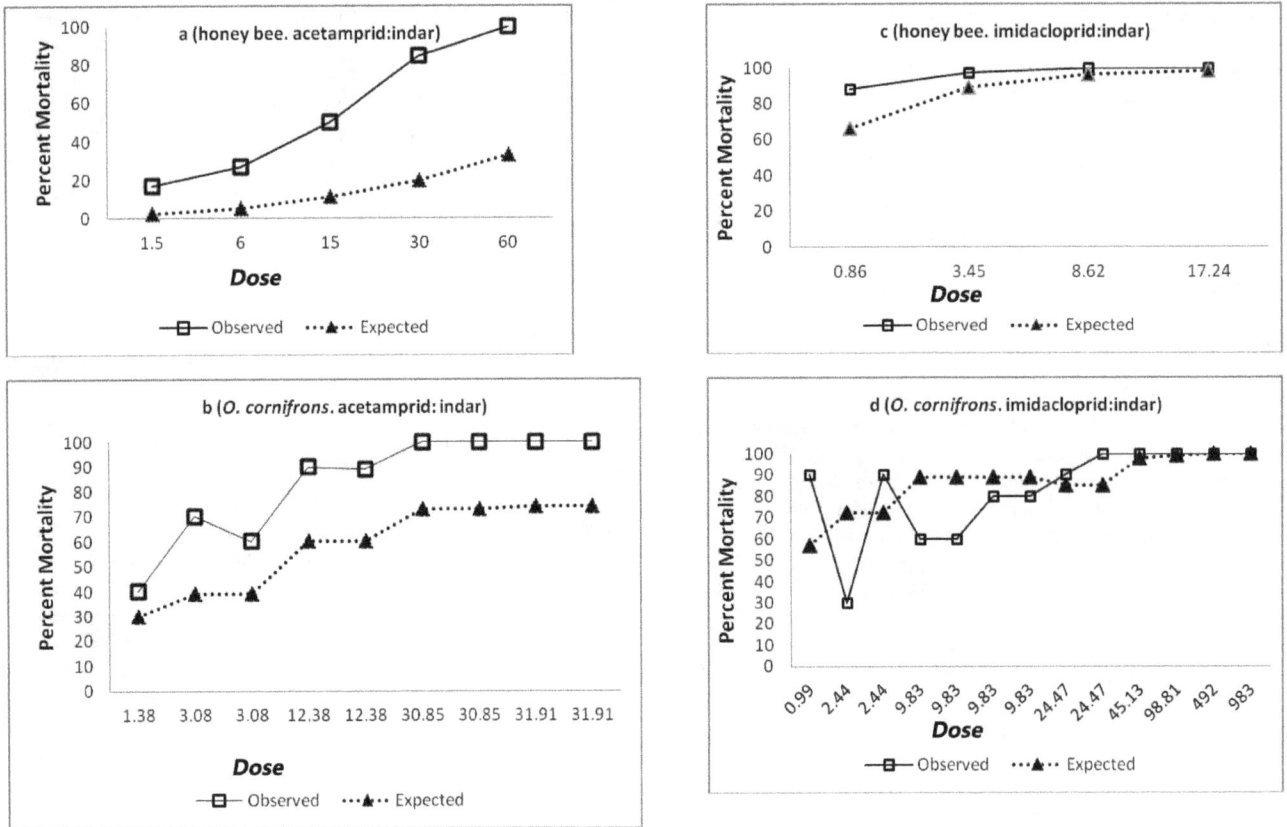

Figure 1. (a–d). Effects of mixtures on *O. cornifrons* **and** *A. mellifera.* Boxes are responses observed and triangles are responses predicted assuming the model of independent joint action (see text). Dose is in units of μg/bee. Figure 1a is for *A. mellifera* treated with acetamiprid:indar; Figure 1b is for *O. cornifrons* treated with acetamiprid:indar; Figure 1c is for *A. mellifera* treated with imidacloprid:indar; Fig. 1d is for *O. cornifrons* treated with imidacloprid:indar.

Table 1. Responses of *Osmia cornifrons* and *A. mellifera* to to acetamiprid mixed with indar.

Chemical	Mixture	Species[a]	n	Slope ± S.E.	LD$_{50}$[b]	95% CL	LDR[c]	95% CI*
Acetamiprid	only	A	245	1.39±0.41	64.6	38.1–252	1.0	–
Indar	only	A	330	Not toxic				
Acetamiprid	Indar (1:1)	A	360	3.13±0.53	14.3	8.5–30.8	4.5	2.5–8.2
Acetamiprid	only	O	272	0.97±0.18	4.0	1.1–7.1	1.0	–
Indar	only	O	60	Not toxic				
Acetamiprid	Indar (1:1)	O	99	1.94±0.42	2.1	1.1–3.2	1.9	0.95–3.9
Imidicloprid	only	A	310	1.26±0.15	0.2	0.1–0.3	1.7	0.92–3.1
Indar	only	A	330	Not toxic				
Imidicloprid	Indar (2:1)	A	300	1.84±0.33	0.3	0.1–0.4	1.0	–
imidacloprid	only	O	310	1.26±0.15	3.8	1.7–12.6	1.7	0.8–3.7
Indar	only	O	90	Not toxic				
Imidicloprid	Indar (2:1)	O	522	3.11±1.09	6.6	1.4–9.6	1.0	–

[a]A is *A.mellifera*, O is *Osmia cornifrons*.
[b]LD is expressed as μg/bee.
[c]LDR is higher LD÷lower LD.
*If the 95% CI of the LDR includes the value 1.0, the LD's are not significantly different.

Our results further suggest that the current practice of using *A. mellifera* as a surrogate species for >90,000 other species of non-target insects, including *O. cornifrons*, in the pesticide registration process as required by the United States EPA is scientifically flawed, and needs significant modifications. Inherent in this registration process is the assumption that the response of *A. mellifera* can be extrapolated to responses of the many different species of predatory and parasitic arthropods that are important to successful Integrated Pest Management (IPM) programs, but which belong to several distantly related orders. This assumption has been shown to be untenable in other studies: many pesticides labeled as "reduced risk" or "organophosphate replacements" by the United States EPA and that have used toxicity tests with *A. mellifera* as part of the criteria for these classifications are in fact toxic to non-target insects and disruptive to IPM programs in tree fruit crops [42], [52], [54], [55]. Use of *A. mellifera* as a surrogate species has also been questioned by members of the European Union [56], which also currently requires toxicity data for multiple insect species [43]. According to Decourtye et al. [46], 42 studies report deleterious side effects on other bee species despite the fact they passed risk assessment on honey bees. Considering the fact that pollinator species significantly differ in their relative toxicity to pesticides, the United States EPA needs to review its current policy on pollinator toxicity that requires data from contact toxicity (from pesticide residues) studies designed to investigate acute effects of a pesticide chemical (under the registration process) on individual bees [57], in addition to adding studies of sublethal effects to these toxicants [58], [59].

Acknowledgments

We thank Eric Mader and Scott Famous for providing some of the *O. cornifrons* for experiments.

Author Contributions

Conceived and designed the experiments: DB ER. Performed the experiments: SA. Analyzed the data: JR. Contributed reagents/materials/analysis tools: CM JF. Wrote the paper: JR DB NJ ER MV.

References

1. Calderone NW (2012) Insect Pollinated Crops, Insect Pollinators and US Agriculture: Trend Analysis of Aggregate Data for the Period 1992–2009. PLoS ONE 7(5): e37235.
2. Losey JE, Vaughan M (2006) The economic value of ecological services provided by insects. Bioscience 56: 311–323.
3. Klein A-M, Vaissière BE, Cane JH, Steffan-Dewenter I, Cunningham SA, et al. (2007) Importance of pollinators in changing landscapes for world crops. Proc R Soc Lond B 274: 303–313.
4. Bosch J, Kemp WP (2002) Developing and establishing bee species as crop pollinators: the example of Osmia spp. (Hymenoptera: Megachilidae) and fruit trees. Bull Entomol Research 92:3–16.
5. Gallai N, Salles J-M, Settele J, VaissiÈre BE (2009) Economic valuation of the vulnerability of world agriculture confronted with pollinator decline. Ecol Econ 68: 810–821.
6. Bosch J, Kemp W, Trostle G (2006) Bee population returns and cherry yields in an orchard pollinated with Osmia lignaria (Hymenoptera: Megachilidae). J Econ Entomol 99: 408–413.
7. Bosch J, Kemp W (2001) How to manage the blue orchard bee. USDA SARE Handbook 5.
8. Mader E, Spivak M, Evans E (2010) Managing alternative pollinators. USDA SARE Handbook. Available: http://palspublishing.cals.cornell.edu/nra_map.html. Accessed 2013 July 31.
9. Maeta V, Kitamura T (1984) How to manage the mame-kobachi, Osmia cornifrons for pollination of the fruit and crops. Ask Co Ltd 1974. 1–16
10. Batra SWT (1979) Osmia cornifrons and Pithitis smaragdula, two Asian bees introduced into the United States for crop pollination. Maryland Agricultural Experiment Station, Special Miscellaneous Publication 1: 307–312.
11. Batra SWT (1998) Hornfaced bees for apple pollination. Am Bee J 138, 364–365.
12. Biddinger D, Frazier J, Frazier M, Rajotte E, Donovall L, et al. (2009) Solitary bees as alternative pollinators in Pennsylvania fruit crops. Penn Fruit News 89: 84–94.
13. Biddinger D, Rajotte E, Joshi NK, Ritz A (2011) Wild bees as alternative pollinators. Fruit Times 30: 1–4.
14. Maeta Y, Kitamura T (1981) Pollinating efficiency by Osmia cornifrons (Radoszkowski) in relation to required number of nesting bees for economic fruit production. Honeybee Science 2:65–72.
15. Park MG, Orr MC, Danforth BN (2010) The role of native bees in apple pollination. NY Fruit Quart 18:21–25.
16. Ritz A, Biddinger D, Rajotte E, Sahli H, Joshi N (2012) Quantifying the efficacy of native bees for orchard pollination in Pennsylvania to offset the increased cost and decreased reliability of honey bees. Penn Fruit News 92: 6–66.
17. Joshi NK, Biddinger D, Rajotte EG (2011) A survey of apple pollination practices, knowledge and attitudes of fruit growers in Pennsylvania. 10th International Pollination Symposium, Puebla, Mexico.
18. Biddinger DJ, Ngugi H, Frazier J, Frazier M, Leslie T, Donovall LR (2010). Development of the mason bee, Osmia cornifrons, as an alternative pollinator to honey bees and as a targeted delivery system for biological control agents in the management of fire blight. Penn Fruit News 90:35–44.
19. Biddinger D, Rajotte E, Joshi NK (2011) Wild bees as alternative pollinators in Pennsylvania apple orchards. Proceedings, the Great Lakes Fruit, Vegetable and Farm Market EXPO and the Michigan Greenhouse Growers Expo, Michigan. Available: http://www.glexpo.com/summaries/2011summaries/webCropPollination.pdf. Accessed 2013 July 31.
20. vanEngelsdorp D, Pettis J, Rennich K, Rose R, Caron D, et al. (2012) Preliminary results: honey bee colonies losses in the U.S., Winter 2011–2012. Available: http://beeinformed.org/2012/05/winter2012. Accessed 2013 July 31.
21. vanEngelsdorp D, Evans JD, Saegerman C, Mullin C, Haubruge E, et al. (2009) Colony Collapse Disorder: A descriptive study. PLoS ONE 4: e6481.
22. vanEngelsdorp D, Hayes J Jr, Underwood RM, Pettis J (2008) A survey of honey bee colony losses in the U.S., fall 2007 to spring 2008. PLoS ONE 3: e4071.
23. National Research Council (2006) Status of pollinators in North America. Washington, DC: National Academies Press. 322 p.
24. Potts SG, Biesmeijer JC, Kremen C, Neumann P, Schweiger O, et al. (2010) Global pollinator declines: trends, impacts and drivers. Trends Ecol Evol 25: 345–353.
25. Henry M, Béguin M, Requier F, Rollin O, Odoux J, et al. (2012) A common pesticide decreases foraging success and survival in honey bees. Science 336:348–350.
26. Whitehorn PR, O'Connor S, Wackers FL, Goulson D (2012) Neonicotinoid pesticide reduces bumblebee colony growth and queen production. Science 336:351–352.
27. Krupke CH, Hunt GJ, Eitzer BD, Andino G, Given K (2012) Multiple routes of pesticide exposure for honey bees living near agricultural fields. PLoS ONE 7:e29268.
28. Laycock I, Lenthall K, Barratt A, Cresswell J (2012) Effects of imidacloprid, a neonicotinoid pesticide, on reproduction in worker bumble bees (Bombus terrestris). Ecotoxicology. Doi: 10.1007/s10646-012-0927-y.
29. Cresswell JE (2011) A meta-analysis of experiments testing the effects of a neonicotinoid insecticide (imidacloprid) on honey bees. Ecotoxicology 20:149.
30. De la Rúa P, Jaffé R, Dall'Olio R, Muñoz I, Serrano J (2009) Biodiversity, conservation and current threats to European honeybees. Apidologie 40:263–284.
31. Chauzat M-P, Faucon J-P, Martel A-C, Lachaize J, Cougoule N, et al. (2006) A survey of pesticide residues in pollen loads collected by honey bees in France. J Econ Entomol 99:253–262.
32. Mullin CA, Frazier M, Frazier JL, Ashcraft S, Simonds R, et al. (2010) High levels of miticides and agrochemicals in North American apiaries: Implications for honey bee health. PLoS ONE 5(3): e9754.
33. Frazier J, Mullin C, Frazier M, Ashcraft S (2011) Managed pollinator coordinated agricultural project: Pesticides and their involvement in colony collapse disorder. Am Bee J 151(8): 779–784.
34. Cox-Foster DL, Conlan S, Holmes EC, Palacios G, Evans JD, et al. (2007) A metagenomic survey of microbes in honey bee colony collapse disorder. Science 318: 283–287.
35. Singh R, Levitt AL, Rajotte EG, Holmes EC, Ostiguy N, et al. (2010) RNA Viruses in Hymenopteran Pollinators: Evidence of Inter-Taxa Virus Transmission

via Pollen and Potential Impact on Non-*Apis* Hymenopteran Species. PLoS ONE 5(12): e14357.

36. Bromenshenk JJ, Henderson CB, Wick CH, Stanford MF, Zulich AW, et al. (2010) Iridovirus and microsporidian linked to honey bee colony decline. PLoS ONE 5: e13181.

37. Ciarlo TJ, Mullin CA, Frazier JL, Schmehl DR (2012) Learning Impairment in Honey Bees Caused by Agricultural Spray Adjuvants. PLoS ONE 7(7): e40848.

38. Evison SEF, Roberts KE, Laurenson L, Pietravalle S, Hui J, et al. (2012) Pervasiveness of Parasites in Pollinators. PLoS ONE 7(1): e30641. doi:10.1371/journal.pone.0030641

39. Andersson GKS, Rundlöf M, Smith HG (2012) Organic Farming Improves Pollination Success in Strawberries. PLoS ONE 7(2): e31599. doi:10.1371/journal.pone.0031599.

40. Anonymous (2012) Pennsylvania Tree Fruit Production Guide, 324 p. Available: http://agsci.psu.edu/tfpg. Accessed 2013 July 31.

41. Hull LA, Joshi NK, Zaman F (2009) Management of internal feeding lepidopteran pests in apple, 2008. Arthropod Mgmt Tests *34*: doi:10.4182/amt.2009.A8

42. Biddinger DJ, Hull LA (1995) Effects of several types of insecticides on the mite predator, Stethorus punctum (Coleoptera: Coccinellidae), including insect growth regulators and abamectin. J Econ Entomol 88: 358–366.

43. Summary of a SETAC Pellston Workshop. Pensacola FL (USA): Society of Environmental Toxicology and Chemistry (SETAC). 45 p.

44. Cresswell JE, Page CJ, Uygun MB, Holmbergh M, Li Y, et al. (2012) Differential sensitivity of honey bees and bumble bees to a dietary insecticide (imidacloprid). Zoology 115: 365–371.

45. Cresswell JE, Laycock I (2012) Towards the comparative ecotoxicology of bees: the response-response relationship. Julius-Kühn-Archiv 437: 55–60

46. Decourtye A, Henry M, Desneux N (2013) Overhaul pesticide testing on bees. Nature 497:168.

47. LeOra Software, PoloPlus (2005) Available: http://www.LeOraSoftware.com. Accessed 2013 July 31.

48. Robertson JL, Russell RM, Preisler HK, Savin NE (2007) Bioassays with Arthropods. Boca Raton: CRC Press. 199 p.

49. LeOra Software, PoloMix (2005) Available: http://www.LeOraSoftware.com. Accessed 2013 July 31.

50. Iwasa T, Motoyama N, Ambrose JT, Roe RM (2004) Mechanism for the differential toxicity of neonicotinoid insecticides in the honey bee, *Apis mellifera*. Crop Prot 23: 371–378.

51. Robertson JL, Boelter LM, Russell RM, Savin NE (1978) Variation in response to insecticides by Douglas-fir tussock moth, *Orgyia pseudotsugata* (Lepidoptera: Lymantriidae), populations. Can Entomol 110: 325–28.

52. Robertson JL, Preisler HK, Ng SS, Hickle L, Gelernter W (1995) Natural variation: A complicating factor in bioassays with chemical and microbial insecticides. J Econ Entomol 88: 1–10.

53. Hopwood J, Vaughan M, Shepherd M, Biddinger D, Mader E, et al. (2012) Xerces Society for Invertebrate Conservation 32 p.

54. Brunner JF, Dunley JE, Doerr M, Beers EH (2001) Effects of pesticides on *Colpoclypeus florus* (Hymenoptera: Eulophidae) and *Trichogramma platneri* (Hymenoptera: Trichogrammatidae), parasitoids of leafro00llers in Washington. J Econ Entomol 94: 1075–1084.

55. Biddinger DJ, Hull LA, LT (2010) Wooly apple aphid control in Pennsylvania apple orchards and impact of Delegate on *Aphelinus mali*. 84th Orchard Pest & Disease Management Conference Abstract, Jan. 14–16, 2010. Portland, OR. Available: http://entomology.tfrec.wsu.edu/wopdmc. Accessed 2013 July 31.

56. Jones VP, Steffan SA, Hull LA, Biddinger DJ (2010) Effects of the loss of organophosphate pesticides in the U.S.: opportunities and needs to improve IPM programs. Outlooks on Pest Management 21:161–66.

57. Roessink I, van der Steen JJM, Kasina M, Gikungu M, Nocelli RCF (2011) Is the European honey bee (*Apis mellifera mellifera*) a good representative for other pollinator species? SETAC Europe 21st Annual Meeting: Ecosystem Protection in a Sustainable World: a Challenge for Science and Regulation. Milan, Italy, 15–19 May 2011.

58. http://www.epa.gov/pesticides/ecosystem/pollinator/then-now.html. Accessed 2013 July 31.

59. Desneux N, Decourtye A, Delpuech JM (2007) The sublethal effects of pesticides on beneficial arthropods. Ann. Rev. Entomology 52:81–106.

Identification of Land-Cover Characteristics Using MODIS Time Series Data: An Application in the Yangtze River Estuary

Mo-Qian Zhang, Hai-Qiang Guo, Xiao Xie, Ting-Ting Zhang, Zu-Tao Ouyang, Bin Zhao*

Coastal Ecosystems Research Station of the Yangtze River Estuary, Ministry of Education Key Laboratory for Biodiversity Science and Ecological Engineering, Institute of Biodiversity Science, Fudan University, Shanghai, P.R. China

Abstract

Land-cover characteristics have been considered in many ecological studies. Methods to identify these characteristics by using remotely sensed time series data have previously been proposed. However, these methods often have a mathematical basis, and more effort is required to better illustrate the ecological meanings of land-cover characteristics. In this study, a method for identifying these characteristics was proposed from the ecological perspective of sustained vegetation growth trend. Improvement was also made in parameter extraction, inspired by a method used for determining the hyperspectral red edge position. Five land-cover types were chosen to represent various ecosystem growth patterns and MODIS time series data were adopted for analysis. The results show that the extracted parameters can reflect ecosystem growth patterns and portray ecosystem traits such as vegetation growth strategy and ecosystem growth situations.

Editor: Gil Bohrer, The Ohio State University, United States of America

Funding: This research was financially supported by the National Basic Research Program of China (No. 2013CB430404), the Natural Science Foundation of China (grant No 31170450), and the National Key Technology R&D Program (No. 2010BAK69B15). The funders had no role in study design, data collection and analysis, decision to publish, or preparation of the manuscript.

Competing Interests: The authors have declared that no competing interests exist.

* E-mail: zhaobin@fudan.edu.cn

Introduction

Land-cover characteristics and their dynamics have captured much attention in the field of ecology, since land-cover exerts a huge influence over ecosystem biodiversity, water budget [1], energy flow [2], and carbon cycling [3]. Remotely sensed time series data provide an opportunity to identify land-cover characteristics at the temporal scale, which often reflect the features of ecosystem growth patterns. Ecosystem growth patterns can be categorized into four types (adapted from [4]): (i) undisturbed ecosystems; (ii) ecosystems that have suffered coverage damage that either lasted the whole growing season or followed by vegetation restoration in the growing season; (iii) ecosystems that have suffered a phenology change that is expressed as either a shift in the growing season or a shortened growing season; and (iv) ecosystems that underwent changes in both coverage and phenology. However, it is challenging to extract desired land-cover characteristics while remaining independent of inter-annual and inter-class variations [1]. Therefore, proper land-cover characteristic identification methods are needed.

Methods that take into account the temporal features of time series data to identify land-cover characteristics have been developed in recent decades; such methods can be roughly classified into two types. The first type is based on signals observed at different temporal scales: vegetation information is often present at seasonal and inter-annual scales, while noise typically has a higher frequency. By decomposing data into different temporal frequencies, noises can be excluded and parameters can be obtained to reflect long-term trends or seasonal patterns. Research

based on this kind of method includes land-cover classification [5] and long-term vegetation dynamic study [6]. However, the ecological meaning of parameters obtained by this kind of method is often limited, and the relations between parameters and land-cover dynamics need further investigation. The second type of methods is based on land surface phenological stages. The phenological stages recognized by time series data include: (i) constant low/no leaf period in winter when the vegetation is dormant, (ii) rapid vegetation growth period in spring, (iii) a period with relatively stable high aboveground biomass in summer, and (iv) rapid senescence period in autumn [7]. Research based on such methods can provide more detailed ecological information (Table 1) that can be applied to study land surface phenology [8], vegetation response to changing climate [9], zoology [10], and so on.

Though methods based on phenological stages have been widely used in ecological studies, phenological stages are often detected based on mathematical criteria such as choosing a certain threshold or detecting curve changes [8,11]. However, it is difficult to choose a mathematically ideal technique [11], and different analysis methods sometimes provide conflicting results on the same research topic (such as the long-term greenup trend in North America [8]). In this study, we propose a method to identify land-cover characteristics from the ecological perspective of sustained vegetation growth. During the analysis, phenological growth stages were first identified based on sustained vegetation growth trends, and parameters designed to reflect land-cover characteristics were extracted accordingly. Improvement was also made in parameter

Table 1. Summary of vegetation metrics used in time series analysis.

Vegetation metric	Interpretation	References
Greenup	Time represents the start of growing season when plant grows and photosynthesis begins	[10,21,22]
Maturity	Time when green leaf area stabilizes with high photosynthesis activity	[22]
Senescence	Time when plant begins senescence either expressed by green biomass decrease or reduced photosynthesis	[22]
Dormancy	Time represents the end of growing season when photosynthesis reaches its minimum and plants become dominant	[10,21,22]
Length of growing season	Time span between greenup and dormancy which represents the duration of photosynthetic activity	[10,21]
Maximum VI	Highest VIs level in growing season	[21]
Timing of maximum VI	Time when VIs reaches its maximum	[10,21]
Seasonal amplitude	VIs value difference between vegetation dormancy and have the highest aboveground biomass	[10,21]
Annual integration	Sum of VIs values in growing season	[21]
Greenup rate	Growth rate during the period between greenup and mature	[10,21]
Senescence rate	Senescence rate during the period between senescence and dormancy	[10,21]

extraction, which was inspired by a technique used for extracting the hyperspectral red edge position.

Materials and Methods

Ethics Statement

As a field survey conducted for remote sensing research, we did not conduct any activities concern field samplings of soil, plants, or animals in the work. All lands where we conducted the survey are non-fenced public areas and accessed to everyone, thus we do not need to ask for any official permission.

Site Description

This study was conducted on the Chongming Island and the Changxing Island, two alluvial islands in the mouth of the Yangtze River, China (121°10′49″ –121°59′10″E, 31°17′4″ –31°54′20″N, Fig. 1). The area is subject to the northern subtropical monsoon climate, with an average annual temperature of 15.3°C and a total annual precipitation around of 1000 mm. Several large land reclamations have taken place since 1960s, the reclaimed areas are much larger than ordinary farmland and neighboring areas are often under the same land management schemes. Diverse land use and a relatively large reclamation area make the study area suitable for identifying land-cover characteristics with remote sensing data.

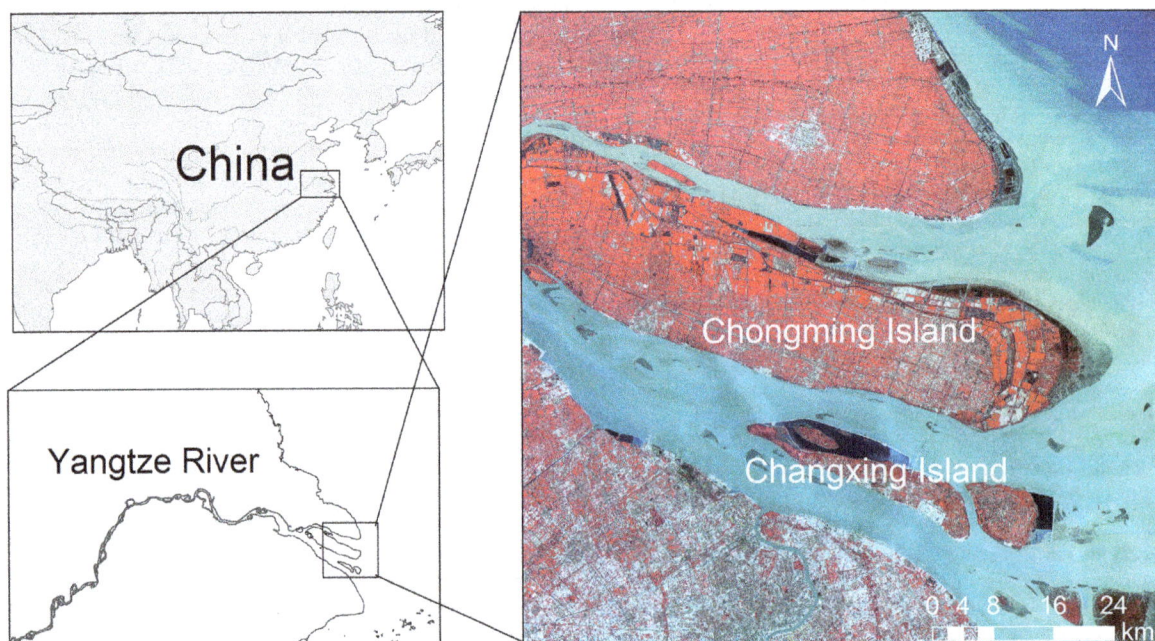

Figure 1. Location of the study area.

Table 2. Descriptions of different land-cover types in study area.

Land-cover types	Description	Vegetation coverage	Disturbance pattern
Urban	Urban area	Low to medium	No
Orchard	Orange tree plantation area	Low to medium	No
Fallow	Farmland where no farming activities conducted, usually covered by natural herbaceous plants such as weed and common reed	Medium to high	No
Cropland-1	Single-cropping farmland with only rice planted from late May to October	High	Yes; Happened early in the year
Cropland-2	Double-cropping farmland with winter wheat planted from late last November to early May and rice planted the same time as cropland1	High	Yes; Happened in the mid year

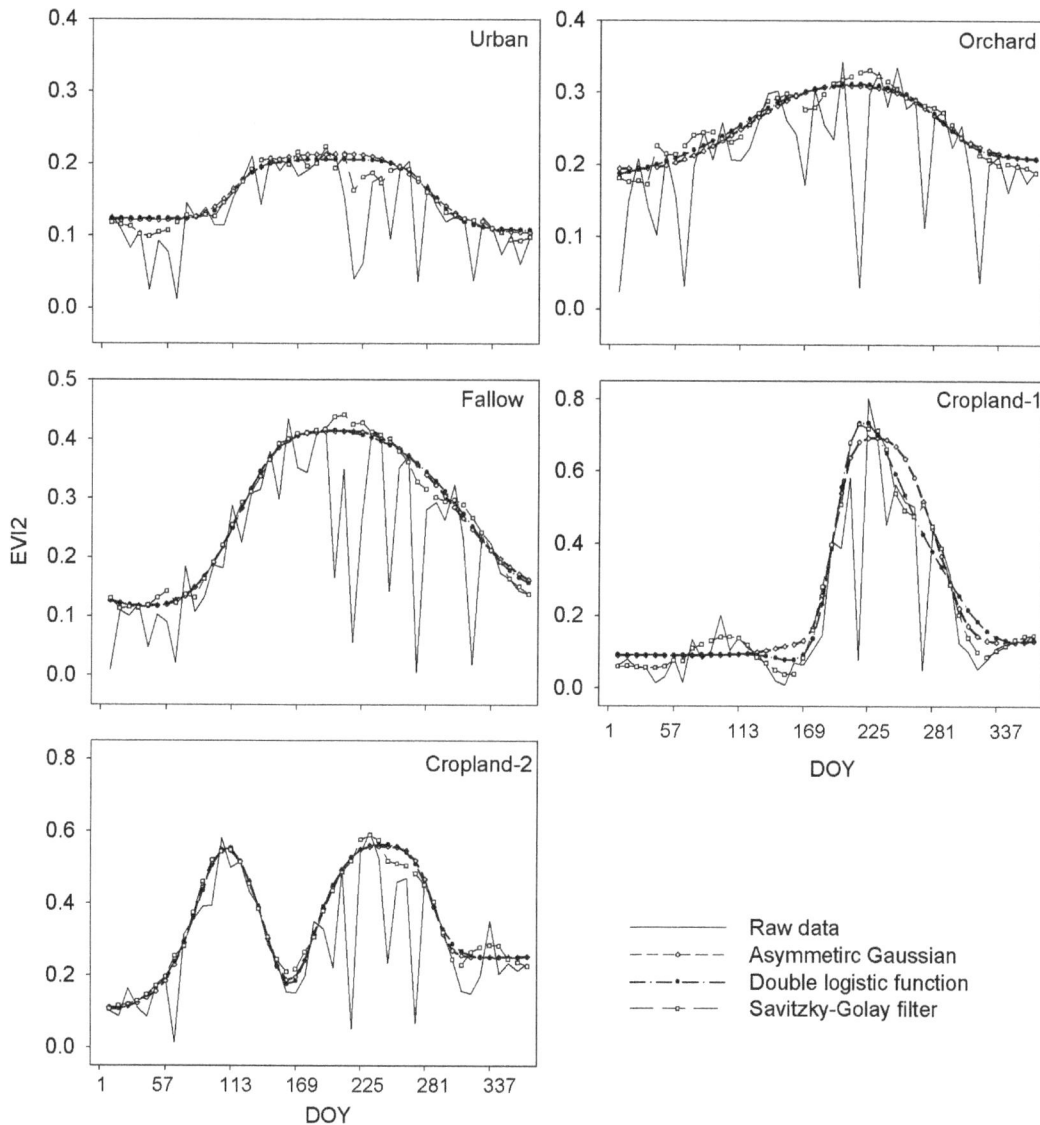

Figure 2. Time series EVI2 data of land cover types processed before and after noise reduction methods.

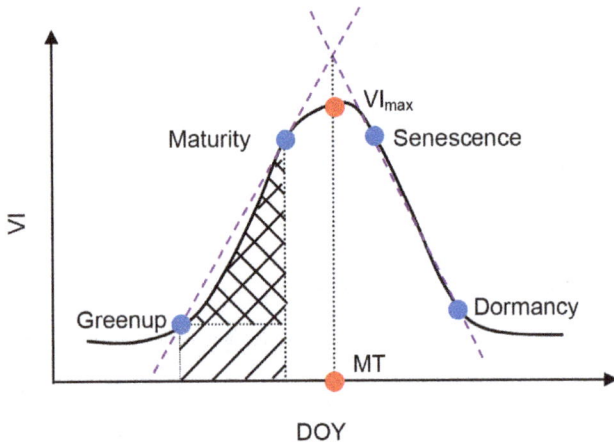

Figure 3. Diagram of parameter extraction from time series vegetation index (VI) data. Blue points represent time points separating different vegetation growth stages, while red points are parts of the extracted parameters.

Analysis Preparation

Remote sensing data. A 250 m 8-day composite surface reflectance data set (MOD09Q1) was used for this study. Satellite quality assurance (QA) data were obtained for further noise reduction, and selected data were derived from MOD09A1 because QA data from MOD09Q1 are insufficient to deliver the actual condition. Remote sensing data for the year 2009, when several field surveys were conducted, were used for analysis. All the remote sensing data used were downloaded from NASA (LP DAAC).

Vegetation indices (VIs) are specially designed indicators that reflect certain properties, such as vegetation coverage (*e.g.*, NDVI, EVI & MSAVI) and land surface water content (*e.g.*, LSWI). A two-band EVI (EVI2) was selected in this study for its superiority over the widely used NDVI [12]. EVI2 is calculated as follows:

$$EVI2 = 2.5 \times (N - R)/(N + 2.4 \times R + 1) \qquad (1)$$

where N and R are reflectance in the near-infrared (NIR) and red bands of MODIS data, respectively.

Field survey. In order to acquire the actual land-cover conditions in different seasons, we conducted three field surveys across the year of 2009. Before the first field survey, historical TM and airborne imageries were studied in the laboratory to identify the relatively homogenous regions for field surveys. During the field surveys, a portable Global Position System (GPS) was used to localize the target ground objects such as cultivated lands, fallow lands, orchards, and buildings. To aid this task, color maps of TM and airborne images were printed beforehand and taken with the investigators for field checks. The field notes were also made and taken to the laboratory for further analysis, such as location check, classification and accuracy assessment.

Land-cover selection. The studied land-cover types were chosen based on ecosystem growth patterns, and five land-cover types (urban, orchard, fallow, and two types of croplands) were chosen for further analysis (Table 2). Among them, urban, orchard, and fallow were used to represent ecosystems that experience a loss in coverage throughout the growing season; cropland-2 was used to represent ecosystems with short-term coverage loss; and cropland-1 was used to represent ecosystems under a growing season shift. Since in the study area no land-cover

type showed the characteristics of ecosystems under a shortened growing season, this ecosystem growth pattern was not included in the present analysis. To better analyze land-cover characteristics, remote sensing pixels that represent only one land-cover type were used in the analysis.

Analysis Techniques

Noise reduction. The asymmetric Gaussian method [13] and double logistic function [14] were chosen for noise reduction in this study, since their ability to maintain the integrity of signals is proven [15]. The Savitzky-Golay filter was also chosen since it could capture detailed variations in time series data and has shown good performance when applied in study related to China [16]. Noise reduction was achieved by using TIMESAT [13,17,18]. Ancillary weights of each data were set according to the QA data. Weights were set at high values for best-quality data (described as *clear* in QA data), at moderate values when data were acquired under less ideal conditions (*cloud shadow* or *mixed*), and at low values when data represent cloudy pixels. Fig. 2 shows the data of different land-cover types represented by EVI2 before and after noise reduction.

Phenological stages discrimination. Though rates of changes in vegetation coverage may vary, the vegetation growth trend inherited in each phenology stage (sustained increase/ decrease, or consistency) remained constant for a certain time; therefore, we propose to discriminate phenological stages based on the sustained vegetation growth trend. The sustained trend was recognized by the following procedure: if the increment/decrement between neighboring data was larger than a certain numerical value (the theoretical increase/decrease threshold), we defined it as an increase or a decrease; and if the increment/ decrement remained constant for some time (for instance more than one month), the period would be identified as showing a sustained increase/decrease trend. Time points (greenup, maturity, senescence, and dormancy; see Table 1) that separate these phenological stages were identified accordingly. Greenup and maturity were identified as the beginning and ending of the period when vegetation showed a sustained increase trend, respectively; the beginning and ending of the sustained decrease trend were termed as senescence and dormancy, respectively.

The theoretical increase/decrease threshold was calculated as:

$$Threshold = (EVI2_{max} - EVI2_{min})/n \qquad (2)$$

where $EVI2_{max}$ represents the maximum value of each time series data. Because the aboveground biomass of evergreen vegetation may vary in winter, when calculating the theoretical increase/ decrease threshold, $EVI2_{min}$ used the minimum value in the first/ second half of the year, respectively. The variable n represents the period when vegetation biomass increases/decreases. The length of this period can be determined from long-term field observations. As the theoretical increase/decrease threshold is not supposed to give a quantitative value, the time period used can be longer than actual value. In this study, we simply assumed that the growing season spans the whole year, with vegetation biomass increase and decrease period accounting for half a year each. Further, the corresponding number of MODIS data was used to represent this period. If there were more than one sustained increase periods, the first period was used to identify greenup and maturity. Senescence and dormancy were identified in a similar way, except that the last sustained decrease period was used for the identification when more than one sustained decrease period existed.

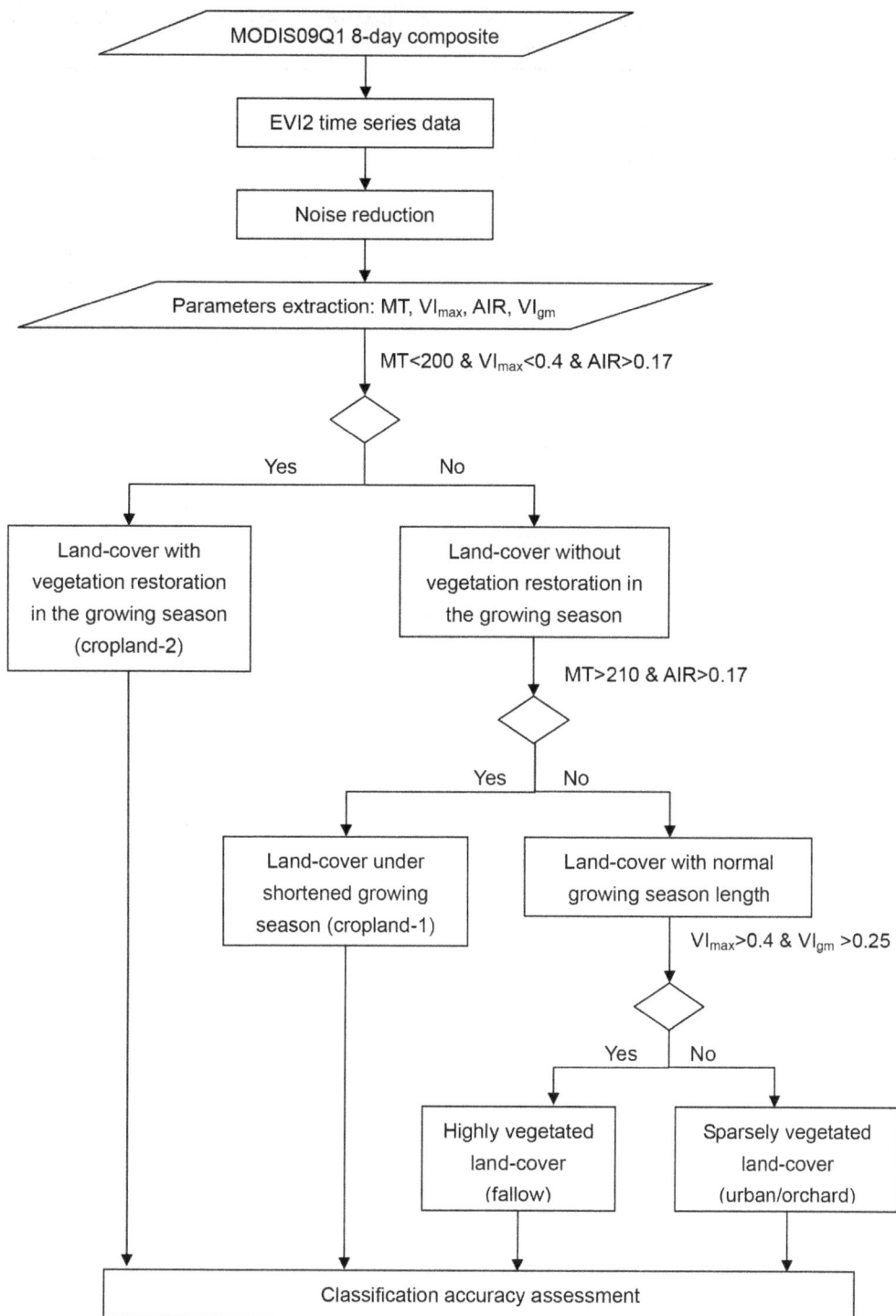

Figure 4. Workflow of hierarchical scheme for land-cover classification. All thresholds that been used in different noise reduction methods were same.

Parameters extraction. The time at which aboveground biomass reaches its maximum (MT, a date) was first identified. MT was extracted by extrapolating two straight lines across the time points that discriminate phenological stages (Fig. 3). This process was inspired by a technique used in hyperspectral analysis, which stabilizes the red edge position when there are multiple peaks in the first derivative curve of hyperspectral data [19]. The

EVI2 extracted on day MT was used to represent the maximum vegetation coverage (VI_{max}, dimensionless). If MOD09Q1 data were missing for that day, VI_{max} was linearly interpolated between the previous and following data.

Two other parameters were further extracted to reflect the vegetation growth status. The average increase rate (AIR, dimensionless) between greenup and maturity was calculated to

Table 3. The mean value and standard deviation (SD) of parameters extracted from time series vegetation index (VI) data with asymmetric Gaussian method (A), double logistic function (B), and Savitzky-Golay filter (C).

A	MT		VI_{max}		AIR		VI_{gm}	
	mean	SD	mean	SD	mean	SD	mean	SD
Urban	200	12.493	0.198	0.035	0.043	0.006	0.149	0.006
Orchard	205	31.842	0.370	0.047	0.059	0.021	0.308	0.041
Fallow	194	10.799	0.496	0.052	0.167	0.025	0.318	0.033
Cropland-1	228	6.799	0.618	0.082	0.250	0.049	0.391	0.035
Cropland-2	166	12.356	0.200	0.070	0.223	0.031	0.327	0.028
B	MT		VI_{max}		AIR		VI_{gm}	
	mean	SD	mean	SD	mean	SD	mean	SD
Urban	204	13.345	0.197	0.037	0.038	0.006	0.149	0.031
Orchard	216	26.139	0.372	0.047	0.053	0.019	0.310	0.043
Fallow	191	10.796	0.504	0.056	0.177	0.029	0.313	0.033
Cropland-1	229	10.878	0.625	0.085	0.252	0.067	0.390	0.040
Cropland-2	168	11.090	0.188	0.081	0.216	0.032	0.327	0.029
C	MT		VI_{max}		AIR		VI_{gm}	
	mean	SD	mean	SD	mean	SD	mean	SD
Urban	189	20.587	0.188	0.081	0.034	0.015	0.140	0.037
Orchard	185	40.105	0.350	0.052	0.063	0.025	0.290	0.055
Fallow	192	10.531	0.495	0.069	0.154	0.041	0.288	0.055
Cropland-1	228	9.770	0.628	0.092	0.285	0.077	0.349	0.047
Cropland-2	169	12.665	0.199	0.076	0.229	0.032	0.331	0.035

reflect how vegetation grows from the minimum vegetation coverage to a relatively stable status. The mean EVI2 between greenup and maturity (VI_{gm}, dimensionless) focuses on the average status in the sustained growing period

$$AIR = \sum_{a}^{b} \Delta VI_i \bigg/ (T_{growth} - 1) \quad (3)$$

$$VI_{gm} = \sum_{a}^{b} VI_i \bigg/ T_{growth} \quad (4)$$

where a and b represent greenup and maturity respectively; $\sum_{a}^{b} \Delta VI_i$ represents the accumulated increments of EVI2 in the sustained growing period (backslash region in Fig. 3); $\sum_{a}^{b} VI_i$ is the EVI2 accumulation in the same period (slash region in Fig. 3); T_{growth} represents the time span between greenup and maturity (Fig. 3), and we used the number of MODIS data to represent this period.

Different ecosystem growth patterns can be expressed by parameter differentiations. Coverage differentiation would be most evident for ecosystems that have suffered vegetation coverage loss lasted the growing season, and the maximum vegetation coverage (hence the values of VI_{max}) would then reduce

accordingly. Since total vegetation coverage increase/decrease status is related to the maximum coverage, the values of AIR and VI_{gm} will also decrease. In ecosystems that are under short-term vegetation loss, VI_{max} will more or less represent the coverage during the period of vegetation damage and not the maximum coverage in the growing season; therefore, VI_{max} value will decrease. However, the changes in phenology or total vegetation coverage (and hence parameters of MT, AIR, and VI_{gm}) depend on the severity and duration of the damage. Phenology differentiation is the most obvious characteristic of an ecosystem undergoing a growing season shift, and the value of MT would change accordingly. Ecosystems with shortened growing seasons exhibit slight shifts of phenology and accelerated vegetation coverage increase/decrease rates. All parameters would change in ecosystems that undergo both coverage and phenological changes.

Land-cover classification. In order to test whether the extracted parameters could be used for actual land-cover change detection, a hierarchical classification scheme was adopted to classify the studied land-cover types (Fig. 4). The parameters used in each classification level were chosen based on the aim of the classification, with each parameter aimed to discriminate only one aspect of the land-cover (sparsely/densely planted, with/without phenological shift, or high/low growth rate). For example, in this study, both coverage and phenology in cropland-2 are distinct from other land-cover types, hence three parameters, MT, VI_{max}, and AIR were chosen in the first classification level. Land-cover types were not sub-classified artificially from each other if no apparent differences in ecosystem features were detected.

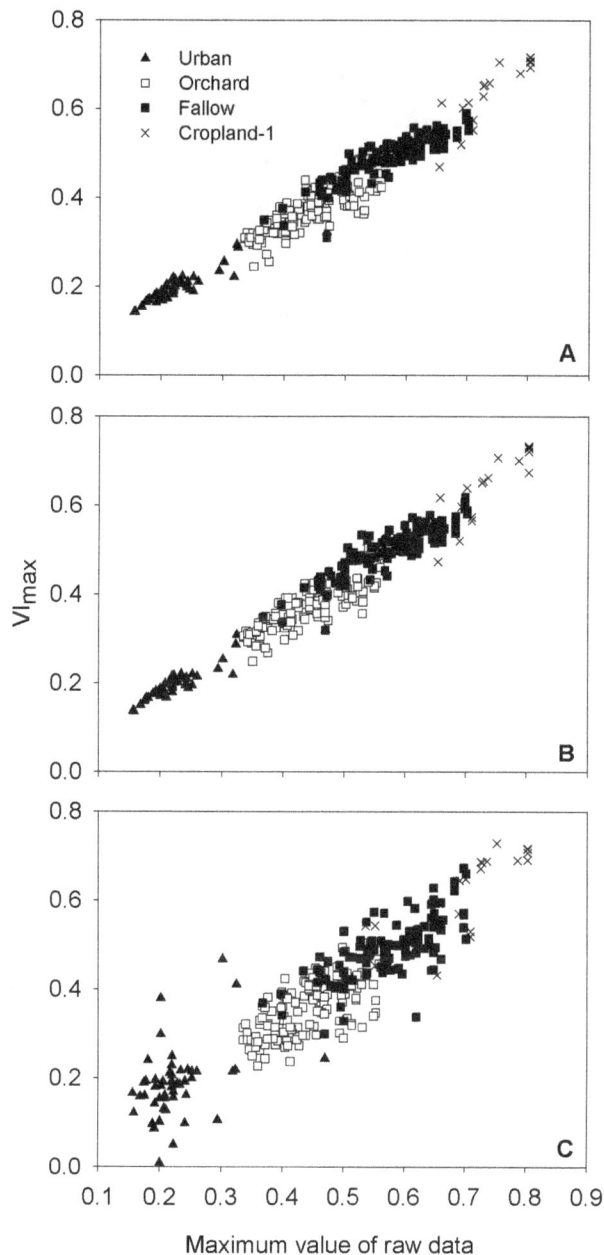

Figure 5. Results evaluation by comparing derived VI$_{max}$ with corresponding maximum values of raw data. Frames showed results that obtained with asymmetric Gaussian method (A) ($R^2 = 0.944$), double logistic function (B) ($R^2 = 0.947$), and Savitzky-Golay filter (C) ($R^2 = 0.837$) respectively.

The thresholds used for classification were set roughly according to predefined criteria rather than on the basis of training data, and hence all original data were used for validation. The thresholds were defined by the following criteria. The threshold of VI$_{max}$/VI$_{gm}$ was set as the arithmetic mean value of soil background and the highest/mean VI value of pixels with the highest vegetation coverage. MT, being representative of phenological information, would change in ecosystems under phenological changes. The threshold of AIR was set as the arithmetic mean value of the observed highest values and the lowest ones. All thresholds used in classification are the same for data processed by different noise

reduction methods. Confusion matrixes were used to evaluate classification accuracies.

Results

Basic Characteristics of Land-cover Types

The extracted parameters can reflect the basic characteristics of different land-cover types (Table 3). MT value changes reflect changes in vegetation phenology. In cropland-1, MT values changes as the growing season has shifted, and these values are the largest among all land-cover types. In cropland-2, since human interference has actually altered vegetation phenology, the MT values have also shifted and are the smallest of all land-cover types. Urban, orchard, and fallow have intermediate MT values, which reflects the fact that the vegetation phenology has not changed here.

VI$_{max}$ reflect the changes in ecosystem coverage. Since cropland-1 did not undergo vegetation coverage loss, this land-cover type has the highest VI$_{max}$ values (Table 3). VI$_{max}$ values of fallow, orchard, and urban decrease with reduced vegetation coverage. Because the tree density in orchard areas is not high, vegetation coverage of orchards is no larger than that of fallow areas (as indicated in Fig. 5); hence, it is understandable that the average VI$_{max}$ values of orchard are lower than those of fallow. In cropland-2, as MT occurs during the time right after rice transplantation at when the land is barely covered, the values of VI$_{max}$ are not high.

AIR and VI$_{gm}$ are parameters that reflect vegetation growth status. On the whole, the change patterns of AIR and VI$_{gm}$ values are similar to those of VI$_{max}$, with cropland-1 having the highest values, followed by fallow and orchard, and urban areas having the smallest values. However, in cropland-2, as VI$_{max}$ values do not reflect the maximum vegetation coverage in the growing season, AIR and VI$_{gm}$ values do not follow the trend exhibited in VI$_{max}$.

MT Results Evaluation

An evaluation was performed on MT results to illustrate the variation in the values (Table 3), because although MT values are quite similar for data processed by different noise reduction methods in fallow, cropland-1, and cropland-2, the results of urban and orchard varied with methods and much lower MT values were obtained when using the Savitzky-Golay filter. As there is no readily available evaluation method, we chose an indirect means of assessment. VI$_{max}$ is based on the position of MT, and a departure of MT from the time of the highest aboveground biomass would result in a decrease in the VI$_{max}$ value. Thus, the VI$_{max}$ value can serve as an indicator for MT evaluation, and comparisons of VI$_{max}$ with the corresponding maximum values of raw data are shown in Fig. 5. The asymmetric Gaussian method and double logistic function provided satisfying results; however, the coefficient of the relationship (R^2) is much lower when using the Savitzky-Golay filter, which indicates greater errors in data processed by this method. Therefore, the observed MT variations in Table 3 should be the result of unstable performance of the Savitzky-Golay filter when it is applied for areas with low vegetation coverage, as vegetation signals are weak and more sensitive to noise in such areas. Cropland-2 was excluded from this evaluation because its VI$_{max}$ values do not reflect the maximum vegetation coverage in the growing season.

Ecosystem Traits Detection

Because the species composition varied among ecosystems, the vegetation growth condition expressed at ecosystem scale differed,

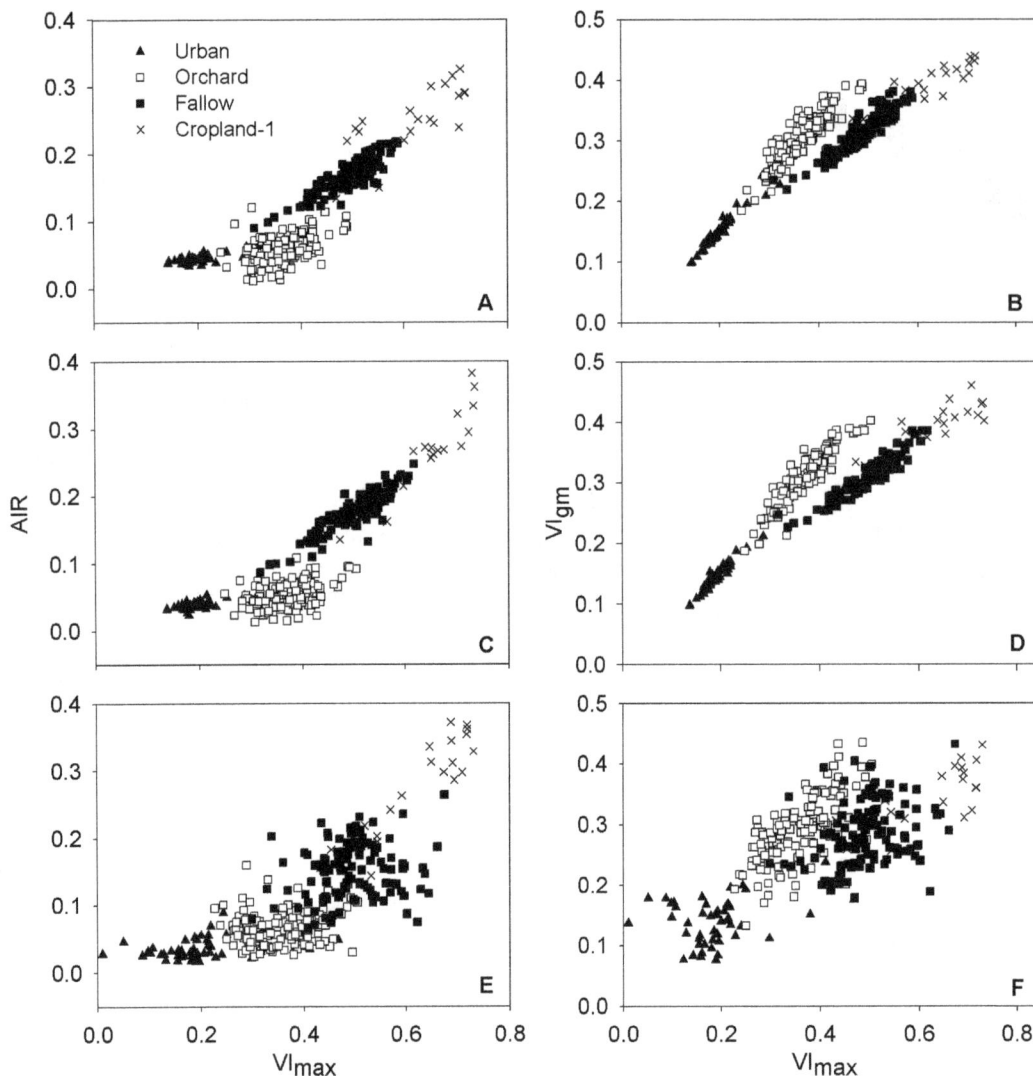

Figure 6. The cross-comparisons among parameters extracted. The left column represent the comparisons between AIR and VI$_{max}$ obtained via asymmetric Gaussian method (A), double logistic function (C), and Savitzky-Golay filter (E), respectively. The right column represent the comparisons of VI$_{gm}$ with VI$_{max}$ by using asymmetric Gaussian method (B), double logistic function (D), and Savitzky-Golay filter (F).

and this trait is inherent in ecosystems. However, AIR and VI$_{gm}$ cannot be used for direct detection of such differences, because the influence of coverage would hide them. By using VI$_{max}$, the coverage differentiation can be partially minimized (Fig. 6), and vegetation growth traits can be conveyed through slope changes.

The differences between vegetation growth rates are shown in Fig. 6A, C, and E. In Fig. 6A and C, slopes of cropland-1 and fallow are larger than those of urban and orchard, indicating that under same coverage, cropland-1 and fallow grow faster. Although intra-class variations in data processed by Savitzky-Golay filter are larger than those in the data processed by the other two methods, a similar pattern could also be observed (Fig. 6E). Since orchard areas comprising woody plants and trees are the dominate urban vegetation in the study area, the slope differences seen in the figures indicate the differences between herbaceous vegetation and woody plants, as herbaceous plants grow faster that woody plants under suitable conditions.

The differences between ecosystem average growing conditions are shown in Fig. 6 B, D, and F. In these figures, the slopes of

orchard are higher than those of cropland-1 and fallow, indicating that the average coverage of orchard is larger at ecosystem level. Since the woody plants in orchard are evergreen and herbaceous plants senescence every year, the aboveground biomasses are different when spring comes, therefore, the coverage of woody plants increased faster. Because some of the trees are deciduous in urban area, the slopes of urban are slightly lower than those of orchard. Cropland-2 was excluded in this part of analysis since its VI$_{max}$ values do not reflect the maximum vegetation coverage in the growing season.

Classification Accuracy Assessment

Confusion matrixes were used to evaluate classification accuracy, (Table 4) and were made for data processed by each different noise reduction method. All noise reduction methods achieve relatively high overall classification accuracy. The user's accuracy of fallow is the lowest in all methods, and errors mainly arise from misclassification of urban/orchard. In our study, the chosen land-cover types are only based on actual land surface situations and we

Table 4. Land classification accuracy assessments of data processed by different smoothing methods (A) asymmetric Gaussian method, (B) double logistic function, and (C) Savitzky–Golay filter.

A	Reference data					
Classification	Urban/orchard	Fallow	Cropland-1	Cropland-2	Total	User's accuracy (%)
Urban/orchard	144	5	0	29	178	80.90
Fallow	36	115	2	9	162	70.99
Cropland-1	0	2	19	0	21	90.48
Cropland-2	0	0	0	439	439	100
Total	180	122	21	477		
Producer's accuracy (%)	80	94.26	90.48	92.03		
Overall accuracy: 89.63%			Kappa: 0.825			

B	Reference data					
Classification	Urban/orchard	Fallow	Cropland-1	Cropland-2	Total	User's accuracy (%)
Urban/orchard	147	5	0	34	186	79.03
Fallow	33	117	2	16	168	69.64
Cropland-1	0	0	19	0	19	100
Cropland-2	0	0	0	427	427	100
Total	180	122	21	477		
Producer's accuracy (%)	81.67	95.90	90.48	89.52		
Overall accuracy: 88.75%			Kappa: 0.811			

C	Reference data					
Classification	Urban/orchard	Fallow	Cropland-1	Cropland-2	Total	User's accuracy (%)
Urban/orchard	144	29	0	18	191	75.39
Fallow	36	91	2	13	142	64.08
Cropland-1	0	1	19	0	20	95
Cropland-2	0	1	0	446	447	99.78
Total	180	122	21	477		
Producer's accuracy (%)	80	74.59	90.48	93.50		
Overall accuracy: 87.5%			Kappa: 0.786			

Integers in tables represent the number of pixels that belongs to a certain classification condition.

did not set any predefined coverage criterion for data selection. Therefore, the actual coverage of fallow, urban, and orchard can overlap (also indicated in Fig. 5), and hence misclassifications are acceptable.

Discussion

Phenological Stages Identification

Remote sensing phenological stages that used to extract land-cover characteristics are often identified by discriminating the time points that separate them. However, it is difficult to choose a mathematically ideal method [11]. Furthermore, as these points are timings that represent dynamic vegetation growth conditions, it is difficult to directly evaluate the results from field phenology observations, because the two kinds of data are not measured at the same spatial scale and often represent different ground phenological events [8]. Therefore, we turned to the sustained vegetation growth trend that phenological stages inherently

exhibited, and time points were thus identified. In this identification process, as the theoretical increase/decrease threshold is not to give a precise quantitative value, the process can be flexible when applied to large scale analysis. Besides, this method can adjust itself according to the maximum and minimum values of the time series data of each pixel. Although time points were only used for later land-cover characteristic identification in this study, they can also be used in land surface phenology research.

Land-cover Characteristics Identification

Parameters were extracted to reflect land-cover characteristics. In this study, the time of highest vegetation biomass (MT) was detected first, and the VI value that represented maximum vegetation coverage (VI$_{max}$) was identified accordingly. Compared with commonly used methods [10], MT was extracted based on temporal features of time series data rather than by using a single maximum value, this makes it more resistant to variations caused by noise. In ecosystems where growth patterns change as a result of disturbances (either in coverage or phenology caused by events such as insect defoliation, windfall, and wildfire), the time points and therefore MT would change consequently; hence, MT can also be used as an indicator of disturbances. This is especially convenient if an irregular growing season was caused by such a disturbance. When MT is combined with VI_{max}, subtle vegetation damages can be more evident. However, the time points of greenup and dormancy are sensitive to the start of spring and the end of autumn, which make these time points vulnerable to inter-annual meteorological variations. In order to obtain a more stable inter-annual result of MT, adjustments such as use of meteorology data or reference area are recommended.

AIR and VI_{gm} can portray ecosystem traits that represent how ecosystems grow. The trait difference between ecosystems with different vegetation composition is especially evident when coverage differences are minimized (Fig. 6). Though effort has been made to discriminate land-cover types that have different species composition by comparing growing season NDVI [20], this method can further explore the temporal features of ecosystems. Therefore, this method has potential for monitoring land-cover changes caused by species variation (such as species invasion and vegetation succession). Similar parameters extracted from vegetation biomass decrease period can also be used for detecting how vegetation senescence. This kind of information could help us to understand ecosystem changes in more detail, and help us to further explore ecosystem processes and functions, as well as the causes of the ecosystem changes.

Land-cover Classification

As the extracted parameters incorporated both spectral and temporal features, land-cover characteristics can be better explored. Results of this study show that this kind of land-cover classification can achieve relatively satisfying results in practice. Classification schemes that include these parameters will facilitate land-cover mapping in complicated situations, such as in regions where the differences between land-cover types are subtle, or in areas with irregular growing seasons.

The Performance of Noise Reduction Methods

Although an 8-day composition scheme is adopted in MODIS products, the presence of cloud remains a problem in retrieving land-cover characteristics in our study area. Therefore, the performance of noise reduction methods affects the ultimate results. Our results show that the asymmetric Gaussian method and double logistic function performed better than the Savitzky-Golay filter, and that some apparent discrepancies exist in the Savitzky-Golay filter. For example, in Fig. 5C some VI_{max} values in urban are obviously larger than the maximum values of raw data (such as 0.380 of VI_{max} corresponds to 0.201 of maximum raw data). This indicates larger errors in the noise reduced data, and indicates that the Savitzky-Golay filter is less robust in areas where vegetation is sparse and noises are frequent. It further confirms a conclusion obtained by [15] that the asymmetric Gaussian method and double logistic function can maintain the integrity of signals and that the Savitzky-Golay filter is sensitive to noise. Our results also give a direct illustration that the Savitzky-Golay filter is not suitable to deal with noise contaminated data at the seashore.

Conclusion and Outlook

In this study, we tried to identify land-cover characteristics based on the consideration of sustained vegetation growth trends. During this process, an improvement was also made by simulating a method used for determining the hyperspectral red edge position. Our results show that this method can capture ecosystem growth patterns and more detailed ecosystem traits such as species growing strategy and ecosystem growth status. This method has a potential in land-cover dynamic studies related to vegetation coverage and composition changes (such as ecosystem damage evaluation, invasive species monitoring, and vegetation succession validation), and also in land surface phenology monitoring. When combined with auxiliary data, such as soil properties, or carbon fluxes between land surface and atmosphere, improvement in the understanding of human-environment interactions and influence of changes in one ecosystem on another can be conceived.

Author Contributions

Conceived and designed the experiments: MQZ BZ. Performed the experiments: MQZ HQG XX TTZ ZTO. Analyzed the data: MQZ BZ. Contributed reagents/materials/analysis tools: XX. Wrote the paper: MQZ HQG BZ TTZ ZTO.

References

1. Turner BL, Lambin EF, Reenberg A (2007) The emergence of land change science for global environmental change and sustainability. Proc Natl Acad Sci U S A 104: 20666–20671.
2. Rotenberg E, Yakir D (2011) Distinct patterns of changes in surface energy budget associated with forestation in the semiarid region. Glob Change Biol 17: 1536–1548.
3. Luo Y, Weng E (2011) Dynamic disequilibrium of the terrestrial carbon cycle under global change. Trends Ecol Evol 26: 96–104.
4. Lupo F, Linderman M, Vanacker V, Bartholome E, Lambin EF (2007) Categorization of land-cover change processes based on phenological indicators extracted from time series of vegetation index data. Int J Remote Sens 28: 2469–2483.
5. Geerken RA (2009) An algorithm to classify and monitor seasonal variations in vegetation phenologies and their inter-annual change. ISPRS J Photogramm Remote Sens 64: 422–431.
6. Martínez B, Gilabert MA (2009) Vegetation dynamics from NDVI time series analysis using the wavelet transform. Remote Sens Environ 113: 1823–1842.
7. Duchemin B, Goubier J, Courrier G (1999) Monitoring phenological key stages and cycle duration of temperate deciduous forest ecosystems with NOAA/AVHRR data. Remote Sens Environ 67: 68–82.
8. White MA, de Beurs KM, Didan K, Inouye DW, Richardson AD et al. (2009) Intercomparison, interpretation, and assessment of spring phenology in North America estimated from remote sensing for 1982–2006. Glob Change Biol 15: 2335–2359.
9. Zhang XY, Tarpley D, Sullivan JT (2007) Diverse responses of vegetation phenology to a warming climate. Geophys Res Lett 34:
10. Pettorelli N, Vik JO, Mysterud A, Gaillard JM, Tucker CJ et al. (2005) Using the satellite-derived NDVI to assess ecological responses to environmental change. Trends Ecol Evol 20: 503–510.
11. Reed BC, White M, Brown JF (2003) Remote sensing phenology. In:Schwartz MD editor Phenology: An integrative environmental science. Dordrecht: Kluwer Academic Publishers. 365–381.
12. Jiang ZY, Huete AR, Didan K, Miura T (2008) Development of a two-band enhanced vegetation index without a blue band. Remote Sens Environ 112: 3833–3845.
13. Jönsson P, Eklundh L (2002) Seasonality extraction by function fitting to time-series of satellite sensor data. IEEE Trans Geosci Remote Sens 40: 1824–1832.
14. Beck P, Atzberger C, Hogda KA, Johansen B, Skidmore AK (2006) Improved monitoring of vegetation dynamics at very high latitudes: A new method using MODIS NDVI. Remote Sens Environ 100: 321–334.
15. Hird JN, McDermid GJ (2009) Noise reduction of NDVI time series: An empirical comparison of selected techniques. Remote Sens Environ 113: 248–258.
16. Chen J, Jönsson P, Tamura M, Gu ZH, Matsushita B et al. (2004) A simple method for reconstructing a high-quality NDVI time-series data set based on the Savitzky-Golay filter. Remote Sens Environ 91: 332–344.
17. Eklundh L, Jönsson P (2009) Timesat 3.0 software manual.
18. Jönsson P, Eklundh L (2004) TIMESAT - a program for analyzing time-series of satellite sensor data. Comput Geosci 30: 833–845.

19. Cho MA, Skidmore AK (2006) A new technique for extracting the red edge position from hyperspectral data: The linear extrapolation method. Remote Sens Environ 101: 181–193.

20. Senay GB, Elliott RL (2002) Capability of AVHRR data in discriminating rangeland cover mixtures. Int J Remote Sens 23: 299–312.

21. Reed BC, Brown JF, Vanderzee D, Loveland TR, Merchant JW et al. (1994) Measuring phenological variability from satellite imagery. J Veg Sci 5: 703–714.

22. Zhang XY, Friedl MA, Schaaf CB, Strahler AH, Hodges J et al. (2003) Monitoring vegetation phenology using MODIS. Remote Sens Environ 84: 471–475.

Ecological and Genetic Differences between *Cacopsylla melanoneura* (Hemiptera, Psyllidae) Populations Reveal Species Host Plant Preference

Valeria Malagnini[1]*, **Federico Pedrazzoli**[1], **Chiara Papetti**[2], **Christian Cainelli**[1], **Rosaly Zasso**[1], **Valeria Gualandri**[1], **Alberto Pozzebon**[3], **Claudio Ioriatti**[1]

1 Centre for Technology Transfer, FEM-IASMA, San Michele all'Adige (TN), Italy, **2** Department of Biology, University of Padua, Padova, Italy, **3** Department of Agronomy, Food, Natural Resources, Animals and Environment, University of Padua, AGRIPOLIS, Legnaro (PD), Italy

Abstract

The psyllid *Cacopsylla melanoneura* is considered one of the vectors of 'Candidatus Phytoplasma mali', the causal agent of apple proliferation disease. In Northern Italy, overwintered *C. melanoneura* adults reach apple and hawthorn around the end of January. Nymph development takes place between March and the end of April. The new generation adults migrate onto conifers around mid-June and come back to the host plant species after overwintering. In this study we investigated behavioural differences, genetic differentiation and gene flow between samples of *C. melanoneura* collected from the two different host plants. Further analyses were performed on some samples collected from conifers. To assess the ecological differences, host-switching experiments were conducted on *C. melanoneura* samples collected from apple and hawthorn. Furthermore, the genetic structure of the samples was studied by genotyping microsatellite markers. The examined *C. melanoneura* samples performed better on their native host plant species. This was verified in terms of oviposition and development of the offspring. Data resulting from microsatellite analysis indicated a low, but statistically significant difference between collected-from-apple and hawthorn samples. In conclusion, both ecological and genetic results indicate a differentiation between *C. melanoneura* samples associated with the two host plants.

Editor: Daniel Doucet, Natural Resources Canada, Canada

Funding: This research was financed by DEMARCATE project, funded by Provincia Autonoma di Trento (Italy). The funders had no role in study design, data collection and analysis, decision to publish, or preparation of the manuscript.

Competing Interests: The authors have declared that no competing interests exist.

* E-mail: valeria.malagnini@iasma.it

Introduction

The agronomic importance of the Hemiptera genus *Cacopsylla* is due to the role that several of its species play in the transmission of phytoplasma diseases belonging to the apple proliferation cluster, including 'Candidatus Phytoplasma mali', 'Ca. P. pyri' and 'Ca. P. prunorum' [1]. 'Ca. P. mali' is the etiological agent of apple proliferation (AP) disease, which is a severe problem in Italian apple (*Malus domestica* Borkh.) orchards. The economic impact of the disease is quite high: besides symptoms on shoots and leaves, such as witches' brooms, enlarged stipules and early leaf reddening, the disease causes a reduction in size (up to 50%), weight (by 63–74%) and, therefore, quality of fruits [2]. In the last ten years, due to the epidemic spread of the AP disease, 6,000 ha of apple orchards were uprooted and replanted in Trentino region.

Cacopsylla melanoneura (Förster), one of the most common psyllids in Italian apple orchards, is known as a vector of AP in Northwestern Italy [3], while it was shown not to transmit this disease in Germany and neighbouring countries [4]. This univoltine species is linked to some *Rosaceae Maloideae*, such as *Crataegus*, *Malus* and *Pyrus* spp. In Italy, the biological cycle of *C. melanoneura* on apple was studied and described by Tedeschi et al. [3] and Mattedi et al. [5]. In Trentino, overwintered adults reach the orchards when the average of the maximum temperatures of

7 days is above 9.5°C [6], which usually corresponds to the end of January. After mating, *C. melanoneura* females oviposit between the beginning of March and the beginning of April. Neanids hatch in the middle of March and complete their development by the end of April. The adults of the next generation (emigrants) leave the orchard around mid-June and move to the overwintering host plants. Conifers have been reported to be shelter plants for the aestivation and overwintering of the new generation [7–9].

Besides apple, the AP agent can also infect other plants, such as other rosaceous fruit trees and woody plants, including hawthorn (*Crataegus monogyna* Jacq.), on which it causes yellowing and/or decline symptoms [10]. For this reason, hawthorn may represent an alternative phytoplasma reservoir for the psyllids, if the insects are able to move from these plants to apple trees. Recently, individuals of *C. melanoneura* collected in nortwestern Italy from hawthorn were found to carry AP-group phytoplasmas, such as 'Ca. P. mali' and 'Ca. P. pyri' [11,12].

Despite the economic importance of this species, little is known about its behavioural aspects, genetic structure, patterns of dispersal at the local and regional scale, and in relation to the host plants. Intrinsic insect characteristics (such as the adult flight capacity), as well as ecological factors related to habitat (i.e. host plant and geographical isolation), may shape the genetic architecture of traits in insect populations [4]. In addition, the existence of

Table 1. *Cacopsylla melanoneura* sampling.

Host plant	Locality	Acronym	Coordinates (lat. and long.)	Altitude (m)	Sample size (N)
Hawthorn	Cles (TN-Italy)	HaCL	N 46°21'E 11°02'	674	22
	Maso Parti (TN-Italy)	HaMP	N 46°11'E 11°06'	204	30
	Rumo (TN-Italy)	HaRU	N 46°26'E 11°01'	953	41
	Chambave (AO-Italy)	HaCH	N 45°44'E 07°33'	723	42
	Neustadt (Germany)	HaNE	N 49°21'E 08°08'	150	48
Apple	Borgo Valsugana (TN-Itay)	ApBO	N 46°02'E 11°28'	481	48
	Oltrecastello (TN-Italy)	ApOL	N 46°04'E 11°09'	377	26
	San Michele (TN-Italy)	ApSM	N 46°11'E 11°08'	291	41
	Vervò (TN-Italy)	ApVE	N 46°18'E 11°07'	766	28
	Vigalzano (TN-Italy)	ApVI	N 46°04'E 11°13'	512	40
	Aosta (AO-Italy)	ApAO	N 45°44'E 07°18'	577	36
	Meckenheim (Germany)	ApME	N 49°24'E 08°14'	116	49
	Stotzheim (France)	ApST	N 48°38'E 07°49'	138	44
Conifers	Sopramonte (TN-Italy)	CoSO	N 46°04'E 11°03'	613	40
	Vason (TN-Italy)	CoVA	N 46°02'E 11°03'	1643	23
	La Grave (Hesperault-France)	CoES	N 43°58'E 03°22'	730	19

Host plant, geographical collection sites, acronyms, coordinates and sample sizes of *Cacopsylla melanoneura* samples are reported. Localities are shown in Figure 4.

host races can also affect gene flow and genetic differentiation in insects [13]. Moreover, especially in agroecosystems, also anthropogenic factors, such as pest management [14], can further contribute to insect population genetic differentiation.

This study was aimed at verifying the hypothesis that the different host plant choice could affect the survival and reproductive performance and shape the genetic structure of *C. melanoneura* samples collected from apple trees and hawthorn bushes, respectively. The ecological effect of the different plant hosting was assessed by a host-switching experiment, while genetic differences between these samples were investigated by genotyping 7 microsatellite markers specifically developed for *C. melanoneura* [15].

Materials and Methods

Ethic Statement

All the insects used in the experiments were collected and treated ethically. The individuals used for the analyses were frozen at −80°C to minimize suffering. This study did not involve endangered or protected species and therefore no specific permissions were required for collecting *C. melanoneura* individuals. The collection of insect specimens in private orchards was carried out after obtaining the permission of the owners.

Sampling

In this study "samples" are defined as groups of individuals of *C. melanoneura* collected from the same plants in a specific locality.

The samples analyzed were collected in apple orchards or hawthorn (*C. monogyna* Jacq.) hedgerows in Italy (Trentino-Alto Adige and Aosta Valley), Southern Germany and France (only from apple plants). Some other psyllids were collected from their shelter plants [conifers such as *Picea abies* (L.) H. Karst. and *Pinus mugo* Turra] in Northeastern Italy and France. Sampling details are reported in Table 1. Samples were collected by sweep-netting between December (from conifers) and the end of March (from apple and hawthorn).

Species Determination

C. melanoneura is often mistaken on hawthorn and conifers for another species, *C. affinis* (Löw), which is morphologically very similar [12]. Only males of the two species can be distinguished by examining terminalia, following Ossiannilsson keys [8], while females are identical. For this reason, at the end of behavioural experiments and before genetic analyses species identifications were verified by specific amplifications with the primers MEL_fw/MEL_rev, which amplify only *C. melanoneura* individuals, and AFF_fw/AFF_rev, specific for *C. affinis*, as described in [16]. These primer pairs amplify a species-specific segment of the control region of the mitochondrial genome;.the PCR products were visualized in an agorose gel (1%), stained with SYBR®Safe (Life Technologies, Grand Island, NY, USA). The total DNA was extracted following Doyle and Doyle method [17], the sequences of the primers and the annealing temperatures are reported in Table 2.

Host-switching Experiments

The effect of different host plants choice on two populations of *C. melanoneura* collected from different host plants was evaluated in terms of survival and reproductive performance. One population of *C. melanoneura* (ApOL) was collected from apple trees in Oltrecastello (Trento - TN) and one from hawthorn bushes (HaCL) in Cles (TN). The distance between the two localities is about 40 km. In two bi-factorial laboratory experiments, the native host plants (apple and hawthorn) and potential host plants (hawthorn and apple, respectively) were considered as experimental factors. The initial experiment involved 80 overwintered adult pairs, 40 collected from apple trees and 40 from hawthorn bushes in Trentino at the end of March 2007. The experimental design consisted of four treatments of 20 replicates. Each treatment was constituted by one population×host plant combination (ApOL× Hawthorn, ApOL×Apple, HaCL×Apple, HaCL×Hawthorn), each replicate corresponding to a shoot. Survival and oviposition of females on different host plants were compared by isolating one

Table 2. Summary data for the microsatellites developed from *Cacopsylla melanoneura.*

Name	Primer sequences (5'-3')	Ta (°C)	GenBank accession no.	Reference
AFF_fw	TTTAACCACCTCAAACTCAA	55		[16]
AFF_rev	CGTAAAATTCTTGGCGA			
MEL_fw	TTTTATCCACTCTTAAAGCTTG	55		[16]
MEL_rev	TGATAGAGCTTTTTGAATTCTC			
Co03	F: TCTGCACGCAATACCAGAAC	60	DQ414790	[15]
	R: CGCTACATGACGTGTTGTCC			
Co04	F: GGATAGCATCCACATTCCAC	60	DQ414791	[15]
	R: CCTCTTTAGGACACGGACTTG			
Co11	F: TTGAATTCTTGAACCTCTGACC	56	DQ631795	[15]
	R: TCACAAATGGAGCTTACAGGTG			
Co12	F: GCTCTTTCTCAATCCGTCCTG	60	DQ414793	[15]
	R: GAGGTGAGAGGGCGGAATAC			
Co13	F: TAAGAAGTTAGAAAGGGAGGGT	56	DQ631796	[15]
	R: GGGTCGGATTTTGGAAACAG			
Co14	F: ACAACACATGGCCCATATTTAC	56	DQ631797	[15]
	R: CTCAGTGGTGTGAATCTGACG			
Co18	F: TTTTGTTTGTTTTAGTGTTCATCCTC	53	DQ414794	[15]
	R: ACTAGGTCGGGGGTGATGTC			

Locus name, primer sequences, annealing temperature (T_a), the GenBank accession number and the reference.

overwintering female and one male on the shoot. Each shoot was placed in a glass tube (diameter 3 cm, height 16 cm), inserted into a green sponge soaked with Murashige-Skoog (MS) nutritive solution [18], and kept in a growth chamber under controlled conditions (20°C with a 16:8 h photoperiod). The shoots were replaced every two days and *C. melanoneura* couples were gently transferred to the new shoots with a thin paintbrush. The survival time of adult females, number of eggs laid and hatching rate were recorded every two to three days. Male survival was not taken into consideration in the analysis because they were let onto the shoot for mating and then removed.

Survival to adulthood was evaluated by transferring with a fine paintbrush six newly emerged nymphs from the shoots used in host-switching experiments to small apple and hawthorn plants. Six replicates for each treatment were considered. Shoots were planted in plastic pots (10 cm diameter) and kept in plexiglas cylindrical vessels (diameter 10.5 cm, height 27 cm) under controlled conditions (20°C with a 16:8 h photoperiod).

Host-switching Data Analysis

The survival rates of the two female populations observed on the different host plants were analyzed applying the LIFEREG procedure of SAS [19] and fitting a Weibull model to survival time. The median insect life spans for the different "Sample×Host plant" combinations (ApOL×Apple, ApOL×Hawthorn, HaCL×Apple, HaCL×Hawthorn) were also estimated. The differences related to population, host plant and their interactions were compared using a Wald chi-square test ($\alpha = 0.05$) [20]. Oviposition was analyzed by fitting the cumulative number of eggs laid during the experiments to a generalized linear Poisson model with a log-

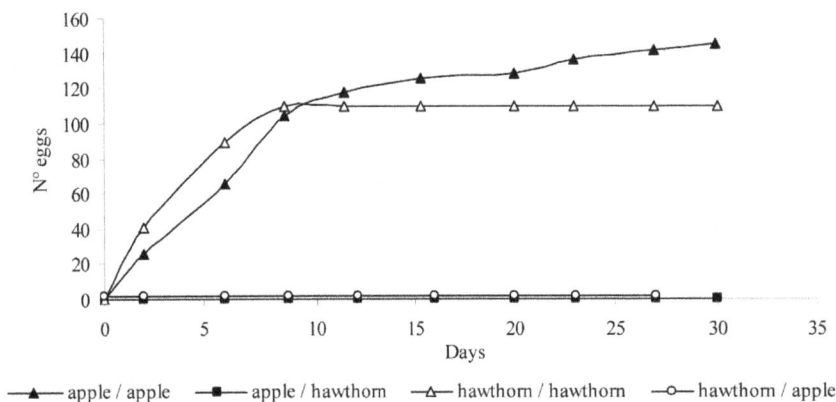

Figure 1. Oviposition of *Cacopsylla melanoneura.* Mean cumulative numbers (untransformed data) of eggs laid by the two samples of *Cacopsylla melanoneura* on the two host plants.

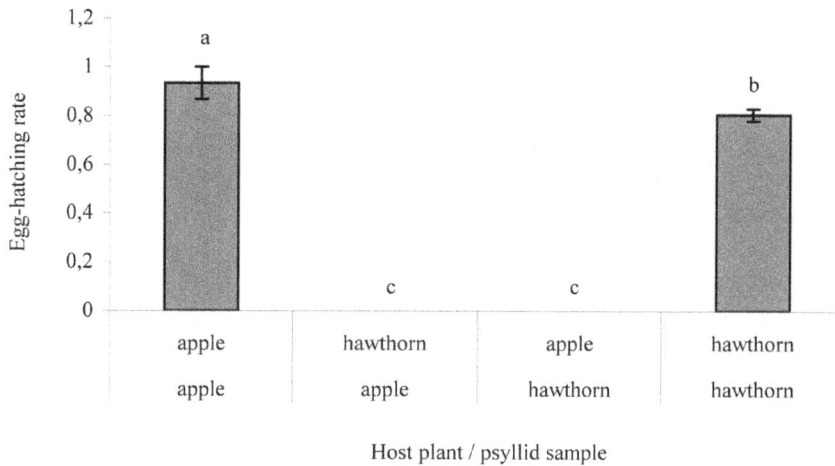

Figure 2. Egg hatching of *Cacopsylla melanoneura*. Egg-hatching rates of the two samples of *Cacopsylla melanoneura* on the different host plants. Different letters indicate significant differences according to the Wald chi-square test ($\alpha = 0.05$).

link function, using the GENMOD procedure of SAS Institute [19] and estimating the least-squares means. The Likelihood ratio chi-square test G^2 ($\alpha = 0.05$) was applied to compare the differences related to population, potential host plant and their interactions, while the differences among the least-squares means were evaluated using a Wald chi-square test ($\alpha = 0.05$). Data on egg-hatching (number of immature insects/number of eggs) and survival to adulthood (number of adults/number of newly emerged nymphs) were analyzed by applying a binomial model with a logit-link function, using the GENMOD procedure of SAS Institute [19]. The Likelihood ratio chi-square test G^2 ($\alpha = 0.05$) was performed to compare the differences related to population, potential host plant and their interactions, while the differences among the least-square means were evaluated with a Wald chi-square test ($\alpha = 0.05$).

Population Genetics Experiments and Statistical Analyses

DNA extraction. Individuals sampled for genetic analyses were immediately frozen at $-80°C$ after collection, lyophilized and homogenized. Samples were then stored at $-80°C$ until the genomic DNA extraction. Total genomic DNA was extracted from single adult specimens following Doyle and Doyle method [17].

Microsatellite genotyping. 577 individuals were genotyped for seven microsatellite loci (shown in Table 2) with the procedure described in Malagnini et al. [15]. Genotypes were obtained using an ABI 3100 sequencer (GeneScan-500 ROX as internal

standard; Applied Biosystems, Foster City, CA, USA). Allele sizing was performed using GENESCAN ver. 3.1.2 and GENEMAPPER ver. 4.0 (both programs from Applied Biosystems, Foster City, CA, USA). Automated binning was performed using FLEXIBIN ver. 2.0 [21], in order to reduce human-related scoring errors [21]. Moreover, the genotype calling was carried out by two independent people. The presence of genotyping artefacts was checked by (i) re-amplifying and scoring a random sub-sample of individuals, (ii) testing for null alleles, stuttering and large allele drop-out using MICROCHECKER ver. 2.2.3 [22] and (iii) subsequently correcting results for loci with null alleles using FREENA [23]. All the analyses described below were performed with both datasets (the original one and the FREENA-corrected one) and the results obtained were compared to assess whether there were significant differences.

Genetic diversity, Hardy-Weinberg equilibrium and linkage disequilibrium. Descriptive statistics such as range of allele sizes (S_R) in a base pair (bp), number of alleles (N_a) and

Table 3. Survival of remigrant *Cacopsylla melanoneura* females of the two samples on the potential host plants.

Sample	Potential host plant	LT$_{50}$±SE (days)	
ApOL	apple	9.11±2.27	a
	hawthorn	1.90±0.45	b
HaCL	apple	4.00±0.97	a
	hawthorn	4.30±1.03	a

Median lethal times (LT$_{50}$) ± standard error (SE) (in days) are reported. Different letters within a sample indicate significant differences according to the Wald chi-square test ($\alpha = 0.05$). For the sample acronyms see Table 1.

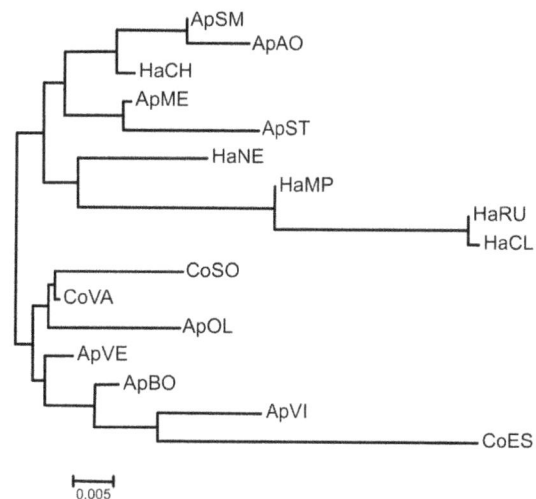

Figure 3. *Cacopsylla melanoneura* **population tree.** Unrooted Neighbour-Joining population tree based on Slatkin linearized F_{ST} pairwise matrix.

Table 4. Genetic differentiation between *Cacopsylla melanoneura* samples.

	ApBO	ApOL	ApSM	ApVE	ApVI	ApAO	ApME	ApST	HaCL	HaMP	HaRU	HaCH	HaNE	CoSO	CoVA	CoES
ApBO		**0.00042**	**0.00042**	0.00292	0.09375	**0.00042**	**0.00042**	**0.00042**	**0.00042**	**0.00042**	**0.00042**	**0.00042**	**0.00042**	**0.00042**	**0.00042**	**0.00042**
ApOL	0.0273		0.00167	0.12167	**0.00042**	**0.00042**	**0.00042**	**0.00042**	**0.00042**	**0.00042**	**0.00042**	**0.00042**	**0.00042**	**0.00042**	0.01042	**0.00042**
ApSM	0.0324	0.0322		0.00292	**0.00042**	0.00125	**0.00042**	**0.00042**	**0.00042**	**0.00042**	**0.00042**	**0.00042**	**0.00042**	**0.00042**	**0.00042**	**0.00042**
ApVE	0.0107	0.0108	0.0159		**0.00042**	**0.00042**	**0.00042**	**0.00042**	**0.00042**	**0.00042**	**0.00042**	**0.00042**	**0.00042**	**0.00042**	**0.00042**	**0.00042**
ApVI	0.0102	0.0499	0.0465	0.0350		**0.00042**	**0.00042**	**0.00042**	**0.00042**	**0.00042**	**0.00042**	**0.00042**	**0.00042**	**0.00042**	**0.00042**	**0.00042**
ApAO	0.0363	0.0445	0.0063	0.0273	0.0526		**0.00042**	**0.00042**	**0.00042**	**0.00042**	**0.00042**	**0.00042**	**0.00042**	**0.00042**	**0.00042**	**0.00042**
ApME	0.0302	0.0414	0.0185	0.0296	0.0430	0.0256		**0.00042**	**0.00042**	**0.00042**	**0.00042**	**0.00042**	**0.00042**	**0.00042**	**0.00042**	**0.00042**
ApST	0.0398	0.0557	0.0266	0.0444	0.0464	0.0309	0.0137		**0.00042**	**0.00042**	**0.00042**	**0.00042**	**0.00042**	**0.00042**	**0.00042**	**0.00042**
HaCL	0.0567	0.0732	0.0692	0.0618	0.0835	0.0833	0.0339	0.0803		0.30417	0.99667	**0.00042**	**0.00042**	**0.00042**	**0.00042**	**0.00042**
HaMP	0.0215	0.0434	0.0369	0.0276	0.0353	0.0459	0.0170	0.0469	0.0101		0.79667	0.00250	**0.00042**	**0.00042**	**0.00042**	**0.00042**
HaRU	0.0617	0.0797	0.0712	0.0649	0.0861	0.0845	0.0368	0.0809	−0.0027	0.0066		**0.00042**	**0.00042**	**0.00042**	**0.00042**	**0.00042**
HaCH	0.0230	0.0421	0.0095	0.0244	0.0398	0.0152	0.0189	0.0329	0.0458	0.0195	0.0451		**0.00042**	**0.00042**	**0.00042**	**0.00042**
HaNE	0.0334	0.0469	0.0335	0.0331	0.0509	0.0382	0.0258	0.0436	0.0555	0.0265	0.0519	0.0269		**0.00042**	**0.00042**	**0.00042**
CoSO	0.0278	0.0394	0.0354	0.0262	0.0478	0.0497	0.0309	0.0463	0.0555	0.0290	0.0576	0.0356	0.0374		0.42708	**0.00042**
CoVA	0.0090	0.0223	0.0176	0.0134	0.0173	0.0259	0.0159	0.0209	0.0614	0.0224	0.0634	0.0212	0.0267	0.0106		0.00583
CoES	0.0492	0.0747	0.0721	0.0577	0.0478	0.0773	0.0664	0.0587	0.1101	0.0610	0.1084	0.0639	0.0736	0.0700	0.0430	

Pair-wise genetic differentiation (F_{ST}) between *Cacopsylla melanoneura* samples (below diagonal) and associated P-values (above diagonal) calculated from the original dataset. Significant P values after Bonferroni correction are in bold.

allelic richness (A_R) were calculated using FSTAT ver. 2.9.3.2 [24]. Observed (H_O) and expected (H_E) heterozygosity were calculated using GENETIX ver. 4.05.2 [25]. Tests for conformity with Hardy-Weinberg equilibrium (HWE) and linkage disequilibrium between pairs of loci in each population were performed using the online version of GENEPOP [26].

Significance levels for multiple comparisons were adjusted the standard Bonferroni technique [27,28].

Population structure. In a first evaluation only samples collected from apple and hawthorn of Trentino region were considered; in a second evaluation all samples were included in the analysis. FSTAT ver. 2.9.3.2 [24] was used to compute the overall and population pair-wise F_{ST} values. The 95% confidence intervals were estimated using 1,000 bootstrap replicates over the loci and probability values were determined using 1,000 permutations. The nominal significance level was set at $\alpha = 0.05$. Statistical significance level for multiple comparisons was adjusted using a standard Bonferroni as described above. To provide a better visualization of population samples relationship, a matrix of Slatkin's linearized F_{ST} values [29] for all population sample pairs was obtained with ARLEQUIN ver. 3.5 software [30] and used to produce a Neighbour-Joining [31] unrooted population tree with MEGA ver. 5.0. [32]. The molecular variance analysis (AMOVA) was performed using the ARLEQUIN ver. 3.5 software [30] to test genetic differentiation among and within groups. In a first evaluation, the analysis was carried out according to a model structure in which the eight samples of Trentino region were divided into two groups: one collected from apple (ApBO, ApOL, ApSM, ApVE and ApVI) and the other collected from hawthorn (HaCL, HaMP, HaRU). In a second analysis, we included also samples from Aosta Valley (ApAO and HaCH), Germany (ApME and HaNE) and France (ApST).

Analysis of the population structure was performed with STRUCTURE ver. 2.3.3 software [30] to infer the most likely number of populations (K), representative of the whole data set, without the use of any *a priori* information. Ten independent runs

of structure were performed for each K value from 1 to 9. Each run consisted of a burn-in period of 100,000 steps, followed by 1,000,000 Monte Carlo Markov Chain replicates, assuming an admixture model and correlated allele frequencies. The most likely K was chosen comparing the average estimates of the likelihood of the data, $\ln[\Pr(X/K)]$, for each value of K [33], as well as calculating *ad hoc* statistics ΔK values (Evanno's method) [34]. The value of $\ln[\Pr(X/K)]$ and ΔK were obtained by Structure Harvester ver. 0.6.1 [35]. The proportions of membership of each individual in each cluster were also calculated. Also in this case, a first analysis was performed using the Trentino data set including only collected-from-apple and hawthorn samples. Ten independent runs of structure were performed for each K value from 1 to 9. A second analysis was carried out using the whole data set. In this case, 10 independent runs were performed for each K value from 1 to 17.

Results

Species Determination

The correct species assessment was confirmed for all male analysed in this study by both morphological and molecular (PCR) analyses. Genetic analyses carried out on females highlighted that only few individuals belonged to the species *C. affinis* and were present within some collected-from-hawthorn samples. These individuals were excluded from the dataset (data not shown).

Ecological Experiments and Data Analysis

Survival analysis detected significant differences between the treatments ApOL×Hawthorn and HaCL×Apple (Wald $\chi^2 = 6.34$; df = 1; $P = 0.012$), but not between the two treatments ApOL×Apple and HaCL×Hawthorn (Wald $\chi^2 = 0.05$; $P = 0.828$). The analysis found a significant "Sample×Host plant" treatment (Wald $\chi^2 = 5.30$; df = 1; $P = 0.021$). The survival rate of the collected-from-apple sample of *C. melanoneura* was higher on apple shoots than on hawthorn shoots (Wald $\chi^2 = 10.35$; df = 1;

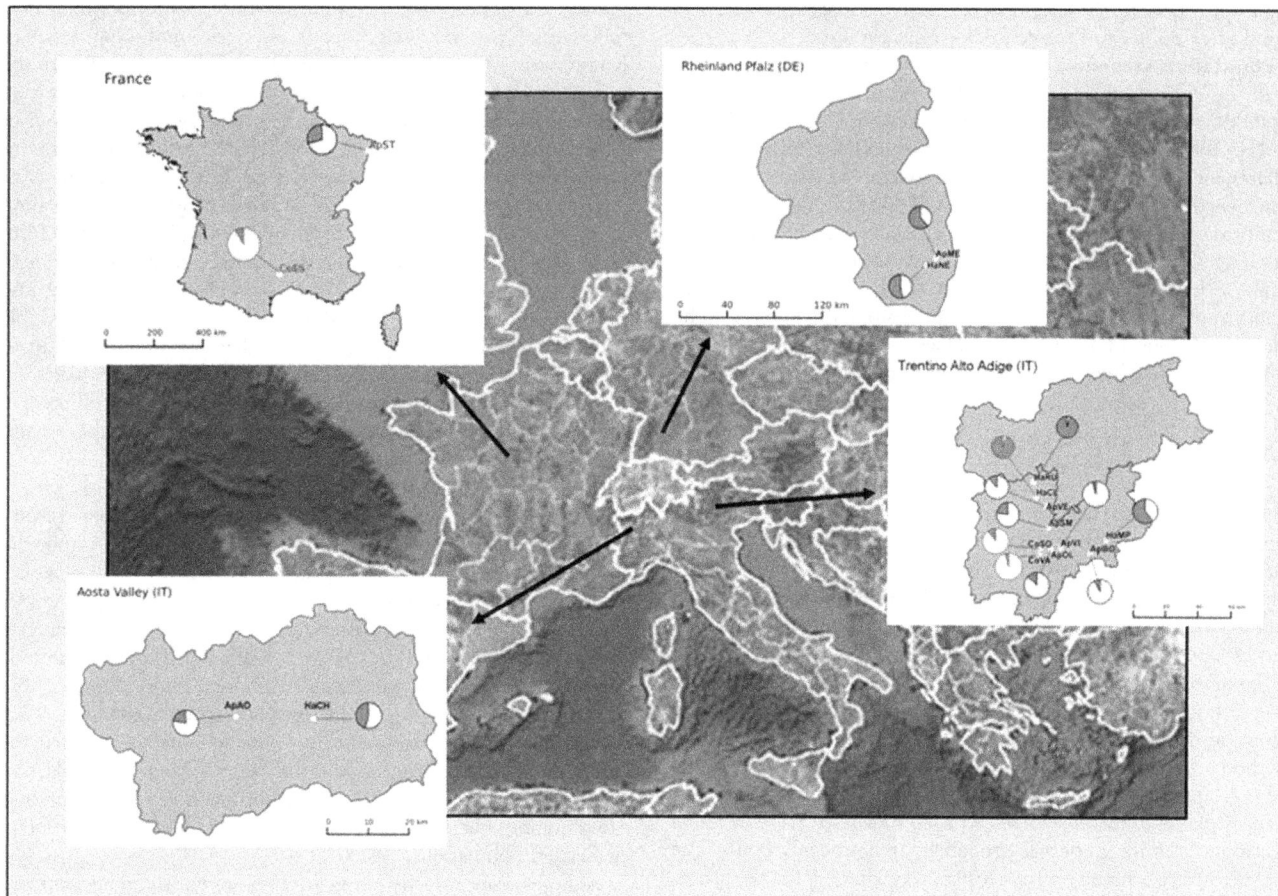

Figure 4. Map of the sampling sites. Insets show detailed maps of Trentino-Alto Adige (Northeastern Italy), Aosta Valley (Northwestern Italy), Rheinland-Pfalz (Southern Germany) and Hesperault (France). The graphs represent the ratio of cluster 1 (white), corresponding to individuals collected from apple with posterior probability greater than 0.70, and cluster 2 (grey), corresponding to individuals collected from hawthorn with posterior probability greater than 0.60, for each sample obtained by STRUCTURE analysis. For samples' acronyms, see Table 1.

$P = 0.001$; Table 3); whereas no host plant effect was observed for the collected-from-hawthorn sample (Wald $\chi^2 = 0.05$; df = 1; $P = 0.832$; Table 3). No difference in the number of eggs laid during the experiment was recorded between the two *C. melanoneura* samples ($G^2 = 0.89$; df = 1; $P = 0.344$; Fig. 1) or between the different host plants ($G^2 = 0.94$; df = 1; $P = 0.332$; Fig. 1). However, the differences between the oviposition of the two samples are significantly related to the native host plants ($G^2 = 13.28$; df = 1; $P < 0.001$; Fig. 1). In particular, the collected-from-apple sample laid more eggs on native shoots (13.74 eggs/day/female) than on hawthorn shoots (0.04 eggs/day/female; Wald $\chi^2 = 48.26$; df = 1; $P < 0.001$) and the collected-from-hawthorn females laid more eggs on hawthorn shoots (13.30 eggs/day/female) than on apple shoots (1 eggs/day/female; Wald $\chi^2 = 20.24$; df = 1; $P < 0.001$). Eggs hatched only on their native host plants as shown in Fig. 2, and no significant differences were found for survival to adulthood of individuals placed on their native host plant shoots ($G^2 = 0.81$; df = 1; $P = 0.36$).

Population Genetics Analyses

Genetic diversity, Hardy-Weinberg equilibrium and linkage disequilibrium. A total of 577 individuals were genotyped for seven microsatellite loci that proved to be polymorphic in the 16 samples analysed. The number of alleles (N_a) per locus varied between 3 for locus Co18 (HaRU) and 43 for locus Co12 (CoSO). Allelic richness (A_R) ranged from a minimum value of 2.905 for locus Co18 (HaRU) to a maximum of 24.383 for locus Co12 (CoVA) (Table S1). The average H_E and H_O ranged from 0.7814 to 0.9575 and from 0.5154 to 0.7094, respectively (Table S1). Loci were not to be in linkage disequilibrium ($P > 0.05$, data not shown).

A significant deviation from Hardy-Weinberg equilibrium was observed for most of the analyzed loci and populations (Table S1). Departure from HWE was due to heterozygotes deficit at most of loci. MICROCHECKER did not provide indications of allele dropouts or stuttering at any marker in any sample, whereas it identified null alleles at all loci. Departure from HWE remained significant after correction for null alleles with FREENA analysis, suggesting the presence of null alleles at all loci. To avoid a possible bias due to the presence of null alleles on F_{ST} estimation, the dataset was corrected using FREENA [23]. However, samples that failed to amplify were <1%, indicating that null homozygotes were not common. For these reasons, we maintained the original dataset for all analyses while cross-checking differentiation results, when appropriate, by comparing outcomes obtained with corrected dataset (with FREENA). Statistics calculated for the original and the corrected dataset provided comparable results (data not reported). Non significant genotypic disequilibrium was found after Bonferroni correction for multiple tests.

Population structure. Overall F_{ST} value was highly significant ($F_{ST}=0.04$; 95% Confidence Interval $C.I. = 0.017 \div 0.039$; $P<0.001$) and pair-wise F_{ST} revealed significant values for 107 out of 120 comparisons, after Bonferroni correction (Table 4). No differences were found among collected-from-hawthorn samples from Northwestern Italy (HaCL, HaMP, HaRU; Table 4), while significant differences emerged among collected-from-apple samples (Fig. 3; Table 4). When the whole data set was considered, the results obtained indicated that most of the comparisons were significantly different, with some exceptions (Fig. 3; Table 4). The matrix of Slatkin's linearized F_{ST} values is reported in Table S2.

The AMOVA analysis of the samples collected from the two host plants revealed that in Trentino region the population is divided into two groups (collected-from-apple and collected-from-hawthorn) and accounts for 5.21% of the whole variability ($F_{TC} = 0.0521$; $P = 0.0169$). Moreover, variation among samples within groups and variation among individuals within samples were highly significant (variation rates of 2.55% and 92.24%; $F_{SC} = 0.02685$ and $F_{ST} = 0.07755$, respectively; both $P<0.001$). In a second analysis, other samples collected from apple and hawthorn in other regions were added to the data set: the subdivision into two groups was confirmed ($F_{TC} = 0.01743$; $P = 0.0386$).

Based on the analysis of data set of Trentino region using structure software, the best model of the number of genetic clusters based on microsatellite variation, as determined by ΔK (Evanno's method), was $K = 2$. All individuals collected from apple plants were assigned to cluster 1 with posterior probability greater than 0.70, while all individuals collected from hawthorn were assigned to cluster 2 with posterior probability major than 0.60. The presence of two main genetic clusters was inferred for the whole dataset with the exception of ApME individuals, which were assigned to the collected-from-hawthorn cluster (Fig. 4).

Discussion

The ecological advantage for insect host races is widely reported in the literature [36–41]. By definition, "host races" are genetically differentiated, sympatric populations of parasites associated with different hosts plant species, on which they feed and reproduce, and between which there is appreciable gene flow [42]. An example of a sympatric host race formation is given in the recent study conducted on *Rhagoletis pomonella* (Walsh), in which a shift from hawthorn to apple is hypothesized [43].

The experiments performed in this study measured the survival and reproductive performance of two *C. melanoneura* groups, showing a significant relationship between this species and its potential host plant. Moreover, a differentiation between the two groups, which are linked to different host plants for their development and reproduction, was suggested by the genetic analyses performed on the different samples.

In the past, *C. melanoneura* was considered to be widely oligophagous on *Crataegus* spp., as in Moravia and Central Europe this species has been observed only on hawthorn [44]. However, there is evidence that *C. melanoneura* lives on, and causes damage to, apple trees (*Malus* spp.) and occasionally to other *Rosaceae* species, such as pear (*Pyrus* spp.) and medlar (*Mespilus germanica* L.) [44]. The presence of different populations of *C. melanoneura* was reported by Lazarev [45,46], who studied and described the so called *C. melanoneura* form *taurica* ("Crimean apple sucker"), which refers to *C. melanoneura* individuals living on apple trees. Members of the Crimean apple sucker do not develop when transferred to hawthorn, and die within a single week after the transfer, without

mating and laying eggs [47]. Furthermore, some morphological differences (e.g. forewing, head, antennae and anal segment length) were observed in *C. melanoneura* samples collected from the two different host plants. Therefore, the presence of different food plants, taken in conjunction with these morphological differences which may correspond to food specialization, enabled to distinguish two forms of the species [46]. In the subsequent years, *C. melanoneura* form *taurica* was considered just as a population living on apple trees and as a synonymous with the typical form [48]. Hence, all the following descriptions gave no taxonomic value to this form and referred to *C. melanoneura* [44]. Anyway, also field observations on *C. melanoneura* naturally occurring in apple orchards and hawthorn hedgerows (Malagnini et al., unpublished data) and preliminary host-switching trials [49] conducted in Northeastern Italy suggested the existence of ecological differences between collected-from-apple and collected-from-hawthorn samples.

The data obtained in this work are in agreement with Lazarev's studies [46], which were focused on the survival of collected-from-apple samples on hawthorn plants. Unfortunately, the Author did not perform the opposite host-switching experiment, so we cannot compare the results. In our trials we used *C. melanoneura* samples collected directly from apple and hawthorn plants after remigration. As hypothesized by Mayer et al. [50], psyllids may be conditioned by the plants which they had started feeding on: feeding and oviposition experiences can induce a host preference switch in females. Nevertheless, genetic results obtained by Trentino data set suggest that the behaviour of the collected-from-apple and collected-from-hawthorn samples is not due to a conditioning effect, but rather to the existence of host-plant associated populations.

Although not in the aims of this study, some preliminary evidences, which deserve further specific experimental investigations, prompted for the two *C. melanoneura* populations to be good candidates for host race definition. In particular, heterozygote deficit and positive F_{IS} values (data not reported) found through most of the samples analyzed and a host-associated trade-off in females' fitness, suggested the hypothesis of coexisting genetically differentiated pools with ecological differences in agreement with what reported for other psyllid species, such as *C. pruni* (Scopoli) [51,52] and *C. chinensis* (Yang and Li) [53]. Therefore, rather than being a result of strong null-alleles pervasiveness, the significant heterozygote deficit observed in our data set may be compatible with the hypothesis of a combined effect of species habits and anthropogenic pressure. In fact, the high inbreeding rates may result from the presence of undetected, differentiated populations with small effective population size (N_e) coexisting at the same sampling area. A condition when populations are subdivided unequally with regard to allele frequencies, so that random mating only involves a portion of the population although gene flow still partially occurs, is also known as Wahlund effect [54]. In this case the heterozygotes deficiency is detected when the differentiated sub-populations are sampled as a single unit [54]. *C. melanoneura* is not a very motile insect: it is passively transported by the wind and, even if it overwinters on conifers at high altitudes, its transfer is not active [55–57]. Therefore it can be hypothesised that individuals form small groups, with low genotype diversity, that are transported by the wind, as a single unit, in a preferential direction. In the same manner, psyllids may be brought back to their native areas by winds blowing in the opposite directions [58]. In addition, if overwintering can take place in apple orchards, as suggested by Mattedi et al. [8], mating choice might be forced and reduced to even fewer individuals.

The phytosanitary measures applied in Northern Italy could have contributed, probably together with other factors (such as climatic conditions), to the dramatic decrease in *C. melanoneura* number recorded in Trentino apple orchards between 2000 and 2006 [5] and therefore to a drop in N_e and genotype diversity of local populations. This reduction, which is relatively recent, might not have been equally severe in all localities, but may have exacerbated the occurring natural conditions, leading to the genetic differences recorded in this study. This differentiation supported also the ecological differences in host plant preferential choice. In fact, the AMOVA analysis performed on microsatellite dataset of Trentino region indicated small, but significant genetic differences between the collected-from-apple and collected-from-hawthorn populations ($F_{TC} = 0.0133$; $P = 0.037$). These results were confirmed when the whole data set was considered. STRUCTURE analysis pointed out the presence of two different clusters corresponding to the two host plants for Trentino region. For the other localities (Germany and Aosta Valley), this subdivision seemed to be not that clear. In Germany, samples collected from apple and from hawthorn resulted to be a mixture of both clusters, with hawthorn cluster as predominant. This result seemed to be in agreement with olfactometer analyses: *C. melanoneura* individuals collected after overwintering on conifers preferred hawthorn to apple [50]. These data reflect the abundance and distribution of the two host plants in Germany, where apple cultivation is not as intensive as in Trentino region, where apple dominates over hawthorn and selective pressure is high and mainly based on insecticides. In Aosta Valley, as in Germany, the two clusters co-occurred in the collected-from-hawthorn samples. The presence of a genetic variability among *C. melanoneura* was already observed also by Tedeschi and Nardi [16]. The Authors isolated two different genetic profiles in the mitochondrial control region of Italian populations collected from apple and from hawthorn. The origin and geographic distribution of such variants is yet to be examined in detail. The two profiles were obtained also in this work when we assessed the identity of *C. melanoneura* (data not reported).

However, this study was mainly focused on Trentino region and only few samples from other regions were considered. Future researches should therefore be carried out, involving measurements of the genetic variation between and within populations of *C. melanoneura*, host association and preferential choice, to confirm these preliminary results. These investigations will help in a better understanding of the entire life cycle of *C. melanoneura*. Moreover, inbreeding trials will be useful also to finally assess all the criteria leading to the definition of host races proposed by Dres and Mallet [42].

Noticeably, our study builds upon the knowledge of one vector of AP phytoplasma since recent investigations pointed out that hawthorn may be an *inoculum* source for the spread of this disease through *C. melanoneura* [12]. Nevertheless, according to ecological and genetic data collected in Trentino, hawthorn and apple clusters are different, indicating no significant exchange between them. If these results were replicated at a larger scale, the actual role of hawthorn as reservoir of 'Ca. P. mali' could be better pinpointed. Furthermore, the significant differences emerged by comparing several collected-from-apple samples could support the different acquisition and transmission efficiencies in *C. melanoneura* samples collected in different areas [4,59].

Acknowledgments

We thank Dr. R. Tedeschi (University of Turin, Italy), Drs. B. and W. Jarausch (AlPlanta-Institute for Plant Research, Neustadt a.d. Weinstrasse, Germany) and Dr. G. Labonne for the insects and their scientific support and Dr. F. Zottele (FEM-IASMA) for his technical support.

Author Contributions

Conceived and designed the experiments: VM FP. Performed the experiments: VM FP CC RZ VG. Analyzed the data: VM CP AP. Contributed reagents/materials/analysis tools: CI. Wrote the paper: VM FP CP.

References

1. Seemüller E, Schneider B (2004) 'Candidatus Phytoplasma mali', 'Candidatus Phytoplasma pyri' and 'Candidatus Phytoplasma prunorum', the causal agents of apple proliferation, pear decline and European stone fruit yellows, respectively. Int J Syst Evol Microbiol 54: 1217–1226.

2. EPPO/CABI (1996) Apple proliferation phytoplasma. In: Quarantine Pests for Europe, 2nd edn. Wallingford: CAB International. 959–962.

3. Tedeschi R, Bosco D, Alma A (2002) Population dynamics of Cacopsylla melanoneura (Homoptera: Psyllidae), a vector of apple proliferation phytoplasma in Northwestern Italy. J Econ Entomol 95: 544–551.

4. Mayer CJ, Jarausch B, Jarausch W, Jelkmann W, Vilcinskas A, et al. (2009) Cacopsylla melanoneura has no relevance as vector of apple proliferation in Germany. Phytopathology 99: 729–738.

5. Mattedi L, Forno F, Varner M (2007) Scopazzi del melo. Conoscenze ed osservazioni di campo. Bolzano: Arti Grafiche La Commerciale-Borgogno.

6. Tedeschi R, Baldessari M, Mazzoni V, Trona F, Angeli G (2012) Population dynamics of Cacopsylla melanoneura (Hemiptera: Psyllidae) in Northeast Italy and its role in apple proliferation epidemiology in apple orchards. J Econ Entomol 105(2): 322–328.

7. Conci C, Rapisarda C, Tamanini L (1992) Annotated catalogue of the Italian Psylloidea. First part (Insecta Homoptera). Atti Accademia Roveretana degli Agiati. Ser. VII, vol. II, B: 33–135.

8. Ossiannilsson F (1992) The Psylloidea (Homoptera) of Fennoscandia and Denmark, vol. 26. Fauna Entomologica Scandinavica. Leiden: ed. Brill EJ.

9. Pizzinat A, Tedeschi R, Alma A (2011) Cacopsylla melanoneura (Foerster): aestivation and overwintering habitats in Northwest Italy. Bull Insectol 64: 135–136.

10. Seemüller E (2002) Apple proliferation: etiology, epidemiology and detection. In: ATTI Giornate Fitopatologiche 1: 3–6.

11. Tedeschi R, Bertignolo L, Alma A (2005) Role of the hawthorn psyllid fauna in relation to the apple proliferation disease. In: Workshop Proceedings of the 3rd National Meeting on Phytoplasma Disease. Petria 15: 47–49.

12. Tedeschi R, Lauterer P, Brusetti L, Tota F, Alma A (2009) Composition, abundance and phytoplasma infection in the hawthorn psyllid fauna of Northwestern Italy. Eur J Plant Pathol 123: 301–310.

13. Miller NJ, Birley AJ, Overall ADJ, Tatchell GM (2003) Population genetic structure of the lettuce root aphid, Pemphigus bursarius (L.), in relation to geographic distance, gene flow and host plant usage. Heredity 91: 217–223.

14. Dorn S, Schumacher P, Abivardi C, Meyhöfer R (1999) Global and regional pest insects and their antagonists in orchards: spatial dynamics. Agr Ecosyst Environ 73: 111–118.

15. Malagnini V, Pedrazzoli F, Forno F, Komjanc M, Ioriatti C (2007) Characterization of microsatellite loci in Cacopsylla melanoneura Föster (Homoptera: Psyllidae). Mol Ecol Notes 7: 495–497.

16. Tedeschi R, Nardi F (2010) DNA-based discrimination and frequency of phytoplasma infection in the two hawthorn-feeding species, Cacopsylla melanoneura and Cacopsylla affinis, in Northwestern Italy. B Entomol Res 100(6): 741–747.

17. Doyle JJ, Doyle JL (1990) Isolation of plant DNA from fresh tissue. Focus 12: 13–15.

18. Murashige T, Skoog F (1962) A revised medium for rapid growth and bioassays with tobacco culture. Physiol Plant 15: 473–497.

19. SAS Institute (1999) SAS/STAT User's Guide, version 8th ed. SAS Institute, Cary, NC.

20. Allison PD (1995) Survival Analysis Using the SAS System: a Practical Guide, 2nd ed. Cary: SAS Institute.
21. Amos W, Hoffman JI, Frodsham A, Zhang L, Best S, et al. (2007) Automated binning of microsatellite alleles: problems and solutions. Mol Ecol Notes 7: 10–14.
22. Van Oosterhout C, Hutchinson WF, Wills DPM, Shipley P (2004) Microchecker: software for identifying and correcting genotyping errors in microsatellite data. Mol Ecol Notes 4: 535–538.
23. Chapuis MP, Estoup A (2007) Microsatellite null alleles and estimation of population differentiation. Mol Biol Evol 24: 621–631.
24. Goudet J (1995) FSTAT (ver. 1.2): a computer program to calculate F-statistics. J Hered 86: 485–486.
25. Belkhir K, Borsa P, Raufaste N, Bonhomme F (2004) GENETIX v. 4.05: logiciel sous WindowsTM pour la génétique des populations. Laboratoire Génome, Populations, Interactions CNRS UMR 5171, Université de Montpellier II, Montpellier, France.
26. Raymond M, Rousset F (1995) GENEPOP (version 1.2), population genetics software for exact tests and ecumenicism. J Hered 86: 248–249.
27. Bonferroni CE (1936) Teoria statistica delle classi e calcolo delle probabilità. Pubblicazioni del R Istituto Superiore di Scienze Economiche e Commerciali di Firenze 8: 3–62.
28. Miller JA, Shanks AL (2004) Evidence for limited larval dispersal in black rock fish (Sebastes melanops): implication for population structure and marine-reverse design. Can J Fish Aquat Sci 61: 1723–1735.
29. Slatkin M (1995) A measure of population subdivision based on microsatellite allele frequencies. Genetics 139: 457–462.
30. Excoffier L, Lischer HEL (2010) Arlequin suite ver 3.5: A new series of programs to perform population genetics analyses under Linux and Windows. Mol Ecol Resour 10: 564–567.
31. Saitou N, Nei M (1987) The neighbor-joining method: a new method for reconstructing phylogenetic trees. Mol Biol Evol 4: 406–425.
32. Tamura K, Peterson D, Peterson N, Stecher G, Nei M, et al. (2011) MEGA5: molecular evolutionary genetics analysis using maximum likelihood, evolutionary distance, and maximum parsimony methods. Mol Biol Evol 28: 2731–2739.
33. Pritchard JK, Stephens M, Donnelly P (2000) Inference of population structure using multilocus genotype data. Genetics 155: 945–959.
34. Evanno G, Regnaut S, Goudet J (2005) Detecting the number of clusters of individuals using the software STRUCTURE: a simulation study. Mol Ecol 14: 2611–2620.
35. Earl DA (2012) STRUCTURE HARVESTER: a website and program for visualizing STRUCTURE output and implementing the Evanno method. Conserv Genet Resour 4: 359–361.
36. Guldemond JA (1990a) Choice of host plant as a factor in reproductive isolation of the aphid genus Cryptomyzus. Ecol Entomol 15: 43–51.
37. Guldemond JA (1990b) Evolutionary genetics of the aphid Cryptomyzus, with a preliminary analysis of host plant preference, reproductive performance and host-alternation. Entomol Exp Appl 57: 65–76.
38. Craig TP, Itami JK, Horner JD, Abrahamson WG (1993) Behavioral evidence for host-race formation in Eurosta solidaginis. Evolution 47: 1696–1710.
39. Feder JL, Opp SB, Wlazlo B, Reynolds K, Go W, et al. (1994) Host fidelity is an effective premating barrier between sympatric races of the apple maggot fly. Proc Natl Acad Sci USA 91: 7990–7994.
40. Filchak KE, Roethele JB, Feder JL (2000) Natural selection and sympatric divergence in the apple maggot Rhagoletis pomonella. Nature 407: 739–742.
41. Via S (2001) Sympatric speciation in animals: the ugly duckling grows up. Trends Ecol Evol 16: 381–390.
42. Dres M, Mallet J (2002) Host races in plant-feeding insects and their importance in sympatric speciation. Philos Trans R Soc Lond B 357: 471–492.
43. Linn CE Jr, Yee WL, Sim SB, Cha DH, Powell THQ, et al. (2012) Behavioral evidence for fruit odor discrimination and sympatric host races of Rhagoletis pomonella flies in the Western United States. Evolution 66–11: 3632–3641.
44. Lauterer P (1999) Results of investigations on Hemiptera in Moravia, made by the Moravian Museum (Psylloidea 2). Acta Musei Moraviae, Scientiae Biologicae (Brno), 84: 71–151.
45. Lazarev MA (1972) Pyylla melanoneura Frst. taurica forma nov. (Homoptera: Psyllidae). An apple tree pest in the Crimea. Proceedings of the All-Union V. I. Lenin Academy of Agricultural Sciences, The State Nikita Botanical Gardens, Yalta, Ukraine, 61: 101–122.
46. Lazarev MA (1974b) The 'Crimean Apple Sucker' Pyylla melanoneura Frst. Forma taurica, nov. (Homoptera, Psylloidea). Nikita State Botanical Gardens, Ukraine. 23–26.
47. Lazarev MA (1974a) Leaf-bugs (Homoptera: Psyllidae) of the apple and pear in the orchards of the Crimea. (Morphology, biology, control). Published degree dissertation, Academy of Sciences of the Moldavian SSR, Kishinyov [in Russian].
48. Gegechkori AM, Loginova MM (1990) Psyllids (Homoptera, Psylloidea) of the USSR (an Annotated List). Metsniereba, Tbilisi, Georgia [in Russian].
49. Malagnini V, Cainelli C, Pedrazzoli F, Ioriatti C (2006) Population diversity within Cacopsylla melanoneura (Förster) based on ecological and molecular studies (Abstract). In: Proceedings of VIII European Congress of Entomology, Izmir, Turkey.
50. Mayer CJ, Vilcinskas A, Gross J (2011) Chemically mediated multitrophic interactions in plant-insect vector-phytoplasma system compared with a partially nonvector species. Agric Forest Entomol 13: 25–35.
51. Sauvion N, Lachenaud O, Genson G, Rasplus JY, Labonne G (2007) Are there several biotypes of Cacopsylla pruni? Bull Insectol 60(2): 185–186.
52. Sauvion N, Lachenaud O, Mondor-Genson G, Easplus JY, Labonne G (2009) Nine polymorphic microsatellite loci from the pysillid Cacopsylla pruni (Scopoli), the vector of European stone fruit yellows. Mol Ecol Res 9(4): 1196–1199.
53. Sun J-R, Li Y, Yan S, Zhang Q-W, Xu H-L (2011) Microsatellite marker analysis of genetic diversities of Cacopsylla chinensis (Yang et Li) (Hemiptera: Psyllidae) population in China. Acta Entomol Sin 54(7): 820–827.
54. Wahlund S (1928) Zusammensetzung von Population und Korrelationserscheinung vom Standpunkt der Vererbungslehre aus betrachtet. Hereditas 11: 65–106.
55. Clark LR (1962) The general biology of Cardiaspina albitextura (Psyllidae) and its abundance in relation to weather and parasitism. Aust J Zool 10: 537–586.
56. Hodkinson ID (1974) The biology of the Psylloidea (Homoptera): a review. Bull Entomol Res 64: 325–339.
57. Hodkinson ID (2009) Life cycle variation and adaptation in jumping plant lice (Insecta: Hemiptera: Psylloidea): a global synthesis. J Nat Hist 43: 65–179.
58. Conci C, Rapisarda C, Tamanini L (1995) Annotated catalogue of the Italian Psylloidea. Second part (Insecta Homoptera) - Atti Accademia Roveretana degli Agiati, ser. VII, vol. V, B: 5–207.
59. Tedeschi R, Alma A (2004) Transmission of apple proliferation phytoplasma by Cacopsylla melanoneura (Homoptera: Psyllidae). J Econ Entomol 97(1): 8–13.

Arbuscular Mycorhizal Fungi Associated with the Olive Crop across the Andalusian Landscape: Factors Driving Community Differentiation

Miguel Montes-Borrego[1], **Madis Metsis**[2], **Blanca B. Landa**[1]*

1 Department of Crop Protection, Institute for Sustainable Agriculture (IAS-CSIC), Cordoba, Spain, 2 Tallinn University, Institute of Mathematics and Natural Sciences, Tallinn, Estonia

Abstract

Background: In the last years, many olive plantations in southern Spain have been mediated by the use of self-rooted planting stocks, which have incorporated commercial AMF during the nursery period to facilitate their establishment. However, this was practised without enough knowledge on the effect of cropping practices and environment on the biodiversity of AMF in olive orchards in Spain.

Methodology/Principal Findings: Two culture-independent molecular methods were used to study the AMF communities associated with olive in a wide-region analysis in southern Spain including 96 olive locations. The use of T-RFLP and pyrosequencing analysis of rDNA sequences provided the first evidence of an effect of agronomic and climatic characteristics, and soil physicochemical properties on AMF community composition associated with olive. Thus, the factors most strongly associated to AMF distribution varied according to the technique but included among the studied agronomic characteristics the cultivar genotype and age of plantation and the irrigation regimen but not the orchard management system or presence of a cover crop to prevent soil erosion. Soil physicochemical properties and climatic characteristics most strongly associated to the AMF community composition included pH, textural components and nutrient contents of soil, and average evapotranspiration, rainfall and minimum temperature of the sampled locations. Pyrosequencing analysis revealed 33 AMF OTUs belonging to five families, with *Archaeospora* spp., *Diversispora* spp. and *Paraglomus* spp., being first records in olive. Interestingly, two of the most frequent OTUs included a diverse group of Claroideoglomeraceae and Glomeraceae sequences, not assigned to any known AMF species commonly used as inoculants in olive during nursery propagation.

Conclusions/Significance: Our data suggests that AMF can exert higher host specificity in olive than previously thought, which may have important implications for redirecting the olive nursery process in the future as well as to take into consideration the specific soils and environments where the mycorrhized olive trees will be established.

Editor: Raffaella Balestrini, Institute for Plant Protection (IPP), CNR, Italy

Funding: This research was supported by grants from Projects AGL2008-00344/AGR from 'Ministerio de Ciencia e Innovación', Project P10-AGR-5908 from 'Consejería de Economía, Innovación y Ciencia' of Junta de Andalucía, Project AGL-2012-37521 from 'Ministerio de Economía y Competitividad' of Spain, and Fondo Europeo de desarrollo regional (FEDER) "Una manera de hacer Europa" from the European Union. The grant 219262 ArimNET_ERANET FP7 2012–2015 Project PESTOLIVE from Instituto Nacional de Investigación y Tecnología Agraria y Alimentaria (INIA), also provided partial financial support. M. Montes-Borrego enjoyed a contract from 'Consejería de Economía, Innovación y Ciencia' of Junta de Andalucía, Spain. The funders had no role in study design, data collection and analysis, decision to publish, or preparation of the manuscript.

Competing Interests: The authors have declared that no competing interests exist.

* E-mail: blanca.landa@csic.es

Introduction

Spain is the world's largest olive oil producer, accounting for more than one-third of global production [1,2]. In Andalusia (Southern Spain), olive orchards dominate the landscape in an impressive monoculture that covers approximately 17% of the total surface of the region (1.5 million ha) [1,3]. In this region, different olive farming systems can be found including: i) *conventional farming* with rain fed orchards of low plant density, intensive tillage, and low inputs in fertilizer, as well as intensive drip-irrigated orchards, grown with higher inputs of pesticides and fertilizers in order to push up olive yields, and ii) *organic farming* using no chemical inputs and mainly non or light-tillage and use of

a vegetative cover to prevent soil erosion [3,4,5]. Additionally, in the last years, many new olive plantations have been mediated by the use of self-rooted planting stocks which have incorporated commercial arbuscular mycorrhizal fungi (AMF) in the potting mixture during the nursery period to facilitate establishment [6,7,8] due to their beneficial effects against biotic and abiotic stresses [9,10,11,12,13].

One of the critical steps for applying AMF to improve crop health is the appropriate selection of effective and well adapted-isolates to be used as inoculants. Although it is well known that olive tree is a mycotrophic plant [7] there is not enough knowledge concerning the effect of cropping practices, olive genotype and environment (soil type and climate) on the biodiversity of AMF in olive orchards in the Mediterranean Region. Knowledge of those factors may be essential to take advantage on the use of AMF in modern oliviculture.

In the present study we have examined the structure and diversity of AMF communities in the rhizosphere of cultivated and wild olives in Andalusia, southern Spain, by using two culture-independent molecular approaches: Fluorescent terminal restriction fragment length polymorphism (T-RFLP) analyses of amplified 28S rDNA sequences and SSU rDNA amplicon parallel 454 pyrosequencing. We also have determined which agronomic or environmental factors associated to the olive orchards sampled are the main drivers of the AMF structure.

Materials and Methods

Ethics Statement

No specific permits were required for the described field studies. Location of organic olive orchards was provided by the Andalusian Committee of Organic Farming (CAAE, Junta de Andalucía). Permission to sample the olive orchards were granted by the landowner. The samples from wild or feral forms of olive are located in public areas or degraded formations and abandoned groves. The specific location of all samples from the study is provided in Table S1. The 96 olive orchards sampled in this study were also included in previous studies [5,14] in which the bacterial communities and functional diversity of the olive rhizosphere was assessed. The sites are not protected in any way. The areas studied do not involve any species endangered or protected in Spain.

Location of Olive Orchards and Rhizosphere Sampling

Soil and roots samples were collected from 90 commercial orchards differing in management system [conventional (49 orchards) vs. organic (41 orchards)] located in the main olive-growing areas of Córdoba (41 orchards), Granada (3 orchards), Jaén (34 orchards), and Sevilla (12 orchards) provinces in Andalusia, southern Spain [5; Table S1]. In addition, six samples (LO, LOBA, BAETICA, MACO, LOMCO, EPCO) from three sites each in Córdoba and Cádiz provinces containing wild or feral forms of olive 'Acebuches' (i.e., secondary sexual derivatives of the cultivated clones or products of hybridization between cultivated trees and nearby oleasters) were included in the study [14; Table S1]. Orchards across all locations sampled differ in climate, soil texture and physicochemical characteristics, soil management system (use of cover crops vs. bare soil), and irrigation regimen (rain-fed vs. drip-irrigated). When possible, we tried to sample a representative distribution of the above considered factors within each orchard management system. Detailed description of soil physicochemical properties, and agronomic and climatic characteristics of the sampled orchards is provided in Table S1. Some of the soil physicochemical and climatic characteristics of the sampled locations were provided recently [5,14].

Root (only young and active) and soil samples were collected in May to July 2009 as described by Aranda et al. [14] and Montes-Borrego et al. [5] in the area of the canopy projection from the upper 5 to 30 cm of soil from three different points around each individual tree. Eight trees per orchard were sampled, and all roots from all trees were thoroughly mixed to obtain a single representative sample per orchard. Intact root systems were shaken gently by hand to remove all but the soil close- and naturally-adhering to the plant root and were kept at 5°C until processing.

Additionally the geographic location and altitude of the sampling sites were determined using a global positioning system (GPS), and climatic variables of each sampling site were obtained from SigMapa, Geographic Information System from the Spanish Ministry of "Medio Ambiente y Medio Rural y Marino" (http://sig.mapa.es/geoportal/) using ArcGIS 10 (ESRI, Redlands, California, EE.UU.) (Table S1).

DNA Extraction from Rhizosphere Samples

Pooled olive root samples were cut into 1-cm pieces with a sterile scissors to get a uniform sample per location. Rhizosphere suspensions (including rhizosphere soil and rhizoplane) were obtained by vigorously shaking 2 g of root segments (four independent replications) suspended in 20 ml of sterile distilled water in an orbital shaker for 10 min and sonicated (Ultrasons, JP Selecta SA, Barcelona, Spain) for 10 minutes. Then, 3 ml of those rhizosphere suspensions were subjected to two consecutive centrifugations at 11,000 rpm for 4 min and the pellet was kept at −20°C until processed. DNA from each of the four rhizosphere soil pellets (approximately 200 mg; four replication per each of the 96 olive orchards) was extracted using the PowerSoil DNA Isolation Kit (MO BIO Laboratories, Inc., Carlsbad, USA) and the FastPrep-24 (MP Biomedicals, Inc., Illkirch, France) instrument run at 6.0 m/s for 40 s as described elsewhere [14].

T-RFLP Analysis

For T-RFLP analysis PCR amplification of partial LSU of rDNA from mycorrhiza were performed using a nested-PCR approach, the first PCR round employing 20 ng of template and the primer pair LR1/FLR2 [15] and the second one the primer pair FLR3/FLR4 [16] following conditions described by Mummey and Rillig [17]. Primer FLR3 was 5′ end-labeled with the fluorescent dye FAM. T-RFLP analysis was performed for all samples using 5 μl of PCR products (about 1000 ng) and TaqI restriction enzyme (Fast Digest, Fermentas, Germany) in a final volume of 10 μl. TaqI restriction enzyme was selected from those (AluI, MboI and TaqI) that were shown to discriminate more AMF groups in a previous study [17] after preliminary testing with a subset of our rhizosphere samples (*data not shown*). Terminal restriction fragments (TRF) were loaded and separated on a 3130XL genetic analyzer (Applied Biosystems, California, USA) at the SCAI-University of Córdoba sequencing facilities. Size of fragments were determined using a ROX500 size standard, and matrices containing incidence as well as peak area data of individual TRFs were generated for all samples with GeneMapper software (Applied Biosystems). Peaks of less than 100 fluorescence units (FU) and shorter than 50 bp were not included in the analysis to eliminate primer dimmers and other small charged molecules. Similarly, molecules that were not present in at least two of the four replicate profiles were disregarded. Also, TRFs that differed by less than 1 bp were clustered, unless individual peaks were detected in a reproducible manner. TRFs profiles were standardized based on methods described previously by Dunbar et al. [18]. The relative abundance of each TRF was calculated as the ratio of the peak area for that TRF to the sum of peak areas for all TRFs in the profile and was expressed as a percentage. Diversity statistics were calculated from standardized profiles of rhizosphere samples by using the number and area of peaks in each profile as representative of the number and relative abundance of OTUs, as defined by Dunbar et al. [19]. Phylotype richness was calculated as the total number of distinct TRF sizes (with length between 50 and 500 bp) in a profile and the Shannon-Wiener diversity index was calculated as described before [14]. Finally, a single standardized T-RFLP profile for each orchard was produced by averaging peak area for each TRF from four replicates.

Pyrosequencing Analysis

For 454-pyrosequencing SSU rRNA Glomeromycota sequences were amplified from a DNA mixture obtained from the four independent rhizosphere DNA extractions per olive orchard. This approach was taken to ideally cover as much biodiversity as possible and to ensure that representative AMF communities from each olive location were sampled [20]. The pyrosequencing was performed as described in Davison et al. [21] using a two-step PCR protocol with the primers NS31 and AML2, which target a ca. 560-bp central fragment of the SSU rRNA gene in Glomeromycota [22], the most widely used marker in AMF surveys to date [23,24]. These primers were linked to partial sequencing primers A and B, respectively. Bar-code sequences, 8 bp in length, were inserted between the A primer and NS31 primer sequences. Thus, the composite forward primer was: 5'-GTCTCCGACTCAG(NNNNNNNN) *TTGGAGGGCAAGTCTG-GTGCC*-3'; and the reverse primer was 5'-TTGGCAGTCT-CAG*GAACCCAAACACTTTGGTTTCC*-3, where partial sequences of A and B primers are underlined, barcode is indicated by N-s in parentheses and specific primers NS31 and AML2 are shown in italic. Then, a 10x dilution of the first PCR product was used in a second PCR where full sequencing adapters were added (Primer A 5'-CCATCTCATCCCTGCGTGTCTCCGAC-3' and Primer B 5'-CCTATCCCCTGTGTGCCTTGGCAGT -3'). The reactions contained 5 µl of Smart-Taq Hot Red 2x PCR Mix (Tartu, Naxo Ltd, Estonia), 1 µl of extracted DNA, and 0.2 µM of each primer in a final volume of 10 µl. The reactions were performed using a Thermal cycler 2720 (Applied Biosystems) under the following conditions: 95°C for 15 min; five cycles of 42°C for 30 s, 72°C for 90 s, 92°C for 45 s; 35 (first PCR) or 20 (second PCR) cycles of 65°C for 30 s, 72°C for 90 s, 92°C for 45 s; followed by 65°C for 30 s and 72°C for 10 min. PCR products were separated by electrophoresis using 1.5% agarose gels in 0.5 x TBE, and the PCR products were purified from the gel using the Qiagen QIAquick Gel Extraction kit (Qiagen Gmbh, Germany) and further purified with AgencourtH AMPureH XP PCR purification system (Agencourt Bioscience Co., Beverly, MA, USA). The 96 quantified samples were finally mixed at equimolar concentrations prior to sequencing. GATC Biotech (Constanz, Germany) performed sequencing procedures as custom service using a Genome Sequencer FLX System and Titanium Series reagents (Roche Applied Science, Indianapolis, IN, USA). Sequencing of 96 samples was performed as a part of a bigger dataset.

Processing of Pyrosequencing Data and phylogenetic analysis

Pyrosequencing data were processed as described by Fierer et al. [25] using the Quantitative Insights Into Microbial Ecology (QIIME) toolkit [26]. In brief, fungal sequences were quality trimmed, assigned to rhizosphere samples based on their barcodes and denoised using default parameters. Chimeras were identified with uclust_ref software [27] and removed, and the remaining sequences were binned into OTUs using a 97% identity threshold with uclust_ref software. Then, to take into account the different number of sequences obtained for each orchard sample in the pyrosequencing analysis we estimated the relative frequency of each OTU in each orchard. Next, the most abundant sequence from each OTU was selected as a representative sequence for that OTU and deposited in the Genbank database under accessions numbers KF831296-KF831328 and the entire dataset of reads in the Sequence Read Archive of Genbank under BioProject ID PRJNA237741. Alpha diversity statistics including Richness (numbers of OTUs) and the Shannon index were also determined for orchard samples with at least five sequences.

Taxonomy was assigned to OTUs by using the Basic Local Alignment Search Tool (BLAST) for each representative sequence against the Silva 108 database (http://www.arb-silva.de/documentation/release-108/) as well as by BLAST search against the Maarj*AM* database (http://maarjam.botany.ut.ee/, [23]). Sequences from the representative set of AMF OTUs obtained in this study, the reference AMF database from Redecker and Raab [28], the blast hits from Silva 108 and the Maarj*AM* databases, and those AMF sequences reported in olive from Calvente et al [7] and present in the GenBank database were aligned using ClustalW software [29] with default parameters. Sequence alignments were manually edited using BioEdit software [30]. Phylogenetic analysis of the sequence data sets was performed based on maximum likelihood (ML) and Bayesian inference (BI) using MrBayes version 3.1.2 software [31]. The best fitted model of DNA evolution was obtained using jModelTest v. 2.1.1 [32] with the Akaike information criterion (AIC). The Akaike-supported model, the base frequency, the proportion of invariable sites, and the gamma distribution shape parameters and substitution rates in the AIC were then used in phylogenetic analyses. BI analysis under a general time reversible of invariable sites and a gamma-shaped distribution (TIM2 +I+G) model for the SSU rRNA, were run with four chains for 1.0×10^6 generations.

Statistical Analysis

The rank-based Kruskall-Wallis test was used to determine differences in the Richness and Shannon diversity indexes in relation to the different agronomic factors of the olive orchard evaluated using the Statistical Analysis System software package (SAS version 9.2; SAS Institute, Cary, NC, USA). Non-metric multidimensional scaling (NMDS) analyses were performed using MetaMDS functions within the vegan package of R software (R Core Development Team, 2005) [33] based on dissimilarities calculated using the Bray–Curtis index obtained for T-RFLP and pyrosequencing results, using 1,000 runs with random starting configurations, and environmental variables (agronomic and climatic characteristics and soil physicochemical properties) were fitted using the envfit routine. For data derived from pyrosequencing analysis only the Glomeromycota sequences were used. Ordinations for the Bray–Curtis dissimilarity derived from relative frequency of OTUs in the pyrosequencing analysis did not reach acceptable [34] stress and stability levels and was not performed. Instead, a Multivariate Regression Tree (MRT) was calculated. MRT are a statistical technique that can be used to explore, describe, and predict relationships between multispecies data and environmental characteristics. MRT forms clusters of sites by repeated splitting of the data, with each split defined by a simple rule based on environmental variables. The splits are chosen to minimize the dissimilarity of sites within clusters [35]. The sums of squares multivariate regression tree was calculated within the mvpart package with the R software, using the one-standard error rule on the cross-validated relative error to determine the number of terminal nodes [35].

Results and Discussion

Diversity of olive AMF communities

i) **T-RFLP analysis.** A total of 36 unique TRFs profiles were consistently identified in the 384 rhizosphere samples analyzed by T-RFLP analysis, with 30 TRFs (83.3%) found in a reproducible manner in 93 of the 96 olive orchards sampled. Mean Richness values ranged from 1 to 15 depending of the rhizosphere sample with an average of 5 TRFs per olive orchard. This translates into an estimated total of 30 different OTUs present across all sampled orchards with 13 and 3 OTUs being common in at least 25% or 50% of olive orchards, respectively. Mean Shannon diversity index values ranged from 0.3 to 2.2 (Figure S1; Table S2). We did not find significant differences ($P>0.120$) in AMF richness or Shannon diversity indexes derived from T-RFLP analysis according to the orchard management system, use of irrigation, presence of vegetative cover, olive tree variety or olive age (Figure S1). It has been shown that although diversity indexes are useful in describing community characteristics, they do not provide information of important compositional features of biodiversity relating to the abundances of shared taxa [36] and statistical analyses that incorporate taxon abundance and identity are more appropriate to specifically assess changes in microbial community composition and to identify the existence, if any, of agronomic or environmental gradients.

ii) **Pyrosequencing analysis.** The pyrosequencing approach of the 96 composite samples yielded a total of 13,772 high-quality reads after denoising, with a length >100 bp and $<$ 550 bp, a mean of 147 sequences per orchard field, and two samples with no sequences (Table S2). Of these, most of sequences 47.2% could not be assigned to any Eukaryota Phyla, 29.6% were assigned to the Phylum Metazoa, whereas 10.2% (1,108 reads) could be assigned to OTUs from fungal families (Figure 1), which indicates that primers NS31 and AML2 are not enough specific for amplifying AMF sequences. Other studies using pyrosequencing analysis have also shown that in spite of the supposed AMF primer specificity, 'contaminant' sequences belonging to taxa different from *Glomeromycota* are detected. For example, Öpik et al. [37] using same NS31 and AML2 primers and Ballestrini et al. [38] and Lumini et al. [20] using NS31 and AMmix primers also found about $>55\%$ of amplified sequences belonging to taxa from non-*Glomeromycota* fungi. In a recent work Kohout et al. [39] also demonstrated that a combination of up to five primer sets specifically designed to amplify *Glomeromycota*, including primer NS31/AML2, co-amplifies to a high extend non-target AMF sequences including plant, Asco- and Basidiomycota (ca. 20 to 50%, depending of the primer set used). In our study, due to the fact that we did not retrieve *Glomeromycota* sequences from some olive orchards our results could be somehow biased. Consequently, to improve the number of sequences and molecular species characterization of AMF, new designed specific AMF-primers should be tested in complex matrixes such as soil or rhizosphere [39,40], or deeper sequencing effort should be done in future studies to face this problem and to capture the total AMF diversity in olive.

The fact that we extracted DNA from the olive rhizosphere (i.e., soil tightly adhered to roots) might have accounted for the low presence of AMF sequences; probably extracting DNA from washed or entire roots may enhance the specificity of AMF amplification. In our study 59.84% of sequences assigned to fungi belonged to *Glomeromycota* (Figure 1), with 38 olive orchard samples retrieving no *Glomeromycota* sequences (Table S2). We did not find any pattern for the lack of *Glomeromycota* sequences in those samples with any of the agronomic, climatic or soil physicochemical properties of the sampled orchards (*data not shown*). When defining an OTU as belonging to AMF on the basis of having at least 97% similarity to sequences classified as AMF in the Silva and GenBank databases we identified 33 OTUs that could be unequivocally assigned to the *Glomeromycota* in 58 out of 94 olive rhizosphere

samples (with a mean of 11.4 sequences per orchard) (Figure 1; Table 1; Figure S2; Table S2). Mean richness values ranged from 1 to 9 with an average of 2.7 OTUs per olive orchard. Mean Shannon diversity index values ranged from 0 (1 single OTU) to 2.7 (Figures S1 and S2; Table S2). We did not find significant differences ($P>0.230$) in AMF richness or Shannon diversity indexes derived from pyrosequencing analysis according to the farm soil management system, use of irrigation, presence of vegetative cover, olive tree variety or olive age (Figure S1).

iii) **Comparison of T-RFLP and pyrosequencing analysis.** Although in the pyrosequencing analysis we got some orchard samples with no *Glomeromycota* sequences and this technique is costly, labour-intensive and allows lower number of samples to be processed, it provides some advantages over the T-RFLP analysis. For example, the latter technique does not provide any information on the taxa identified, different taxa (species) may share similar TRFs in electropherograms, and multiple TRFs profiles can exist within a single species [41]. Furthermore, the lack of specifity of the primers used for T-RFLP have also been shown in other studies and may also be a source of errors when PCR products serve as basis for those fingerprinting approaches [39,41]. Consequently, data dervived from T-RFLP analysis should be complemented with other techniques that provide taxa identity information such as library cloning or pyrosequencing as was the case of this study.

Species identity of olive AMF communities

In our study the 33 OTUs identified represented most of the major AMF lineages, including *Paraglomus* spp. (family Paraglomeraceae; two OTUs, comprising 4.20% of reads and 6.90% of fields), *Glomus* group A (family Glomeraceae; 22 OTUs, comprising 59.73% of reads and 81% of fields), *Glomus* group B (family Claroideoglomaceae; four OTUs, 26.40% of reads and 38% of fields), *Glomus* group C (family Diversisporaceae; three OTUs, 6.03% of reads and 14% of fields), and *Archaeospora* (two OTUs, 3.62% and 12% of fields) (Figures 1 and 2; Table 1); with *Archaeospora* spp., *Diversispora* spp. and *Paraglomus* spp. being first records in olive. It should be noted that OTU OAMF127 clustered with the virtual taxon sequence AF131054 that has been recently proposed as a potential new taxon (new family or even order) within Glomeromycota [37].

The fact that most sequences from our study belonged to the *Glomeraceae* family (*Glomus* group A) that contains several cryptic taxa with differences in ecological properties agrees with other studies that have found many isolates of this group in different locations through the world, including Mediterranean-type environments, on both natural woodlands to high input managed agro-ecosystems [20,42,43,44] suggesting that these taxa have a generalist ruderal style with tolerance to disturbance such as in agricultural ecosystems.

Interestingly, we found that those AMF species commonly used as olive inoculants or previously isolated from olive roots (i.e., *Claroideglomus claroideum*, *Funneliformis mosseae*, *Rhizophagus clarus*, *R. intraradices*, and *Septoglomus viscosum*; see [7,8,9,12,13]) showed low abundance since they were present in 1.7 to 18.97% of fields. This is in agreement with the fact that AMF belonging to Glomeraceae family colonize preferentially the roots and might be present in lower densities in the rhizosphere soil [45]. On the contrary, two AMF sequences including OAMF216 belonging to Claroideoglomeraceae (18.6% of sequences), and OAMF91246, a new

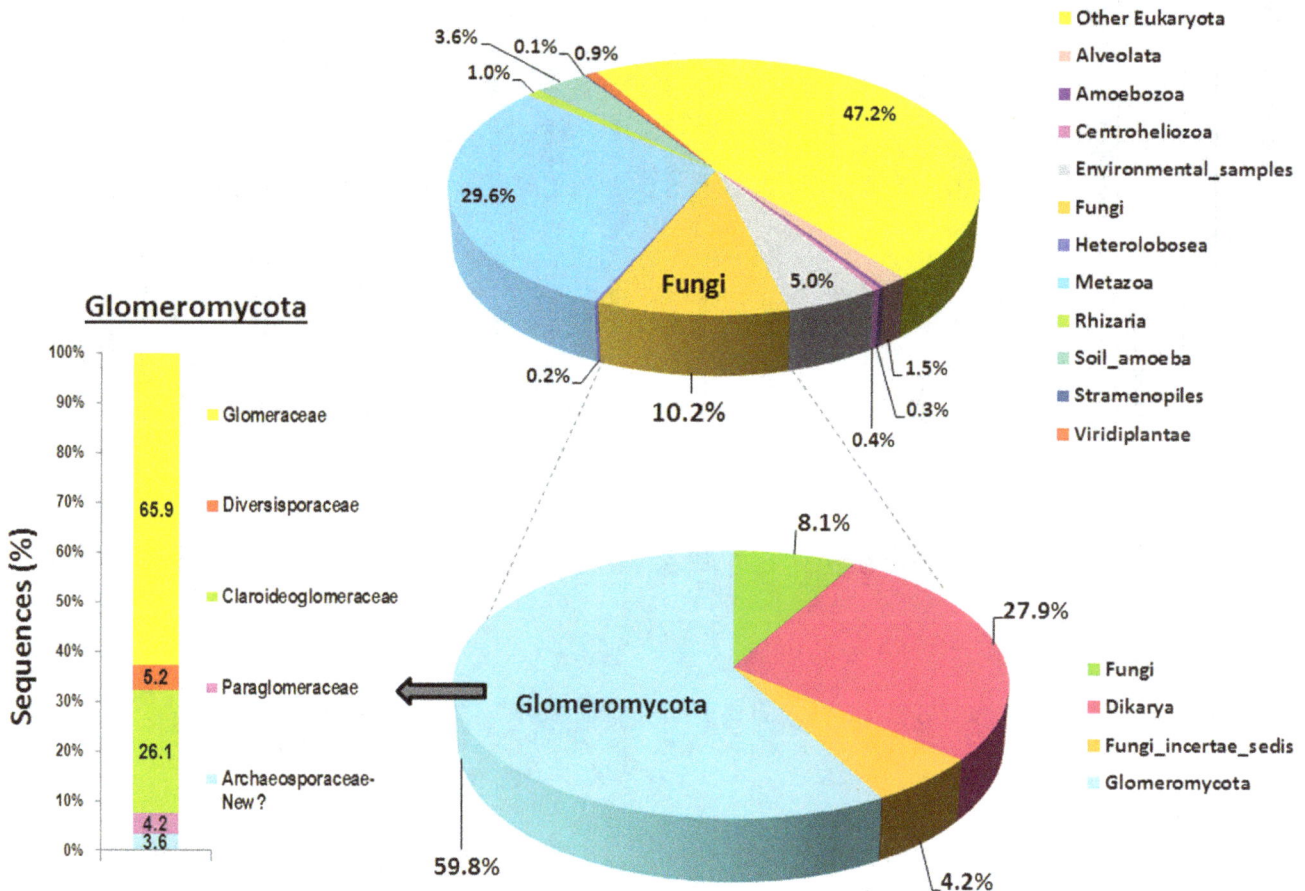

Figure 1. Proportion of overall phyla and disectioning of the fungal and Glomeromycota phyla detected by pyrosequencing analysis with primers NS31/AML2 from rhizosphere samples obtained from 96 olive orchards in Andalusia, southern Spain.

unidentified Glomeraceae (18.9% of sequences) that formed a separate cluster from other well-known AMF Glomeraceae taxa, were identified in 27.6% and 36.21% of olive orchards, respectively (Figure 2; Table 1). This might indicate that AMF can exert higher host specificity in olive than previously thought which may have implications for the olive nursery process. Thus, some authors have reported a differential growth response of olive cultivars to AMF inoculation where this responsiveness to mycorrhization has been found to depend on both the AMF species and the plant genotype [7,46]. In our study, we did not find any clear differences between the AMF sequences detected in the rhizosphere of wild olives and those found in the cultivated ones (Figure S2). This could be due to the small number of sequences that we sampled from wild olives which deserves further studies since wild olives have been shown to be a potential reservoir for discovering microbial species of diverse biotechnological and commercial interest [14,47].

Factors shaping the structure of AMF communities in olive rhizosphere

It has been shown that although diversity indexes (such as Richness and Shannon used in our study) are useful in describing community characteristics they do not provide information of important compositional features of microbial diversity related to the abundances of shared taxa [36] which migth explain that we did not find an effect of the environmental and agronomic

variables on the estimated alpha-diversity indexes. Consequently, in a second approach to specifically assess changes in AMF community composition (incorporating taxon abundance (frequency) and identity), we used NMDS ordination to represent, in two dimensions, the pairwise Bray-Curtis dissimilarities between AMF communities derived from T-RFLP analyses. Then, we projected each of the environmental and agronomic variables independently onto the NMDS ordination to identify hypotetical gradients likely related to the differentiation in AMF composition (Figure 3; Table 2). In relation to agronomic variables AMF communities were differentiated according to the cultivar genotype and age of plantation and the irrigation regimen of the olive orchard, in that order, whereas the grouping according to the orchard management system or presence of a vegetative cover was not significant (Table 2). Thus, there was a tendency to locate rhizosphere samples in the NMDS ordination from olive orchards <15 year old at the bottom quandrant of Y = 0 (with only two exceptions), whereas olive orchards of 15 to 30 year old were all located on the left cuadrant of X = 0 (Figure 3). The effect of olive gentoype in affecting soil biota has also been shown in a recent study [48] which demostrated that olive genotypes significantly influenced the nematode assemblages present in their rhizospheric soil.

We also identified C:N ratio, soil C and organic matter content and pH as the environmental variables better explaining ($P <$ 0.001; $0.2836 > r^2 > 0.1301$) the AMF community composition among the olive orchards, in that order (Figure 3, Table 2). Other

Table 1. Glomeromycota taxa detected in olive rhizosphere samples, taxonomic affiliation, number of sequences from each taxa, frequency of occurrence in the sampled olive orchards, and the closest related AMF sequence.

Family	OTU identification[a] Phylogenetic group	Code	Acc. Number	Number of sequences	Frequency of sequences (%)	Frequency of orchards (%)	Closest taxa[b] Silva 108	MaarjAM
Archaeosporaceae	Ia	OAMF127	KF831299	23	3.47	10.34	Glomeromycota sp.	Archaeospora sp. VT5
	Ib	OAMF64639	KF831323	1	0.15	1.72	Archaeospora trappei	Archaeospora trappei VT245
Paraglomeraceae	IIa	OAMF131	KF831300	1	4.07	1.72	Fungi	Paraglomus laccatum VT281
	IIb	OAMF60009	KF831322	27	0.15	6.90	Fungi	Paraglomus brasilianum VTX239
Claroideoglomeraceae	IIIa	OAMF26258	KF831310	52	7.84	18.97	Glomus etunicatum	Glomus sp. group B VT193
(Glomus group B)	IIIb	OAMF216	KF831306	121	18.25	27.59	Glomus etunicatum	Glomus geosporum VT65
		OAMF71	KF831324	1	0.15	1.72	Glomus etunicatum	Glomus sp. group B VT56
		OAMF333	KF831315	1	0.15	1.72	Glomus etunicatum	No match
		Total		123	18.55	27.59		
Diversisporaceae	IVa	OAMF79857	KF831327	19	2.87	3.45	Glomus eburneum	Glomus sp. MO-GC1 VT60
(Glomus group C)	IVb	OAMF264	KF831311	20	3.02	12.07	Glomus eburneum	Glomus sp. Wirsel OTU20 VT62
		OAMF156	KF831302	1	0.15	1.72	Glomus eburneum	Diversispora sp. VT62
		Total		21	3.31	13.79		
Glomeraceae	V	OAMF19034	KF831303	16	2.41	1.72	Glomeromycota	Glomus sp. Dictamnus2 VT163
(Glomus group A)		OAMF443	KF831318	11	1.66	5.17	Uncultured Glomus sp.	Glomus sp. Douhan6 VT143
		Total		27	4.07			
	VI	OAMF43	KF831317	23	3.47	5.17	Uncultured Glomus sp.	Glomus sp. VT145
	VIIa	OAMF359	KF831316	1	0.15	1.72	Glomeromycota	Glomus sp. VT153
		OAMF91246	KF831328	124	18.70	36.21	Uncultured Glomus sp.	Glomus sp. VT118
		Total		125	18.85	36.21		
	VIIb	OAMF73	KF831325	59	8.90	15.52	Uncultured Glomus sp.	Glomus Wirsel VT137
	VIIc	OAMF499	KF831320	9	1.36	1.72	Uncultured Glomus sp.	Glomus JP3 VT128
	VIIIa	OAMF213	KF831305	12	1.81	3.45	Glomeromycota	Glomus sp. Alguacil09b VT109
	VIIIb	OAMF452	KF831319	11	1.66	8.62	Uncultured Glomus sp.	Glomus sp. Glo45 VT109
	IXa	OAMF22696	KF831307	14	2.11	6.90	Uncultured Glomus sp.	Glomus sp. Glo24 VT105
		OAMF15	KF831301	2	0.30	3.45	Rhizophagus intraradices	Glomus intraradices VT113
		OAMF22729	KF831308	29	4.37	8.62	Rhizophagus intraradices	Glomus intraradices VT113
		Total		45	6.79	18.97		

Table 1. Cont.

Family	OTU identification[a]		Acc. Number	Number of sequences	Frequency of sequences (%)	Frequency of orchards (%)	Closest taxa[b]	
	Phylogenetic group	Code					Silva 108	MaarjAM
	IXb	OAMF293	KF831312	8	1.21	6.90	Uncultured Glomus sp.	Glomus sp. VT94
	IXc	OAMF123	KF831297	2	0.30	3.45	Uncultured Glomus sp.	No match
	IXd	OAMF521	KF831321	17	2.56	8.62	Uncultured Glomus sp.	Glomus sp. Glo7 VT214
	Xa	OAMF102182	KF831296	4	0.60	1.72	Glomus sp.	Funneliformis caledonium VT65
	Xb	OAMF77556	KF831326	5	0.75	1.72	Glomus sp.	Funneliformis mosseae VT67
	Xc	OAMF311	KF831314	13	1.96	10.34	Uncultured Glomus sp.	Septoglomus viscosum VT63
	XI	OAMF3	KF831313	1	0.15	1.72	Uncultured Glomus sp.	Glomus sp. VT301
		OAMF20	KF831304	4	0.60	3.45	Uncultured Glomus sp.	Glomus sp. VT301
		Total		5	0.75	3.45		
	XII	OAMF25566	KF831309	30	4.52	18.97	Glomeromycota	Glomus Alguacil09c Glo4 VT166
		OAMF125	KF831298	1	0.15	18.97	Glomeromycota	Glomus Alguacil09c Glo4 VT166
		Total		31	4.68	18.97		

[a]The phylogenetic groups were arbitrarily named according to their position in the Bayesian analysis shown in Figure 2. Each AMF OTU sequence found in the study was assigned a Code (OAMF# S#) where OAMF refers to 'olive arbuscular mycorrhizal fungi' and # to the number assigned to each representative AMF OTUs derived from uclust_ref analysis with the QIIME software.
[b]Closest taxa assigned by BLAST analysis using the Silva 108 or MaarjAM database. Numerical codes of 'virtual taxa' VT as appear in the MaarjAM database are shown as in Figure 2.

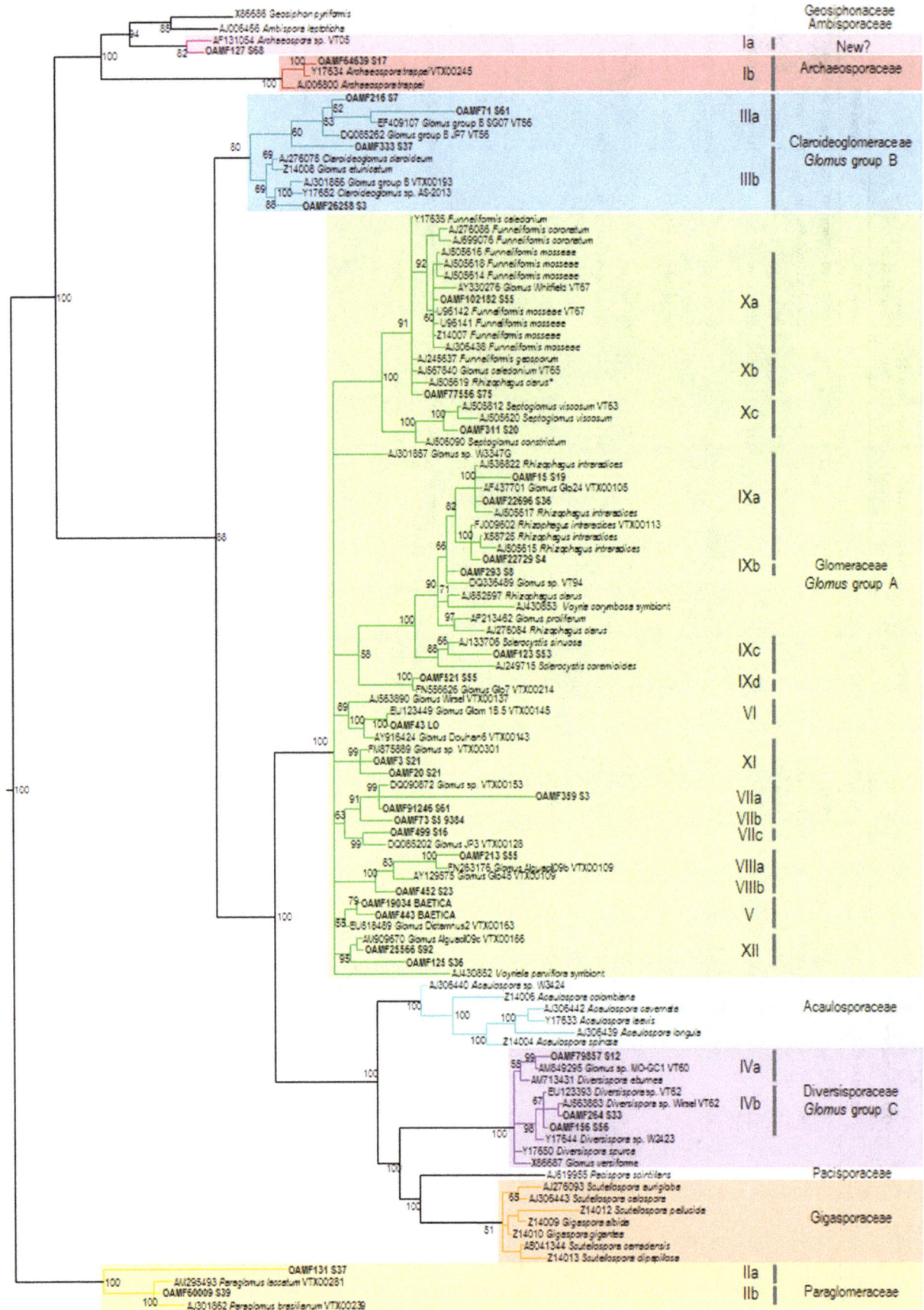

0.2

Figure 2. Phylogenetic relationships of nuclear small subunit ribosomal RNA (SSU rRNA) gene sequences of *Glomeromycota* **reference sequences derived from uclust_ref search with those that matched the silva_108 and Maarj***AM* **databases, the reference AMF database from Redecker and Raab [28] and those reported in olive from Calvente et al [7] and present in the GenBank.** Bayesian 50% majority rule consensus tree as inferred from nSSU rRNA sequences alignments under the general time reversible + G + I model. Numbers on the nodes indicate Bayesian posterior probabilities (>50%). The phylogram was rooted with Paraglomeraceae sequences. Numerical codes in bold name each representative AMF OTUs from olive rhizosphere derived from uclust_ref analysis with the QIIME software and are labelled (OAMF# S#) where OAMF refers to 'olive arbuscular mycorrhizal fungi' and # to the number assigned and groups identified and the remaining code refers to the soil sample. Phylogenetic groups (I to Xd) were arbitrarily described and are shown in Table 1. (*) Although this sequence was originally identified as belonging to *R. clarus* by Calvente et al. [7] its closest taxonomic affiliation is to *Funneliformis* sp. and clearly differs from sequences AJ276084 and AJ852597 of *R. clarus*.

environmental factors showing a significant (*P*<0.034) but lower effect included clay, sand and N content, extractable P and annual evapotranspitation and minimum temperature of the sampled locatios (Table 2).

A multivariate regression tree was also calculated to summarize the relationships between AMF community composition derived from pyrosequencing analysis and environmental and agronomic variables with the most informative variables in each split shown in Figure 4. The tree explained >30% of the variability in AMF profiles, much of which were accounted by the first split based on exchangeable K and in a lower extend by altitude and sand and clay content (Figure 4). Then, climatic variables from sampled locations including total rainfall and evapotranspiration (ETP) followed by soil pH were the next best predictors for the second-order split. Two climatic variables (altitude and average rainfall), and nutrient contents of soil samples (including OM, C and extractable P and the C:N ratio) allowed differentiating five groups of soils that included three groups of soils showing high richness in

OTUs and two groups of soils characterized by two specific OTUs each (Figure 4).

In our study we were able to identify some agronomic and environmental gradients driving the AMF community differenti-ation; however, we found some differences when using data derived from T-RFLP or pyrosequencing analysis. This could be due to the unequal sample size included in each data set (93 vs. 58 olive orchards, respectively). Another factor that is likely to have contributed to those differences is the usage of two different rRNA regions and methods, SSU for pyrosequencing and LSU for T-RFLP. These regions differ in their phylogenetic resolutions, and the methods of T-RFLP versus full amplicon sequencing differ as well in this respect as found in previous studies [49]. Consequently, in our study results from the pyrosequencing analysis should be interpreted with caution due to the smaller data set analysed and the possibility of introducing some bias due to the fact that from some olive orchards we did not amplified any *Glomeromycota* sequences. Nevertheless, we retrieved consistent results with both techniques. Thus, soil pH, textural characteristics, nutrient

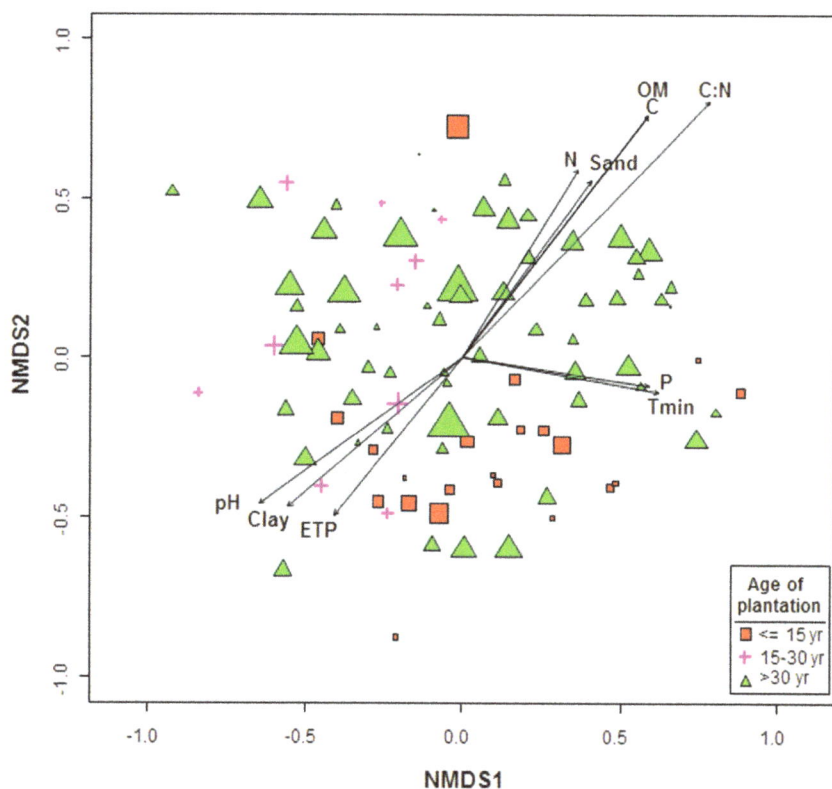

Figure 3. NMDS biplot of a Bray-Curtis dissimilarity matrix of T-RFLP analysis. The fitted vectors of environmental and physicochemical soil variables and the agronomic variable age of plantation (indicated with different symbols) most significantly and strongly associated (*P*<0.05) with the ordination and shown in Table 2 are also represented. Size of symbols is proportional to AMF richness in those olive orchards.

Table 2. Summary of relationships[a] between agronomic, soil and environmental factors and AMF communities assessed by T-RFLP analysis.

Factors [b]	r^2	P	
Soil physicochemical properties			
Clay (%)	**0.1087**	**0.003996**	**
Sand (%)	**0.0987**	**0.007992**	**
Organic C (%)	**0.1915**	**0.000999**	***
Organic N (%)	0.1006	0.010989	*
Extractable P (ppm)	0.0735	0.02997	*
Exchangeable K (ppm)	0.0315	0.228771	
CEC	0.0005	0.976024	
C:N ratio	**0.2623**	**0.000999**	***
pH(KCl)	**0.1301**	**0.000999**	***
SOM (%)	**0.1914**	**0.000999**	***
Climatic characteristics			
Total Rainfall	0.0183	0.438561	
Average Rainfall	0.0045	0.811189	
ETP	0.0849	0.026973	*
Tmax	0.0486	0.117882	
Tmin	0.0815	0.016983	*
Tmean	0.0412	0.154845	
Altitude	0.0274	0.306693	
Agronomic characteristics			
Olive variety	**0.2836**	**0.000999**	***
Presence of vegetative cover	0.0032	0.738262	
Age of plantation	**0.1233**	**0.001998**	**
Irrigation regimen	0.0401	0.033966	*
Orchard management system	0.002	0.986014	

[a]Correlations with all environmental variables (r^2) were obtained by fitting linear trends to the NMDS ordination obtained with each restriction enzyme and significance (P) was determined by permutation (nperm = 1000).
'***' = $P<0.001$;
'**' = $P<0.01$;
'*' = $P<0.05$. Variables with highest significant weight are shown in bold.
[b]Orchard agronomic and climatic characteristics, and soil physicochemical properties are shown in Table S1 and some of them were reported before [5,14]. Climatic variables were obtained from SigMapa, Geographic Information System from the Spanish Ministry of "Medio Ambiente y Medio Rural y Marino" (http://sig.mapa.es/geoportal/) using ArcGIS 10 (ESRI, Redlands, California, EE.UU.).

contents, and some climatic variables appear as the most important vectors driving the AMF community differentiation in olive. Soil texture has typically not been identified as being of great importance on AMF community composition until recently [43,50]. Landis et al. [51] identified a texture effect in an oak savannah ecosystem, but the effects of texture could not be separated from plant community composition and soil N. A separate study [52] did find closely influences of clay, moisture, and pH in AMF composition in a maize agricultural system in Zimbabwe, along with strong effects of soil organic C and total N. On the other hand, Moebius-Clune et al. [51] studied AMF communities in an assemblage of maize fields across an eastern New York State landscape and found soil textural components as the most strongly related to AMF community differences followed by nutrient concentrations, particularly Mg, whereas soil P or pH, were less important.

These results demonstrate that when there are small differences in soil physicochemical characteristics, the composition of the AMF communities might be similar with an overlapping in the AMF assemblages among different agronomic or soil use practises [20]. All the data obtained in this work, reinforce the concept that the general AMF assemblage structure and composition in olive might be influenced primarily by soil type and climate and at less extent by host plant features (age, vegetative stages, host genotype) or agricultural practices as it has been shown in other woody crop such as vineyards [20,53]. To our best knowledge this study provides the first evidence of a specific effect of such factors on AMF community composition in olive. Further research using a deeper pyrosequencing effort or more specific primers should be conducted to determine how this specific selection of AMF communities by the different olive varieties may be related to olive resilience to mycorrhization during the olive nursery process or to the successful establishment of those mycorrhized planting stocks when transplanted to soils in the different biogeographical areas (as identified by climatic and soil physicochemical properties) present in southern Spain.

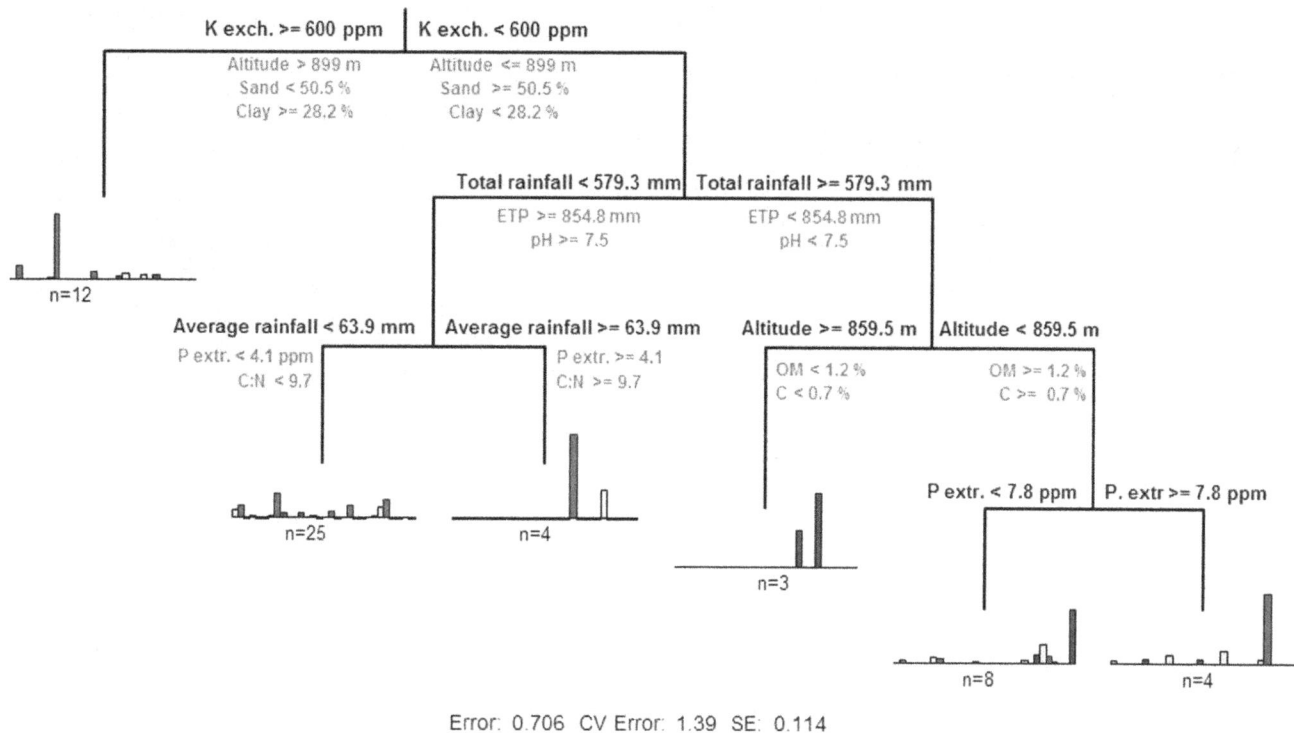

Figure 4. Sums of squares multivariate regression tree summarizing olive AMF community–agronomic, environmental and soil factors relationships. The tree was calculated using frequency of AMF OTUs derived from pyrosequencing analysis (Figure S1). For each split a rule is selected based on the predictors to minimize the dissimilarity within the AMF OTUs profiles in the resulting two nodes (main rule is shown above the node, and second rules are shown below the node). At each terminal node, the mean relative abundances of each AMF OTU are shown, together with the number of olive orchards for each group.

Supporting Information

Figure S1 Summary box-plots of Richness and Shannon diversity indexes derived from T-RFLP (93 olive orchards) and pyrosequencing analysis (43 olive orchards) grouped by the agronomic characteristics of the olive orchards sampled (Table S1).

Figure S2 Frequency of occurrence of the different *Glomeromycota* OTUs detected with primers NS31/AML2 and listed in Table 1 in 56 rhizosphere samples from 96 olive orchards sampled in Andalusia, southern Spain.

Table S1 Datasets, location and characteristics of the olive orchards sampled.

Table S2 Number of sequences and diversity indexes values obtained in the T-RFLP and pyrosequencing analysis in each orchard sampled.

Acknowledgments

We thank C. Cantalapiedra, F. Durán and G. Contreras for technical assistance and P. Castillo and J.A. Navas-Cortés for suggestions to improve the manuscript.

Author Contributions

Conceived and designed the experiments: MMB BBL. Performed the experiments: MMB BBL. Analyzed the data: BBL. Contributed reagents/materials/analysis tools: MM BBL MM. Wrote the paper: MM BBL MM.

References

1. CAP (2011) Estadísticas agrarias [Agricultural statistics]. Consejería de Agricultura y Pesca, Junta de Andalucía, Sevilla, Spain. Available at Web site http://www.juntadeandalucia.es/agriculturaypesca/portal/servicios/estadisticas/estadisticas/agrarias/index.html (verified October 21, 2013).
2. IOC (2011) World Olive Oil Figures. International Olive Council. Available at Web site http://www.internationaloliveoil.org/estaticos/view/131-world-olive-oil-figures (verified October 21, 2013).
3. Soriano MA, Álvarez S, Landa BB, Gómez JA (2013) Soil properties in organic olive orchards following different weed management in a rolling landscape of Andalusia, Spain. Renew Agr Food Syst: in press, doi:10.1017/S1742170512000361.
4. Milgroom J, Soriano MA, Garrido JM, Gómez JA, Fereres E (2007) The influence of a shift from conventional to organic olive farming on soil management and erosion risk in southern Spain. Renew Agr Food Syst 22: 1–10.

5. Montes-Borrego M, Navas-Cortés JA, Landa BB (2013) Linking microbial functional diversity of olive rhizosphere soil to management systems in commercial orchards in southern Spain. Agric Ecosyst Environ 181: 169–178.
6. Binet MN, Lemoine MC, Martin C, Chambon C, Gianinazzi S (2007) Micropropagation of olive (*Olea europaea* L.) and application of mycorrhiza to improve plantlet establishment. In Vitro Cell Dev Biol Plant 43: 473–478.
7. Calvente R, Cano C, Ferrol N, Azcon-Aguilar C, Barea JM (2004) Analysing natural diversity of arbuscular mycorrhizal fungi in olive tree (*Olea europaea* L.) plantations and assessment of the effectiveness of native fungal isolates as inoculants for commercial cultivars of olive plantlets. Appl Soil Ecol 26: 11–9.
8. Estaún V, Camprubi A, Calvet C, Pinochet J (2003) Nursery and field response of olive trees inoculated with two arbuscular mycorrhizal fungi, *Glomus intraradices* and *Glomus mosseae*. J Am Soc Horticult Sci 128:767–775.

9. Castillo P, Nico AI, Azcón-Aguilar C, Del Río Rincón C, Calvet C, et al. (2006). Protection of olive planting stocks against parasitism of root-knot nematodes by arbuscular mycorrhizal fungi. Plant Pathol 55: 705–713.

10. Castillo P, Nico A, Navas-Cortés JA, Landa BB, Jiménez-Díaz RM, et al. (2010) Plant-parasitic nematodes attacking olive trees and their management. Plant Dis 94: 148–162.

11. Dag A, Yermiyahu U, Ben-Gal A, Zipori I, Kapulnik Y (2009) Nursery and post-transplant field response of olive trees to arbuscular mycorrhizal fungi in an arid region. Crop Pasture Sci 60: 427–433.

12. Meddad-Hamza A, Beddiar A, Gollotte A, Lemoine MC, Kuszala C, et al. (2010) Arbuscular mycorrhizal fungi improve the growth of olive trees and their resistance to transplantation stress. Afr J Biotechnol 9: 1159–1167.

13. Porras-Soriano A, Soriano-Martín ML, Porras-Piedra A, Azcón R (2009). Arbuscular mycorrhizal fungi increased growth, nutrient uptake and tolerance to salinity in olive trees under nursery conditions. J Plant Physiol 166:1350–1359.

14. Aranda S, Montes-Borrego M, Jiménez-Díaz RM, Landa BB (2011) Microbial communities associated with the root system of wild olives (*Olea europaea* L. subsp. *europaea* var. *sylvestris*) are good reservoirs of bacteria with antagonistic potential against *Verticillium dahliae*. Plant Soil 343: 329–345.

15. Van Tuinen D, Jacquot E, Zhao B, Gollotte A, Gianinazzi-Pearson V (1998) Characterization of root colonization profiles by a microcosm community of arbuscular mycorrhizal fungi using 25S rDNA-targeted nested PCR. Mol Ecol 7: 879–887.

16. Gollotte A, van Tuinen D, Atkinson D (2004) Diversity of arbuscular mycorrhizal fungi colonizing roots of the grass species *Agrostis capillaries* and *Lolium perenne* in a field experiment. Mycorrhiza 14: 111–117.

17. Mummey DL, Rillig MC (2007) Evaluation of LSU rRNA-gene PCR primers for analysis of arbuscular mycorrhizal fungal communities via terminal restriction fragment length polymorphism analysis. J Microbiol Methods 70: 200–204.

18. Dunbar J, Ticknor LO, Kuske CR (2001) Phylogenetic specificity and reproducibility and new method for analysis of terminal restriction fragment profiles of 16S rRNA genes from bacterial communities. Appl Environ Microbiol 67: 190–197

19. Dunbar J, Ticknor LO, Kuske CR (2000) Assessment of microbial diversity in four southwestern United States soils by 16S rRNA gene terminal restriction fragment analysis. Appl Environ Microbiol 66: 2943–2950.

20. Lumini E, Orgiazzi A, Borriello R, Bonfante P, Bianciotto V (2010) Disclosing arbuscular mycorrhizal fungal biodiversity in soil through a land-use gradient using a pyrosequencing approach. Environ Microbiol 12: 2165–2179.

21. Davison J, Öpik M, Zobel M, Vasar M, Metsis M, et al. (2012) Communities of arbuscular mycorrhizal fungi detecting in forest soil are spatially heterogeneous but do not vary throughout the growing season. PLoS One 7:e41938.

22. Lee J, Lee S, Young JPW (2008) Improved PCR primers for the detection and identification of arbuscular mycorrhizal fungi. FEMS Microbiol Ecol 65: 339–349.

23. Öpik M, Vanatoa A, Vanatoa E, Moora M, Davison J, et al. (2010) The online database MaarjAM reveals global and ecosystemic distribution patterns in arbuscular mycorrhizal fungi (Glomeromycota). New Phytol 188:223–241.

24. Kivlin SN, Hawkes CV, Treseder KK (2011) Global diversity and distribution of arbuscular mycorrhizal fungi. Soil Biol Biochem 43: 2294–2303.

25. Fierer N, Hamady M, Lauber CL, Knight R (2008) The influence of sex, handedness, and washing on the diversity of hand surface bacteria. Proc Natl Acad Sci USA 105: 17994–17999.

26. Caporaso JG, Kuczynski J, Stombaugh J, Bittinger K, Bushman FD, et al. (2010). QIIME allows analysis of high-throughput community sequencing data. Nature Methods 7, 335 – 336.

27. Edgar RC (2010) Search and clustering orders of magnitude faster than BLAST, Bioinformatics 26(19): 2460-2461. doi: 10.1093/bioinformatics/btq461.

28. Redecker D, Raab P (2006) Phylogeny of the Glomeromycota (arbuscular mycorrhizal fungi): recent developments and new gene markers. Mycologia 98: 885–895.

29. Thompson JD, Gibson TJ, Plewniak F, Jeanmougin F, Higgins DG (1997) The CLUSTAL_X windows interface: flexible strategies for multiple sequence alignment aided by quality analysis tools. Nucleic Acids Res 25: 4876–4882.

30. Hall TA (1999). BioEdit: a user-friendly biological sequence alignment editor and analysis program for windows 95/98/NT. Nucleic Acids Symp Ser 41: 95–98.

31. Huelsenbeck JP, Ronquist F (2001) MrBAYES: Bayesian inference of phylogenetic trees. Bioinformatics 17: 754–755.

32. Darriba D, Taboada GL, Doallo R, Posada D (2012) jModelTest 2: more models, new heuristics and parallel computing. Nat Methods 9: 772.

33. Oksanen J, Blanchet FG, Kindt R, Legendre P, O'Hara RG, et al. (2011) Vegan: community ecology package. R package version 1.17-6. Available at: http://CRAN.R-project.org/package = vegan.; accessed 08/06/2013.

34. Clarke KR (1993) Non-parametric multivariate analyses of changes in community structure. Aust J Ecol 18: 117–143.

35. De'ath G (2002) Multivariate regression trees: a new technique for modeling species – environment relationships. Ecology 83: 1105–1117.

36. Griffiths RI, Thomson BC, James P, Bell T, Bailey MJ, et al. (2011) The bacterial biogeography of British soils. Environ Microbiol 13: 1642–1654.

37. Öpik M, Davison J, Moora M, Zobel M (2013) DNA-based detection and identification of Glomeromycota: the virtual taxonomy of environmental sequences. Botany 10.1139/cjb-2013-0110.

38. Balestrini R, Magurno F, Walker C, Lumini E, Bianciotto V (2013) Cohorts of arbuscular mycorrhizal fungi (AMF) in *Vitis vinifera*, a typical Mediterranean fruit crop. Environ Microbiol Rep 2: 594–604.

39. Kohout P, Sudováa R, Janouškováaa M, Čtvrtlíkováac M, Hejdaa M, et al. (2014) Comparison of commonly used primer sets for evaluating arbuscular mycorrhizal fungal communities: Is there a universal solution?. Soil Biol Biochem 68: 482–493.

40. Krüger M, Stockinger H, Krüger C, Schüßler A (2009) DNA-based species level detection of Glomeromycota: one PCR primer set for all arbuscular mycorrhizal fungi. New Phytol 183: 212–223.

41. Dickie IA, FitzJohn RG (2007) Using terminal restriction fragment length polymorphism (T-RFLP) to identify mycorrhizal fungi: a methods review. Mycorrhiza 17:259–270.42.

42. Öpik M, Zobel M, Cantero JJ, Davison J, Facelli JM, et al. (2013) Global sampling of plant roots expands the described molecular diversity of arbuscular mycorrhizal fungi. Mycorrhiza 23:411–430.

43. Oehl F, Laczko E, Bogenrieder A, Stahr K, Bosch R, et al. (2010) Soil type and land use intensity determine the composition of arbuscular mycorrhizal fungal communities. Soil Biol Biochem 42: 724–738.

44. Öpik M, Moora M, Liira J, Zobel M (2006) Composition of root-colonizing arbuscular mycorrhizal fungal communities in different ecosystems around the globe. J Ecol 94: 778–790.

45. Hart MH, Reader RJ (2002) Taxonomic basis for variation in the colonization strategy of arbuscular mycorrhizal fungi. New Phytol 153:335–344.

46. Estaún V, Calvet C, Campubrí A (2010) Effect of Differences among crop species and cultivars on the arbuscular mycorrhizal symbiosis. In: Arbuscular Mycorrhizas: Physiology and Function. Hinanit Koltai and Yoram Kapulnik, Editors. Springer pp. 279–295.

47. Aranda S, Montes-Borrego M, Landa BB (2011) Purple-pigmented violacein-producing *Duganella* spp. inhabit the rhizosphere of wild and cultivated olives in Southern Spain. Microb Ecol 62:446–459.

48. Palomares-Rius JE, Castillo P, Montes-Borrego M, Müller H, Landa BB (2012) Nematode community populations in the rhizosphere of cultivated olive differs according to the plant genotype. Soil Biol Biochem 45:168–171.

49. Verbruggen E, Kuramae EE, Hillekens R, de Hollander M, Kiers ET, et al. (2012). Testing potential effects of maize expressing the Bacillus thuringiensis Cry1Ab endotoxin (Bt maize) on mycorrhizal fungal communities via DNA- and RNA-based pyrosequencing and molecular fingerprinting. Appl Environ Microbiol 78: 7384–7392.

50. Moebius-Clune DJ, Moebius-Clune BN, van Es HM, Pawlowska TE (2013) Arbuscular mycorrhizal fungi associated with a single agronomic plant host across the landscape: Community differentiation along a soil textural gradient. Soil Biol Biochem 64: 191–199.

51. Landis FC, Gargas A, Givnish TJ (2004) Relationships among arbuscular mycorrhizal fungi, vascular plants and environmental conditions in oak savannas. New Phytol 164: 493–504.

52. Lekberg Y, Koide RT, Rohr JR, Aldrich-Wolfe L, Morton JB (2007) Role of niche restrictions and dispersal in the composition of arbuscular mycorrhizal fungal communities. J Ecol 95: 95–105.

53. Schreiner RP, Mihara K (2009) The diversity of arbuscular mycorrhizal fungi amplified from grapevine roots (*Vitis vinifera* L.) in Oregon vineyards is seasonally stable and influenced by soil and vine age. Mycologia 101: 599–611.

10

Manure Refinement Affects Apple Rhizosphere Bacterial Community Structure: A Study in Sandy Soil

Qiang Zhang, Jian Sun, Songzhong Liu, Qinping Wei*

Institute of Forestry and Pomology, Beijing Academy of Agriculture & Forestry Sciences, Beijing, China

Abstract

We used DNA-based pyrosequencing to characterize the bacterial community structure of the sandy soil of an apple orchard with different manure ratios. Five manure percentages (5%, 10%, 15%, 20% and 25%) were examined. More than 10,000 valid reads were obtained for each replicate. The communities were composed of five dominant groups (Proteobacteria, Actinobacteria, Chloroflexi, Acidobacteria and Bacteroidetes), of which Proteobacteria content gradually decreased from 41.38% to 37.29% as manure ratio increased from 0% to 25%, respectively. Redundancy analysis showed that 37 classes were highly correlated with manure ratio, 18 of which were positively correlated. Clustering revealed that the rhizosphere samples were grouped into three components: low manure (control, 5%) treatment, medium manure (10%, 15%) treatment and high manure (20%, 25%) treatment. Venn analysis of species types of these three groups revealed that the bacteria community difference was primarily reflected by quantity ratio rather than species variety. Although greater manure content led to higher soil organic matter content, the medium manure improved soil showed the highest urease activity and saccharase activity, while 5% to 20% manure ratio improvement also resulted in higher bacteria diversity than control and 25% manure ratio treatment. Our experimental results suggest that the use of a proper manure ratio results in significantly higher soil enzyme activity and different bacteria community patterns, whereas the use of excessive manure amounts has negative effect on soil quality.

Editor: Jose Luis Balcazar, Catalan Institute for Water Research (ICRA), Spain

Funding: China Agriculture Research System, CARS-28, Minister of Agriculture (PRC), http://english.agri.gov.cn/. The funders had no role in study design, data collection and analysis, decision to publish, or preparation of the manuscript.

Competing Interests: The authors have declared that no competing interests exist.

* E-mail: qinpingwei@gmail.com

Introduction

China is the largest apple producer in the world, of which apple orchard cultivation area and yield were 2.06×10^6 ha and 3.33×10^7 tons, accounting for 43.78% and 47.86% of the world's production, respectively (FAOstat, 2010). However, 74% of China's total apple orchard area is distributed in hills, mountains and dry highland region, where limited soil organic matter (SOM) sustains crop productivity [1]. SOM can facilitate water entry [2], resist surface structure degradation and adsorb minerals and nutrients, all of which significantly impact fruit yield and quality. Manure improvement has been used in farmlands in China for thousands of years, and it is now the most effective manual way to increase SOM within apple orchards. Meanwhile, microbes within orchard soil is an important component of soil function [3] that are related with soil mineral nutrient release and plant disease resistance, and bacteria mineralization [4] is an important natural source of SOM. The evaluation of effects of manure application on the structure and functionality of soil microbial community is crucial to gaining a more holistic understanding of soil status under manure refinement.

Inspecting rhizosphere soil microbial community structure shifts after manure application is an important method for determining the proper way to improve ecosystem services and soil function. Bacteria are the most abundant and diverse group of soil organisms [5], and it is estimated that 1 g of soil contains more than 1,000,000 bacteria from thousands of different species. Since

no more than 1% of these microbes can be cultured in the laboratory [6], the culture-based method cannot be used. The recent development of the microbial ecology base makes this investigation possible using partial ribosomal amplification and pyrosequencing techniques [7] to examine microbial diversity depth in virtually any sample type and detect changes within soil microbial groups. Soil profiles can then be compared using multivariate statistical techniques to reveal differences among microbial communities. Studies have shown that soil microbial communities were sensitive to management [8], seasonal changes [9], rhizosphere effects [10], pollution and the addition of composts and pesticides [11]. Qiu et al. [12] reported that bio-organic fertilizer changed the soil microbe community and reduced *Fusarium oxysporum* counts in cucumber plants. Sugiyama et al. studied [13] the soil bacteria communities of organic and conventional potato farms. Manter et al. [14] reported a highly diverse and cultivar-specific bacterial endophyte community in potato roots. Stéphane characterized rhizosphere soil bacteria profiles of oak [15] and found some rare phylogenetic groups. Lin et al. [16] found that different fertilizers had various impacts on soil bacteria. However, little was discovered about the apple rhizosphere microbial profile, and although manure treatment was employed in an earlier study, little was learned about the impact of different doses.

Our objectives were to determine the effects of manure improvement on apple rhizosphere soil and evaluate whether

Table 1. Soil properties under different manure refinement levels.

Manure Ratio (%)	Total N (g/kg)	Organic Matter (g/kg)	Available N (mg/kg)	Available P (mg/kg)	Available K (mg/kg)	Available Ca (g/kg)	Available Fe (mg/kg)	Available Zn (mg/kg)	Available B (mg/kg)	Cation Exchange Capacity (mmol/kg)	pH
0	0.994±0.009c	2.09±0.04a	29.8±0.3a	30.8±2.0a	78±3a	2.77±0.06d	4.98±0.10a	4.77±0.06b	0.21±0.02a	160±7a	8.63±0.01c
5	0.876±0.004b	4.38±0.04b	64.3±0.8e	67.7±2.5b	120±5b	2.61±0.02c	9.59±0.12b	3.97±0.02a	0.25±0.02a	151±6b	8.64±0.02c
10	1.97±0.11d	6.63±0.13c	35.7±1.1b	85.5±2.8c	200±9c	2.36±0.09b	10.3±0.2c	5.32±0.11d	0.49±0.04b	154±7a	8.43±0.02a
15	0.718±0.011a	8.21±0.04d	46.4±0.9d	150.0±2.0d	342±8d	2.16±0.04a	13.0±0.2d	5.15±0.06c	0.66±0.03c	181±6b	8.52±0.08b
20	0.975±0.010c	17.30±0.44f	40.5±1.4c	161.1±10.6e	334±5d	2.33±0.03b	14.3±0.5e	8.95±0.09f	0.63±0.02c	144±6a	8.50±0.02b
25	0.775±0.016a	12.20±0.26e	47.6±0.9f	196.2±4.3f	420±6e	2.35±0.04b	16.1±0.8f	7.31±0.03e	0.86±0.02d	152±10a	8.59±0.03c

Averages of replicates ± standard error; means followed by different letters are significantly different at $P<0.05$.

microbial community composition changes will directly affect microbial functionality as indicated by enzyme activities over a three-year study period. We hypothesized that soil bacteria community structures and bioactivity are affected by manure ratio.

Materials and Methods

Ethics statement

The experiment was carried out in our scientific research field for pomology studies which is owned by our institute, therefore, no specific permissions were required for these locations/activities, the field studies did not involve endangered or protected species.

Soil sampling sites and collection

In spring 2009, sandy soil was purchased from a local construction materials market and sieved through a 5-mm screen. Different ratios of manure (0%, 5%, 10%, 15%, 20%, 25%) were then premixed into the sandy soil prior to nursery stock transplantation. One-year-old "Fuji" apple ("Red Delicious" × "Ralls Genet") nursery with SH40 (*Malus honanensis*) as a dwarfing interstock and *Malus hupehensis* Var. Pingyiensis Jiang as a base stock was planted into 64 liter cubic Plexiglas boxes in triplicate. Soil samples were collected from depths of 0–40 cm in three different locations at 20-cm distances from the center of a 64 liter box using a 5-cm-diameter soil auger and transferred on ice to the laboratory at the beginning of autumn 2012. The soil samples were sieved through a 2-mm screen and homogenized prior to the analysis. One portion of the composite soil was stored in polythene bags and kept with dry ice for the molecular analysis, while another portion was used for the total N and SOM content and enzyme activity measurements.

Soil enzyme activity characterization

Soil urease activity was detected using improved sodium phenate and sodium hypochlorite colorimetry. A total of 5 g of soil and 1 mL of methylbenzene were added to a 50-mL triangular flask and mixed. Fifteen minutes later, 10 mL of a 10% urea solution and 20 mL of citrate buffer (pH 6.7) were added to the mixture, the flask was shaken and the mixture was incubated at 37°C for 24 h. The solution was then filtered, 1 mL of the liquid was transferred to a 50-mL volumetric flask, 4 mL of 1.35 M sodium phenate and 3 mL sodium hypochlorite solution (0.9%

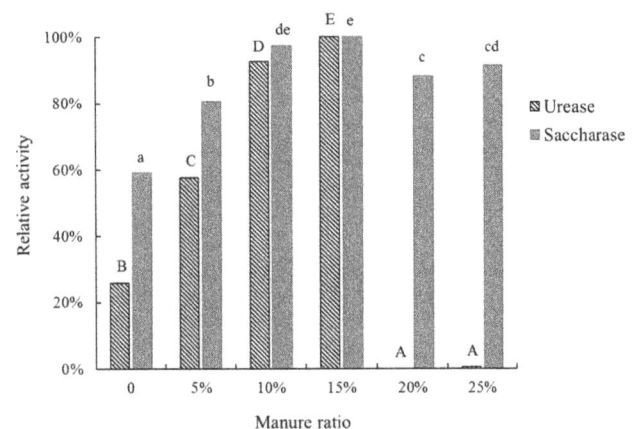

Figure 1. Urease and saccharase activities of soil with different manure ratios. Each enzyme's highest activity levels was defined as 100%. Different letters means significantly different at P<0.05, urease is represented as capital letters.

Table 2. Comparison of the estimated operational taxonomic unit (OTU) richness and diversity indexes of the 16S rRNA gene libraries for clustering at 97% identity as obtained from the pyrosequencing analysis.

Manure ratio	Reads	OTU	ACE	Chao	Coverage	Shannon	Simpson
0%	9015±2997[a]	3533±812[a]	9581±2336[a]	6816±1559[a]	0.767±0.024[a]	7.58±0.112[a]	0.0010±5.7735[b]
5%	9485±2305[a]	4032±878[a]	1101±2941[a]	7754±1732[a]	0.749±0.005[a]	7.776±0.179[a]	0.0007±5.7735[a]
10%	8501±1153[a]	3643±478[a]	9814±1751[a]	7204±1217[a]	0.747±0.026[a]	7.706±0.117[a]	0.0008±5.7735[a]
15%	9138±1096[a]	3871±213[a]	1069±835[a]	7648±393[a]	0.749±0.019[a]	7.766±0.005[a]	0.0007±0.0001[a]
20%	9916±1429[a]	4136±496[a]	1105±1744[a]	8133±1134[a]	0.755±0.032[a]	7.836±0.116[a]	0.0007±0.0001[a]
25%	7991±768[a]	3354±393[a]	9651±2464[a]	6903±1290[a]	0.749±0.019[a]	7.573±0.081[a]	0.0010±5.7735[b]

Averages of replicates ± standard error; means followed by different letters are significantly different at $P<0.05$.

active chlorine) were added and the flask was shaken slightly at constant volume. After 20 minutes, the solution was read at $A_{578\ nm}$ and the urease activity was calculated according to a working curve based on the nitrogen concentration of ammonium sulfate. The control solution contained no substrate. During the experiment, a control with no soil was also processed to rule out any system errors.

Saccharase and cellulase activities were estimated using the 3,5-dinitrosalicylic acid method according to previous report [17], all enzyme activity experiment was placed in triplicate.

DNA extraction

Soil genomic DNA for the polymerase chain reaction (PCR) amplification was extracted from 0.5 g of soil using an E.Z.N.A

Soil DNA Kit (Omega Biotek Inc., Norcross, GA, USA) following the manufacturer's instructions. DNA purity and concentration were analyzed spectrophotometrically using the e-Spect ES-2 (Malcom, Tokyo, Japan). The extracted DNA was stored at $-20°C$ for up to 10 days before use.

PCR amplification, quantitation, pyrosequencing

Bacterial 16 s rDNA was amplified by PCR to construct a community library using tag pyrosequencing. The bar-coded broadly conserved primers 27F and 533R containing the A and B sequencing adaptors (454 Life Sciences, Branford, CT, USA) were used to amplify this region. The forward primer (B-27F) was 5′-CCTATCCCCTGTGTGCCTTGGCAGTCTCAGAGAGTTT GATCCTGGCTCAG-3′, in which the B adaptor is underlined.

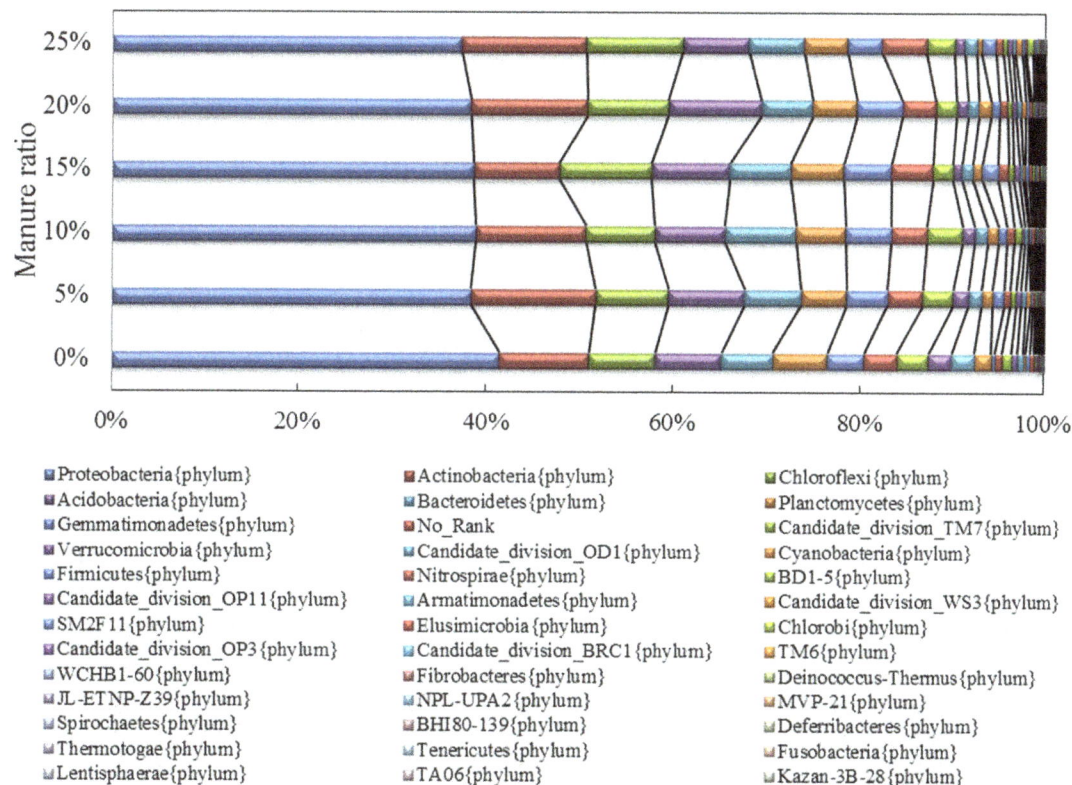

Figure 2. Comparison of the bacterial communities at the phylum level. Relative read abundance of different bacterial phyla within the different communities. Sequences that could not be classified into any known group were labeled "No_Rank".

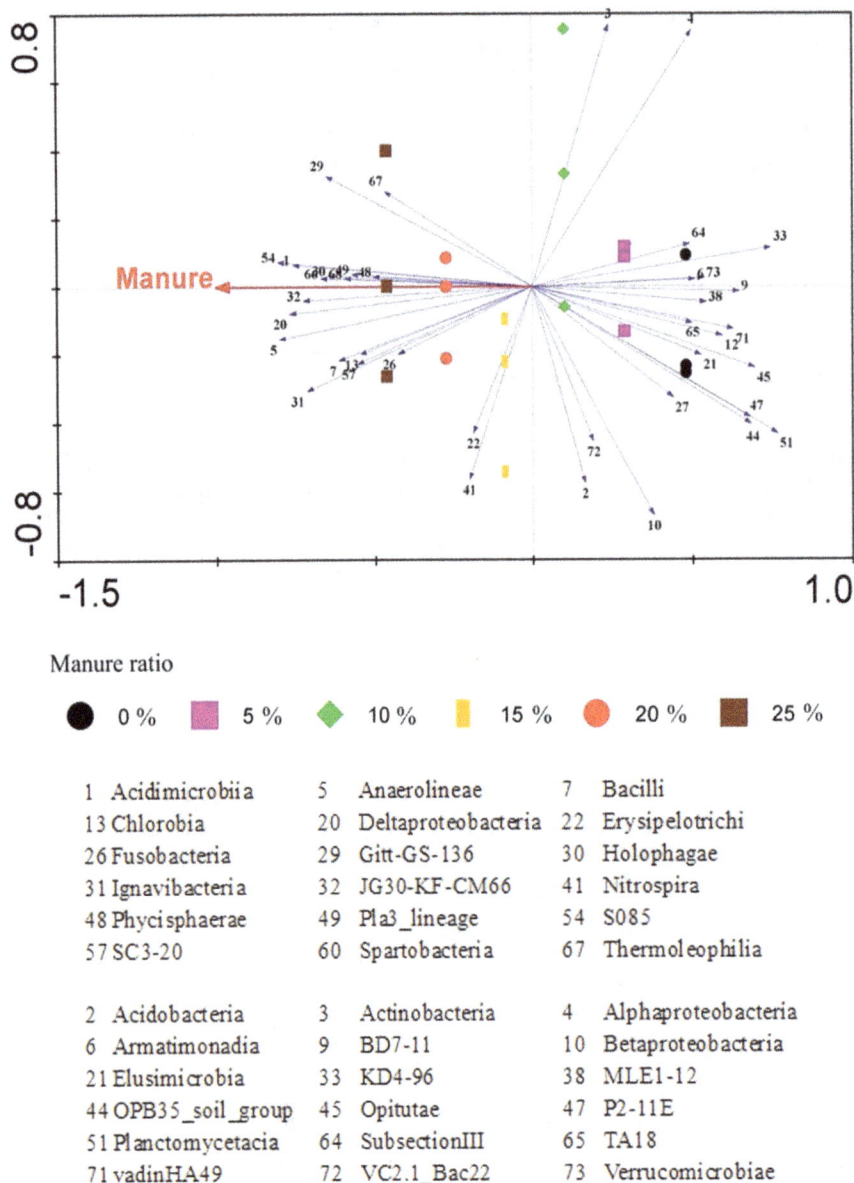

Figure 3. Redundancy analysis on the relative abundance of bacterial class using Canoco 4.5. The canonical axis (horizontal axis) shows that with 21.5% total variability, the correlation with manure ratio is 0.909. The 37 classes that passed the criterion of having >21.5% variability are shown on the horizontal axis.

The reverse primer (A-533R) was 5′-CCATCTCATCCCTG CGTGTCTCCGACTCAGNNNNNNNNNNTTACCGCGGC TGCTGGCAC-3′, in which the sequence of A adaptor is underlined and the string of Ns represents an eight-base sample-specific barcode sequence. Including the barcode and 454 primers, the length of the amplicon was approximately 596 nt. The PCR procedures were carried out in triplicate 20-μL reactions using 0.4 mM of each primer, 5 ng of template DNA, 1× PCR reaction buffer and 1.5 U of *TransStart FastPfu* DNA Polymerase (TransGen Biotech, Beijing, China). The amplification program consisted of the following: initial denaturation, 95°C for 2 min; 25 cycles of 95°C for 30 s (denaturation), 55°C for 30 s (annealing) and 72°C for 30 s (extension); and a final extension of 72°C for 5 min; and 10°C forever. PCR products of the same sample were assembled and visualized on 2% agarose gels (TBE buffer) using ethidium bromide and further purified using a gel extraction kit (TransGen). The DNA concentration of each PCR product was determined using a Quant-iT PicoGreen double-stranded DNA assay (Invitrogen, Darmstadt, Germany) and was quality controlled on an Agilent 2100 Bioanalyzer (Agilent Technologies, Palo Alto, CA, USA). The PCR products from each reaction mixture were pooled in equimolar ratios and then subjected to emulsion PCR to generate amplicon libraries. Amplicon pyrosequencing was performed from the A-end using a 454/Roche A sequencing primer kit on a Roche Genome Sequencer GS FLX Titanium platform at Majorbio Bio-Pharm Technology Co., Ltd. (Shanghai, China). Complete data sets are submitted to the NCBI Short Read Archive under accession no. SRX337490.

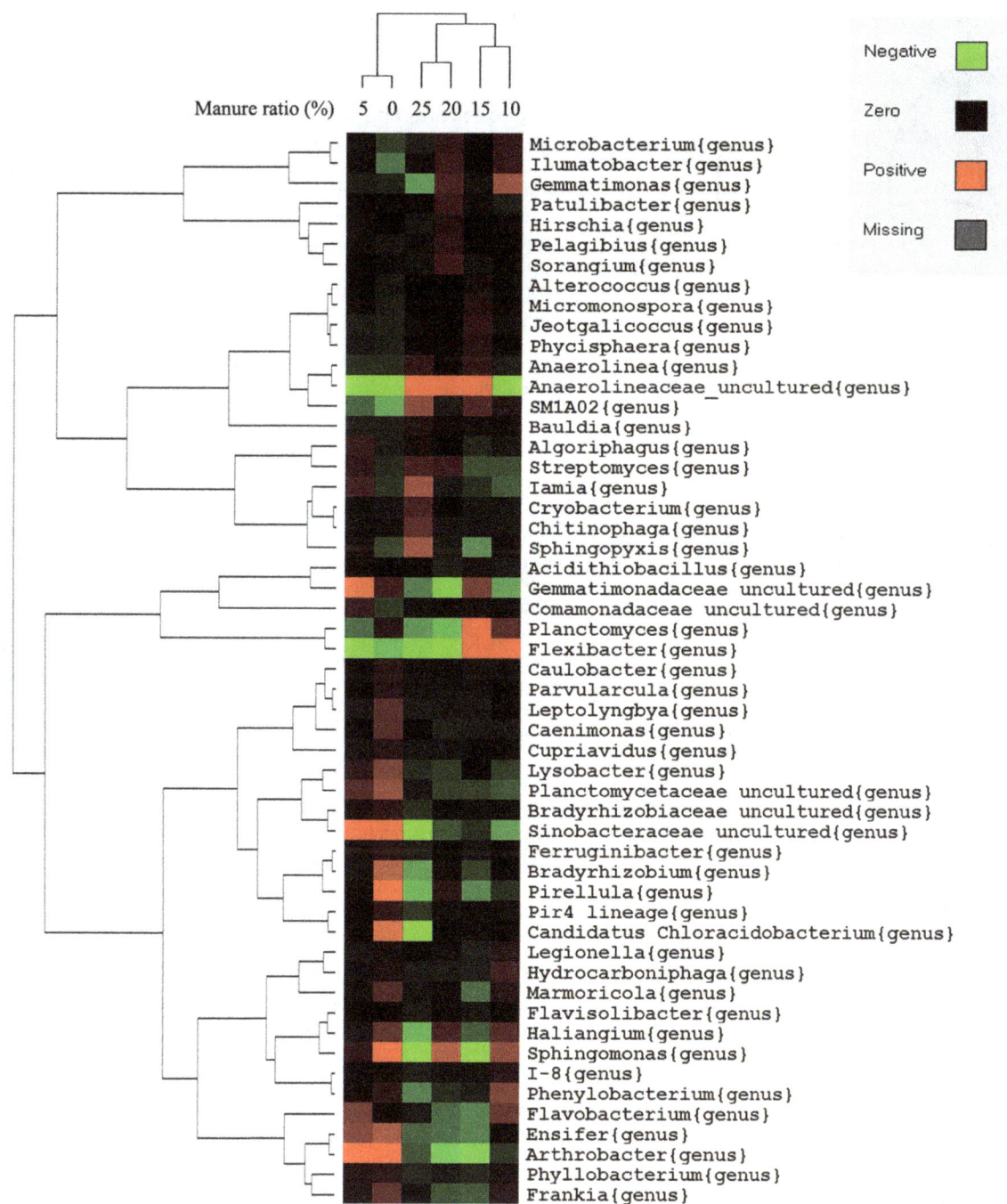

Figure 4. Genus level bacterial distribution among the six samples. The phylogenetic tree was calculated using the neighbor-joining method, while the relationship among samples was determined using Bray distance and the complete clustering method. The heatmap plot was generated by Gene cluster and Tree View (written by Michael Eisen, Stanford University). Genuses with >0.1% relative abundances are quoted and the species were mean centered before clustering.

Statistical and bioinformatics analysis

Rarefaction analysis and Good's coverage for the libraries were determined using custom scripts on R version 2.15.2 (R Foundation for Statistical Computing). Heatmap figures were generated using Gene cluster and Tree View (written by Michael Eisen, Stanford University). Venn diagrams curves were created with the online tool Venny (Juan Carlos Oliveros; http://bioinfogp.cnb.csic.es/tools/venny/index.html) and Canoco 4.5

was used to run a redundancy analysis (RDA). Analysis of variance was performed using SPSS Statistics 18 (IBM, Armonk, NY, USA). Community richness index, community diversity index, data preprocessing, operational taxonomic unit-based analysis and hypothesis tests were performed using mothur (http://www.mothur.org/). The histogram was created using Microsoft Excel 2007 (Microsoft, Redmond, WA, USA).

0% and 5%   10% and 15%

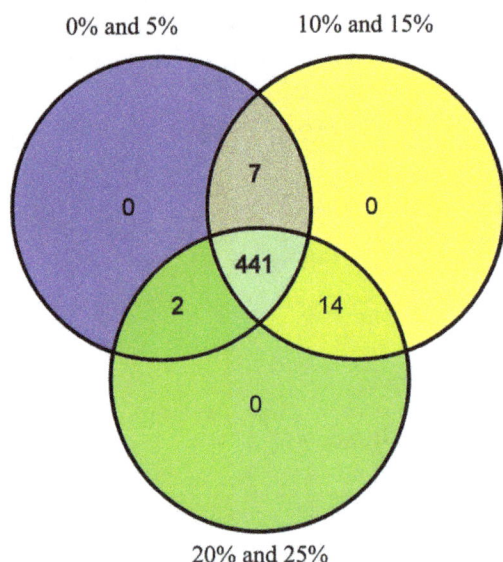

20% and 25%

Figure 5. Shared species analysis of the different treatment groups. The Venn diagram shows the unique and shared species among the different treatments.

Results and Discussion

Soil properties and enzyme activity

Three years after planting, soil was collected and physicochemical properties were tested (Table 1). The SOM contents of the apple rhizosphere soil correlated with manure ratio except that the 20% manure treatment was 17.30 g/kg, which is higher than that of the 25% manure treatment. Available contents of the soil nutrients P, K, Fe, Zn and B content were increased by manure application, while that of Ca was not. Ca is known to be an important factor affecting apple root and shoot growth [18], and all manure treatments decreased available Ca content compared to control and increased plant mass growth (data not shown), probably because manure application enhanced plant growth and increased Ca consumption. Cation exchange capacity (CEC), the number of positive charges that a soil can contain, was mainly influenced by SOM and pH [19]. In our experiment, although SOM content increased stepwise with manure ratio and the pH of all treatment soils was 8.43–8.64, the 15% manure treatment had the highest CEC, indicating that overdose levels of manure might have a negative impact on soil fertility. This is in accordance with a former study which showed higher ratio ash admixture to compost changed apple rhizosphere microbe communities but did not enhance C utilization as lower dose [20].

Hydrolytic enzymes regulate the rate at which organic materials are degraded and nutrients become available to plants, thus reflecting soil fertility [21]. Soil urease, saccharase and cellulase activity were tested (Fig. 1). Urease activity was increased by manure application, but once the manure ratio was ≥20%, urease activity was heavily inhibited. In an earlier study, urease content was reportedly increased by the use of both organic and inorganic fertilizers [22,23], while it was significantly correlated with SOM and total N content. However, we found that the use of overdose manure levels significantly inhibited urease activity. Saccharase has been shown to be an important factor affecting soil alkali-hydrolyzable N [24], while our experimental results showed a similar trend to that of available N in that, 15% manure treatment resulted in the highest value (relatively high soil fertility). High

manure doses might change the soil microbe community structure and reduce soil enzyme activity. Cellulase was not detected in any of our treatments.

Pyrosequencing and sequence analysis

A total of 162,141 valid reads and 67,714 operational taxonomic units (OTUs) were obtained from the 18 samples through 454 pyrosequencing analysis (Table 2). After excluding non-ribsomal DNA, low quality (containing fuzzy bases) or short sequences (reads <200 bp long), mothur was used to assign OTUs at the 97% sequence similarity level and to calculate Shannon and Simpson's diversity indexes, Chao, Ace and Good's coverage. Total reads and OTUs did not differ significantly among treatments, so the sequencing depth was unique for each sample. The bacterial diversity richness indexes ACE and Chao at the 3% dissimilarity level tended to be higher under manure treatment than the control; however, it is important to emphasize that 95% confidence interval calculations for the two indexes showed that none of these trends were significant. On the other hand, the Simpson indexes for 5%, 10%, 15% and 20% manure treatment were lower than those for the 25% manure treatment and the control, indicating that these treatments led to higher bacterial diversity. Good's coverage was around 0.75 for all samples, indicating a 75% species detection rate which was normal for pyrosequencing of soil microbes [25].

Nacke et al. [26] reported that for forest soil, with >20,000 reads per sample, the rarefaction curve was not saturated until genetic distances of 25% rather than 3% were used. In contrast Kunin et al. [27] suggested that a 97% clustering threshold is necessary for reliable estimation of community diversity. However, the pyrosequencing data gave much more information than the traditional microbe culture method or the molecular cloning library, and it effectively showed differences between treatments and revealed that manure application changed the soil microbial diversity in this study.

Taxonomic composition analysis

All of the sequences were classified to 39 phyla or groups by the mothur program. The predominant bacterial components of the different treatments were similar, while the distribution among phyla and groups varied (Fig. 2). Proteobacteria, Actinobacteria, Chloroflexi, Acidobacteria and Bacteroidetes were the five most dominant phyla in all treatments, accounting for >70% of the total reads. Proteobacteria presented the highest percentage in all treatments, a finding that supports those of other reports [25,28]. Most Proteobacteria bacteria are Gram-negative and many are responsible for legume symbionts nitrogen fixation [29] and polycyclic aromatic hydrocarbons [30], Proteobacteria bacteria are widely distributed in soil samples including contaminated soil [31].

A wide variety of pathogens are also Proteobacteria, such as Agrobacterium, Escherichia, Salmonella and Helicobacter. In the present study, Proteobacteria content gradually decreased from 41.38% to 37.29% as manure ratio increased from 0% to 25%, respectively. Relatively low Proteobacteria distribution (23%) was observed in an area of environmentally protected soil through metatranscriptome [32], while an earlier study of a cucumber field [12] suggested that the use of manure composts with antagonistic microorganisms decreased Proteobacteria distribution from 40% to 20%, suppressed disease incidence by 83% and reduced yield losses. Hence, we presumed that using manure treatment to reduce the Proteobacteria ratio was beneficial for apple plants.

Actinobacteria are Gram-positive bacteria that are widely distributed throughout the soil and water ecosystem. In many studies, Actinobacteria ratios are less than Proteobacteria ratios in soil samples [12,33]; however, Actinobacteria was the most abundant phylum in sludge [34] and dry land [35], probably because the secondary metabolites produced by Actinobacteria enhanced its adaptability. The lowest Actinobacteria content (9.09%) was detected after 15% manure treatment, so we assume that 15% manure application supplied a better nutrition environment for most of the bacteria.

The proportion of Chloroflexi was not more than 5% of the soil bacteria community in many pyrosequencing studies [33], whereas it increased under heat-exposed conditions [36]. In the present study, it was approximately 10% in all treatments, probably because Chloroflexi includes many aerobic thermophile members while the sandy soil led to better oxygen supply.

Acidobacteria is another abundant bacterial phylum that was recently discovered within soils. The majority had not been cultured but studies have shown that the phylum was negatively correlated with environmental pH [37]. In the present study, the Acidobacteria ratio did not differ significantly among the different treatments, probably since the soil pH values was similar.

Due to the relative abundance of bacterial classes, RDA was performed to check the effect of manure ratio using Canoco 4.5 (Fig. 3). The canonical axis (horizontal axis) explains 21.5% of the total variability and a correlation with manure ratio of 0.909; thus, the horizontal axis highly represents the influence of manure ratio. A total of 37 classes passed the criterion of having >21.5% variability in its values explained by the horizontal axis. Of these, Acidimicrobiia, Anaerolineae, Bacilli, Chlorobia, Deltaproteobacteria, Erysipelotrichi, Fusobacteria, Gitt-GS-136, Holophagae, Ignavibacteria, JG30-KF-CM66, Nitrospira, Phycisphaerae, Pla3_lineage, S085, SC3-20, Spartobacteria and Thermoleophilia were positively correlated with manure ratio, while Acidobacteria, Actinobacteria, Alphaproteobacteria, Armatimonadia, BD7-11, Betaproteobacteria, Elusimicrobia, KD4-96, MLE1-12, OPB35_soil_group, Opitutae, P2-11E, Planctomycetacia, SubsectionIII, TA18, vadinHA49, VC2.1_Bac22 and Verrucomicrobiae were negatively correlated.

Overall, 231 genera were found across the samples. All of the genera data underwent ANOVA by SPSS, and 82 genera showed significant difference among treatments, of which 53 genera with abundance >0.1% were used to cluster the samples and a heatmap was generated using Treeview (Fig. 4). The phylogenetic tree was calculated using the neighbor-joining method, while relationships among samples were determined by Bray distance and the complete clustering method. Samples were clustered into three groups, the low manure group (LM; 0% and 5% manure), medium manure group (MM; 10% and 15% manure) and high manure group (HM; 20% and 25% manure).

Anaerolineaceae was reported to be found in many samples including soil, sludge and water samples [38], and little was known about this genus except for its anaerobism, probably because the difficulty of culturing anaerobic bacteria. One earlier pyrosequencing revealed that it was more frequently encountered in anaerobic environments [39]. In this study, *Anaerolineaceae* increased with manure ratio from 1.56% (0% manure) to 3.83% (25% manure). We presume that this was due to the sandy soil and that the higher SOM led to lower oxygen supply, thus increasing the *Anaerolineaceae* ratio.

Ratios of *Sinobacteraceae*, *Ensifer* and *Arthrobacter* were higher in LM treatment. *Sinobacteraceae* consisted of Gram-negative, non-motile, rod-shaped bacteria, and a strain was isolated from polluted soil sample by enriched culture. The genus, which is represented in many contaminated soil samples [40,41], decreased from 1.65% (0% manure) to 0.66% (25% manure). *Arthrobacter* is also commonly found in soil. All species in this genus are Gram-positive obligate aerobes that are rod-shaped during exponential growth and cocci in the stationary phase. This genus was also involved in the soil bioremediation of polluted environments. In an earlier study, Mazzola [42] reported that the proportion of *Arthrobacter* in apple rhizosphere decreased naturally as the plant grows. One explanation could be that plant growth benefited under manure application, which impacted microbial diversity.

Ensifer, also called *Sinorhizobium*, is a genus of N-fixing bacteria (rhizobia), and the decreasing *Ensifer* content reflective of high nutrient levels reduced the N fixing capacity of the soil microbes. *Flexibacter* is well known for its yellow-hued species that are identified as fish pathogens [43]; however, in soil, *Flexibacter* species such as *Flexibacter canadensis* have a denitrification [44] function in which they denitrify NO_3^- and NO_2^- to gaseous forms with increased oxygen tolerance. The higher content of *Flexibacter* in MM compared to LM and HM was in accordance with soil urease activity and indicative of a higher nitrous oxide reductase level. It has been reported that integrated apple farms maintained higher level of *Flexibacter* than organic farms [45] since denitrification causes nutrient loss, so the dynamic trend of *Flexibacter* indicates that the impact of manure application on apple rhizosphere soil was not linear. The profile differences of LM, MM and HM also fit the soil enzyme results (Fig. 1) in which MM showed highest urease and saccharase activities.

A Venn diagram based on species level comparing the LM, MM and HM groups revealed that most species were seen in each group; thus the bacterial community differences were mainly reflected by ratio rather than species variety (Fig. 5). However, most species showing different distributions among groups were uncultured. This finding indicated that pyrosequencing could be used to discover many uncultured bacterial species.

In the present study, we employed high throughput pyrosequencing to characterize apple rhizosphere soil microbial communities under different doses of manure applied after three-year culturing. The study results indicated that different manure ratios resulted in different changes in soil enzyme activity and bacterial communities, and that the use of overdose levels of manure maintained different bacteria community structures and decreased soil enzyme activity.

In conclusion, our study findings support the hypothesis that manure improvement exerts significant effect on soil enzyme activity and bacterial communities. Mid-level manure improvement increased soil enzyme activity and bacterial diversity, whereas over-dose levels of manure had an adverse impact.

Author Contributions

Conceived and designed the experiments: JS QPW. Performed the experiments: QZ JS SL. Analyzed the data: JS QPW SL. Contributed reagents/materials/analysis tools: QZ. Wrote the paper: QZ JS.

References

1. Wang J, Fu B, Qiu Y, Chen L (2001) Soil nutrients in relation to land use and landscape position in the semi-arid small catchment on the loess plateau in China. J Arid Environ 48: 537–550.

2. Chenu C, Le Bissonnais Y, Arrouays D (2000) Organic matter influence on clay wettability and soil aggregate stability. Soil Sci Soc Am J 64: 1479–1486.

3. Torsvik V, Øvreås L (2002) Microbial diversity and function in soil: from genes to ecosystems. Curr Opin Microbiol 5: 240–245.

4. Kindler R, Miltner A, Richnow HH, Kästner M (2006) Fate of gram-negative bacterial biomass in soil—mineralization and contribution to SOM. Soil Biology and Biochemistry 38: 2860–2870.

5. Plassart P, Terrat S, Thomson B, Griffiths R, Dequiedt S, et al. (2012) Evaluation of the ISO Standard 11063 DNA extraction procedure for assessing soil microbial abundance and community structure. PloS one 7: e44279.

6. Torsvik V, Goksøyr J, Daae FL (1990) High diversity in DNA of soil bacteria. Appl Environ Microb 56: 782–787.

7. Jackson SA, Kennedy J, Morrissey JP, O Gara F, Dobson AD (2012) Pyrosequencing reveals diverse and distinct sponge-specific microbial communities in sponges from a single geographical location in Irish Waters. Microbial Ecol 64: 105–116.

8. Acosta-Martinez V, Dowd S, Sun Y, Allen V (2008) Tag-encoded pyrosequencing analysis of bacterial diversity in a single soil type as affected by management and land use. Soil Biology and Biochemistry 40: 2762–2770.

9. Dumbrell AJ, Ashton PD, Aziz N, Feng G, Nelson M, et al. (2011) Distinct seasonal assemblages of arbuscular mycorrhizal fungi revealed by massively parallel pyrosequencing. New Phytol 190: 794–804.

10. Uroz S, Buée M, Murat C, Frey Klett P, Martin F (2010) Pyrosequencing reveals a contrasted bacterial diversity between oak rhizosphere and surrounding soil. Environmental Microbiology Reports 2: 281–288.

11. Imfeld G, Vuilleumier S (2012) Measuring the effects of pesticides on bacterial communities in soil: a critical review. Eur J Soil Biol 49: 22–30.

12. Qiu M, Zhang R, Xue C, Zhang S, Li S, et al. (2012) Application of bio-organic fertilizer can control *Fusarium wilt* of cucumber plants by regulating microbial community of rhizosphere soil. Biol Fert Soils 48: 807–816.

13. Sugiyama A, Vivanco JM, Jayanty SS, Manter DK (2010) Pyrosequencing assessment of soil microbial communities in organic and conventional potato farms. Plant Dis 94: 1329–1335.

14. Manter DK, Delgado JA, Holm DG, Stong RA (2010) Pyrosequencing reveals a highly diverse and cultivar-specific bacterial endophyte community in potato roots. Microbial Ecol 60: 157–166.

15. Uroz S, Buée M, Murat C, Frey Klett P, Martin F (2010) Pyrosequencing reveals a contrasted bacterial diversity between oak rhizosphere and surrounding soil. Environmental Microbiology Reports 2: 281–288.

16. Lin X, Feng Y, Zhang H, Chen R, Wang J, et al. (2012) Long-Term Balanced Fertilization Decreases Arbuscular Mycorrhizal Fungal Diversity in an Arable Soil in North China Revealed by 454 Pyrosequencing. Environ Sci Technol 46: 5764–5771.

17. Wang B, Liu GB, Xue S, Zhu B (2011) Changes in soil physico-chemical and microbiological properties during natural succession on abandoned farmland in the Loess Plateau. Environmental Earth Sciences 62: 915–925.

18. Miller SS (2002) Prohexadione-calcium controls vegetative shoot growth in apple. Journal of tree fruit production 3: 11–28.

19. Guckland A, Jacob M, Flessa H, Thomas FM, Leuschner C (2009) Acidity, nutrient stocks, and organic-matter content in soils of a temperate deciduous forest with different abundance of European beech (*Fagus sylvatica* L.). Journal of Plant Nutrition and Soil Science 172: 500–511.

20. Bougnom BP, Insam H (2009) Ash additives to compost affect soil microbial communities and apple seedling growth. Die Bodenkultur 60: 5–15.

21. Marx MC, Kandeler E, Wood M, Wermbter N, Jarvis SC (2005) Exploring the enzymatic landscape: distribution and kinetics of hydrolytic enzymes in soil particle-size fractions. Soil Biology and Biochemistry 37: 35–48.

22. Ge G, Li Z, Fan F, Chu G, Hou Z, et al. (2010) Soil biological activity and their seasonal variations in response to long-term application of organic and inorganic fertilizers. Plant Soil 326: 31–44.

23. San Francisco S, Urrutia O, Martin V, Peristeropoulos A, Garcia Mina JM (2011) Efficiency of urease and nitrification inhibitors in reducing ammonia volatilization from diverse nitrogen fertilizers applied to different soil types and wheat straw mulching. J Sci Food Agr 91: 1569–1575.

24. Alkorta I, Aizpurua A, Riga P, Albizu I, Amézaga I, et al. (2003) Soil enzyme activities as biological indicators of soil health. Reviews on environmental health 18: 65–73.

25. Lauber CL, Hamady M, Knight R, Fierer N (2009) Pyrosequencing-based assessment of soil pH as a predictor of soil bacterial community structure at the continental scale. Appl Environ Microb 75: 5111–5120.

26. Nacke H, Thürmer A, Wollherr A, Will C, Hodac L, et al. (2011) Pyrosequencing-based assessment of bacterial community structure along different management types in German forest and grassland soils. PLoS One 6: e17000.

27. Kunin V, Engelbrektson A, Ochman H, Hugenholtz P (2009) Wrinkles in the rare biosphere: pyrosequencing errors can lead to artificial inflation of diversity estimates. Environ Microbiol 12: 118–123.

28. Sugiyama A, Vivanco JM, Jayanty SS, Manter DK (2010) Pyrosequencing assessment of soil microbial communities in organic and conventional potato farms. Plant Dis 94: 1329–1335.

29. Raymond J, Siefert JL, Staples CR, Blankenship RE (2004) The natural history of nitrogen fixation. Mol Biol Evol 21: 541–554.

30. DeBruyn JM, Mead TJ, Wilhelm SW, Sayler GS (2009) PAH biodegradative genotypes in Lake Erie sediments: Evidence for broad geographical distribution of pyrene-degrading mycobacteria. Environ Sci Technol 43: 3467–3473.

31. Feris K, Ramsey P, Frazar C, Moore JN, Gannon JE, et al. (2003) Differences in hyporheic-zone microbial community structure along a heavy-metal contamination gradient. Appl Environ Microb 69: 5563–5573.

32. Urich T, Lanzén A, Qi J, Huson DH, Schleper C, et al. (2008) Simultaneous assessment of soil microbial community structure and function through analysis of the meta-transcriptome. PLoS One 3: e2527.

33. İnceoğlu Ö, Al-Soud WA, Salles JF, Semenov AV, van Elsas JD (2011) Comparative analysis of bacterial communities in a potato field as determined by pyrosequencing. PLoS One 6: e23321.

34. Kwon S, Kim TS, Yu GH, Jung JH, Park HD (2010) Bacterial community composition and diversity of a full-scale integrated fixed-film activated sludge system as investigated by pyrosequencing. J Microbiol Biotechnol 20: 1717–1723.

35. Makhalanyane TP, Valverde A, Lacap DC, Pointing SB, Tuffin MI, et al. (2012) Evidence of species recruitment and development of hot desert hypolithic communities. Environmental microbiology reports.

36. Nocker A, Richter-Heitmann T, Montijn R, Schuren F, Kort R (2010) Discrimination between live and dead cells in bacterial communities from environmental water samples analyzed by 454 pyrosequencing. International Microbiology 13: 59–65.

37. Jones RT, Robeson MS, Lauber CL, Hamady M, Knight R, et al. (2009) A comprehensive survey of soil acidobacterial diversity using pyrosequencing and clone library analyses. The ISME journal 3: 442–453.

38. Tekere M, Prinsloo A, Olivier J, Jonker N, Venter S (2012) An evaluation of the bacterial diversity at Tshipise, Mphephu and Sagole hot water springs, Limpopo Province, South Africa. African Journal of Microbiology Research 6: 4993–5004.

39. Sherry A, Gray ND, Ditchfield AK, Aitken CM, Jones DM, et al. (2013) Anaerobic biodegradation of crude oil under sulphate-reducing conditions leads to only modest enrichment of recognized sulphate-reducing taxa. Int Biodeter Biodegr 81: 105–113.

40. Bell TH, Yergeau E, Martineau C, Juck D, Whyte LG, et al. (2011) Identification of nitrogen-incorporating bacteria in petroleum-contaminated arctic soils by using [15N] DNA-based stable isotope probing and pyrosequencing. Appl Environ Microb 77: 4163–4171.

41. Militon C, Boucher D, Vachelard C, Perchet G, Barra V, et al. (2010) Bacterial community changes during bioremediation of aliphatic hydrocarbon-contaminated soil. Fems Microbiol Ecol 74: 669–681.

42. Mazzola M (1999) Transformation of soil microbial community structure and Rhizoctonia-suppressive potential in response to apple roots. Phytopathology 89: 920–927.

43. Wakabayashi H, Hikida M, Masumura K (1986) *Flexibacter maritimus* sp. nov., a pathogen of marine fishes. International Journal of Systematic Bacteriology 36: 396–398.

44. Wu Q, Knowles R, Niven DF (1994) O_2 regulation of denitrification in *Flexibacter canadensis*. Can J Microbiol 40: 916–921.

45. Bougnom BP, Greber B, Franke-Whittle IH, Casera C, Insam H (2012) Soil microbial dynamics in organic (biodynamic) and integrated apple orchards. Organic Agriculture 2: 1–11.

11

A Simple Model to Predict the Probability of a Peach (*Prunus persicae*) Tree Bud to Develop as a Long or Short Shoot as a Consequence of Winter Pruning Intensity and Previous Year Growth

Daniele Bevacqua*, Michel Génard, Françoise Lescourret

INRA, UR1115 PSH, Avignon, France

Abstract

In many woody plants, shoots emerging from buds can develop as short or long shoots. The probability of a bud to develop as a long or short shoot relies upon genetic, environmental and management factors and controlling it is an important issue in commercial orchard. We use peach (*Prunus persicae*) trees, subjected to different winter pruning levels and monitored for two years, to develop and calibrate a model linking the probability of a bud to develop as a long shoot to winter pruning intensity and previous year vegetative growth. Eventually we show how our model can be used to adjust pruning intensity to obtain a desired proportion of long and short shoots.

Editor: Randall P. Niedz, United States Department of Agriculture, United States of America

Funding: The authors have no support or funding to report.

Competing Interests: The authors have declared that no competing interests exist

* E-mail: daniele.bevacqua@avignon.inra.fr

Introduction

Two morphologically distinct shoots, commonly referred to as short and long shoots (SS and LS, respectively), occur in many woody plants. In SS the rib meristems fails to become active after opening of the buds so that little or no intermodal elongation occurs. The putative long and short shoots buds are generally identical and differences emerge during growing season. Both type of shoots bear foliage and contribute to photosynthesis. Short shoots generally do not exceed 2 cm length and are important providers of photosynthate in the first weeks following bud breaking, whereas LSs have elongated stems and constitute tree architecture [1]. The probability of a bud to develop as LS (P_{LS}) is controlled by both genetic, environmental and, when present, management factors. For example in genera *Pinus* and *Larix* it is almost constant. A wider range in the proportion between SSs and LSs seems to exist in deciduous rather than coniferous trees. In some genera (*e.g. Fagus*, *Betula* and *Acer*) it is less predictable than in others (*e.g. Malus*, *Prunus*, *Pyrus*, *Cytrus*, *Cratageus etc.*), although P_{LS} generally decreases with age.

In commercial orchards, the quantity of 1-yr old LS in trees is artificially regulated by winter pruning intended to shape trees and adjust crop load in order to *i*) improve or maintaining tree vigor and *ii*) increase the yield or quality of fruits [2]. In fact, pruning alters the shoot : root ratio, by removing shoot biomass, and forces the plant to increase new shoot growth, according to the functional-balance concept [3] which states that new biomass is partitioned between roots and shoots in favor of the organ that capture the limiting resource (e.g. carbon or nitrogen, respectively captured by shoots and roots). In the growing season, new shoots will then emerge from remaining 1-yr old LSs (i.e. those LSs that

have not been cut). Consequences of winter pruning on fruit production are not straightforward since it generally increases the fraction of buds that develop into LS, but it also eliminates 1-yr old shoots bearing flower and vegetative buds. Moreover, some trees bear most of fruits on SSs (e.g. cherry trees, apple trees) while others such as peach trees bear fruits on LSs [2].

Quantitative relationship between cultural practices and shoots development should be explicitly considered in mathematical crop models. These models can then be used to predict crop dynamics under different cultural practices, for which direct field observations would be extremely difficult and/or expensive [4]. Despite winter pruning is one of the most common cultural practice influencing P_{LS}, there are no dedicated studies to quantify its effect on P_{LS} (but see Grechi et al. [5] who studied the effect of winter pruning on peach tree-aphid interactions and also provided a first estimate of the effect of winter pruning on P_{LS}). Bussi et al. [6] focused on probability of sprouts emergency as response to pruning intensity. Stephan et al. [7] analyzed the effect of pruning intensity on apple tree architecture and Gordon & Dejong [8] studied the effect of sprouts removal and fruit crop on P_{LS}. Fumey et al. [9], in a comprehensive experimental study, analyzed the consequences of different pruning practices on tree branching in apple trees. They found that pruning enhanced vegetative growth and decreased flowering, yet they did not provide a quantitative model to predict the consequences of different practices. In present work we use peach (*Prunus persicae*) trees, subjected to different winter pruning levels, and monitored for two consecutive years, to develop and calibrate a model linking P_{LS} to intensity of winter pruning and overall length of 1-yr old shoots before pruning. A high capacity for neoformation determines high plastic adaptation in response to branch removal and makes peach a good model to study the effects of winter pruning [8]. We use the calibrated model to assess the number of long shoots N_{LS} relevant to different combinations of winter pruning intensities and 1-yr old wood present before pruning. Non-linearity of interactions between considered variables gives rise to a maximum value of N_{LS} at different values of pruning intensity, depending on the overall length of 1-yr old shoots before pruning. Eventually, we show how our model can be used to adjust winter pruning intensity to get an optimal production of LSs.

Materials and Methods

Available Data

Data were collected in 2005 and 2006 from an experimental peach orchard planted in 1998 with 20 late maturing trees (cv Suncrest/GF677) (see Grechi et al. [5] for a full description of the experiment). Trees were pruned in winter with a pruning intensity (PI) (i.e. fraction of mass of 1-yr old wood pruned on total 1-yr old wood) varying from 0 to 0.8. Each year, for each tree, we measured tree length of 1-yr old wood before pruning (L_{W1}), PI and the fraction of shoots that developed as long shoots (i.e. P_{LS}). Data are reported in table 1.

No specific permits were required for the described field studies. The location is part of our public institute (INRA) domain. The field studies did not involve endangered or protected species.

Table 1. Characterization of the 20 peach trees monitored in 2005 and 2006: total length of 1-yr old wood before winter pruning (L_{W1}), fraction of L_{W1} pruned (PI) and fraction of buds developing as long shoots (P_{LS}).

Tree no.	L_{W1} (m)	PI	P_{LS}	L_{W1} (m)	PI	P_{LS}
	2005			**2006**		
1	190	0.4	0.9	157	0.11	0.14
2	170	0.24	0.23	152	0.36	0.14
3	164	0.56	0.23	247	0.63	0.36
4	182	0.39	0.25	309	0.49	0.42
5	185	0.32	0.9	259	0.49	0.24
6	158	0.49	0.25	250	0.56	0.36
7	138	0.5	0.13	131	0.14	0.9
8	215	0.28	0.8	143	0.36	0.25
9	230	0.64	0.36	354	0.66	0.52
10	186	0.65	0.36	357	0.77	0.50
11	265	0.48	0.20	184	0.54	0.34
12	269	0.15	0.16	106	0.21	0.22
13	265	0.65	0.23	190	0.68	0.61
14	219	0.61	0.24	182	0.73	0.50
15	298	0.26	0.7	101	0.38	0.17
16	264	0.34	0.9	137	0.46	0.24
17	220	0.39	0.16	178	0.49	0.23
18	282	0.19	0.11	170	0.34	0.24
19	226	0.54	0.19	157	0.56	0.41
20	222	0.10	0.8	144	0.25	0.18

The Models

Being P_{LS} a probability (i.e. it ranges between 0–1) and assuming that its variability can depend on PI, L_{W1} and their interaction PI×L_{W1} (abundance of pruned 1-yr old wood), the full model is:

$$P_{LS} = 1/(1 + a \cdot \exp(bPI + cL_{W1} + ePI \times L_{W1})) \qquad (1)$$

where $1/(1+a)$ represents a possible constant value for P_{LS} and b, c, e are coefficients respectively accounting for the effect of winter pruning intensity, vegetative growth of last year and amount of pruned biomass.

To estimate the unknown parameters a, b, c and e, we linearized the model as follow:

$$\log(1/P_{LS} - 1) = \log a + bPI + cL_{W1} + ePI \times L_{W1} \qquad (2)$$

We selected the model providing the best fit to observed data through a backward stepwise selection procedure based on the Akaike information criterion (AIC).

After having checked for constancy of variance and normality of errors of the selected model, we estimated uncertainty associated with parameters of the best model by bootstrapping [10]. We resampled 10,000 times the original data, generating empirical probability distribution for each parameter.

We computed N_{LS} on a virtual tree, where L_{W1} varied between 0–600 m, and subjected to PI between 0–1, as:

$$N_{LS} = L_{W1}(1 - PI)N_S P_{LS} \qquad (3)$$

where N_s is the constant number of shoots emerging per unit of 1-yr old wood left after pruning ($N_S = 45.55$ m^{-1} according to Grechi et al. [11]). Finally, we searched for the minimum value of PI that maximizes N_{LS}, given different values of L_{W1}. This is equivalent of finding the optimal level of PI in the case of a farmer that wanted to maximize N_{LS} in each tree. This is reasonable in most commercial peach orchards being LSs the most fruitful shoots [12].

Results and Discussion

The best model explained 68% of observed P_{LS} variability and included L_{W1} and interaction L_{W1}×PI as explanatory variables, while it excluded the sole effect of PI. Estimated model parameters and their empirical distributions are reported in table 2. The derivate of $P_{LS} = 1/(1 + a \cdot \exp(cL_{W1} + ePI \times L_{W1}))$ (i.e. the selected model to estimate P_{LS}) with respect of PI is positive for $e \cdot a < 0$ (see Supporting Information for details). Parameter estimates (i.e. $a = 2.75$ and $e = -1.51 \times 10^{-2}$) indicate thus that, in our considered model system, P_{LS} is a monotone increasing function of PI. In other words, the probability of a bud to develop into a LS is always increased by higher values of PI. On the other hand, the derivate of P_{LS} with respect of L_{W1} is positive for $PI > -c/e$ (see Supporting Information for details). Parameters estimates indicate thus that there is a critical value PI* = 0.52 over which trees with higher values of L_{W1} (referred to as "bigger trees" in the following) have higher P_{LS}. Under this value of PI*, bigger trees are expected to have lower P_{LS}. Note that L_{W1} is a proxy of previous tree growth and that with the term "bigger trees" we refer to those trees that produced more shoot biomass in the previous growing season and not necessarily to those that cumulated more biomass over the entire life span.

Table 2. Basic statistics of models parameters (see equations 3 and 4), as obtained by bootstrapping the 2005 and 2006.

Parameter	Mean	St.dev	Median	5th percentile	95th percentile
log a	1.01	0.25	1.02	0.59	1.41
C	7.85×10^{-3}	1.83×10^{-3}	7.78×10^{-3}	4.83×10^{-3}	10.8×10^{-3}
E	-1.51×10^{-2}	0.20×10^{-2}	-1.51×10^{-2}	-1.85×10^{-2}	-1.2×10^{-2}

The fraction of LSs and the overall number of LSs, predicted by the model for different scenarios of PI and L_{W1}, are shown in fig. 1 and fig. 2 respectively. As discussed above, it is evident in fig. 1 that P_{LS} increases for increasing values of PI while the effect of L_{W1} depends on the PI level. According to our model, in order to maximize N_{LS}, one should start to prune peach trees only when $L_{W1} > 80$ m and winter pruning should never cut more than 70% of 1-yr old wood (fig. 3). Figures of observed versus estimated values of P_{LS} and empirical parameters distributions are reported in figures S1 and S2.

The fact that P_{LS} increases with PI is coherent with previous studies (e.g. [5]) and with the functional balance theory [3]. Our results also suggest that, for low levels or absence of winter pruning (i.e. PI < 0.52 in *P. persicae*), bigger trees produce a lower fraction of LSs. Similarly, Wilson [13] found that, on red maple (*Acer rubrum*)

trees over 30-yr old, less than 10% of the shoots developed as LSs, and Greenwood et al. [14] found that old red spruce (*Picea rubens*) showed reduced LSs elongation. Such mechanism would allow the plant to maintain a fairly constant N_{LS} as bigger trees produce more shoots yet a lower fraction of long ones. Assuming N_{LS} as a proxy of peach tree vegetative growth potential in a growing year, our model suggests the existence of overcompensation in response to winter pruning only for plants with $L_{W1} > 80$ m. In fact, whenever winter pruning determines a higher N_{LS} with respect of undisturbed situation (i.e. PI = 0), it is the case of overcompensation i.e. the plant responds to a stress by increasing its ability to grow, and finally grows more than in undisturbed conditions [15]. Figure 3 shows that "disturbing" pruning practices become efficient only if $L_{W1} > 80$ m. Capacity for overcompensation in plants is likely to have evolved as a response

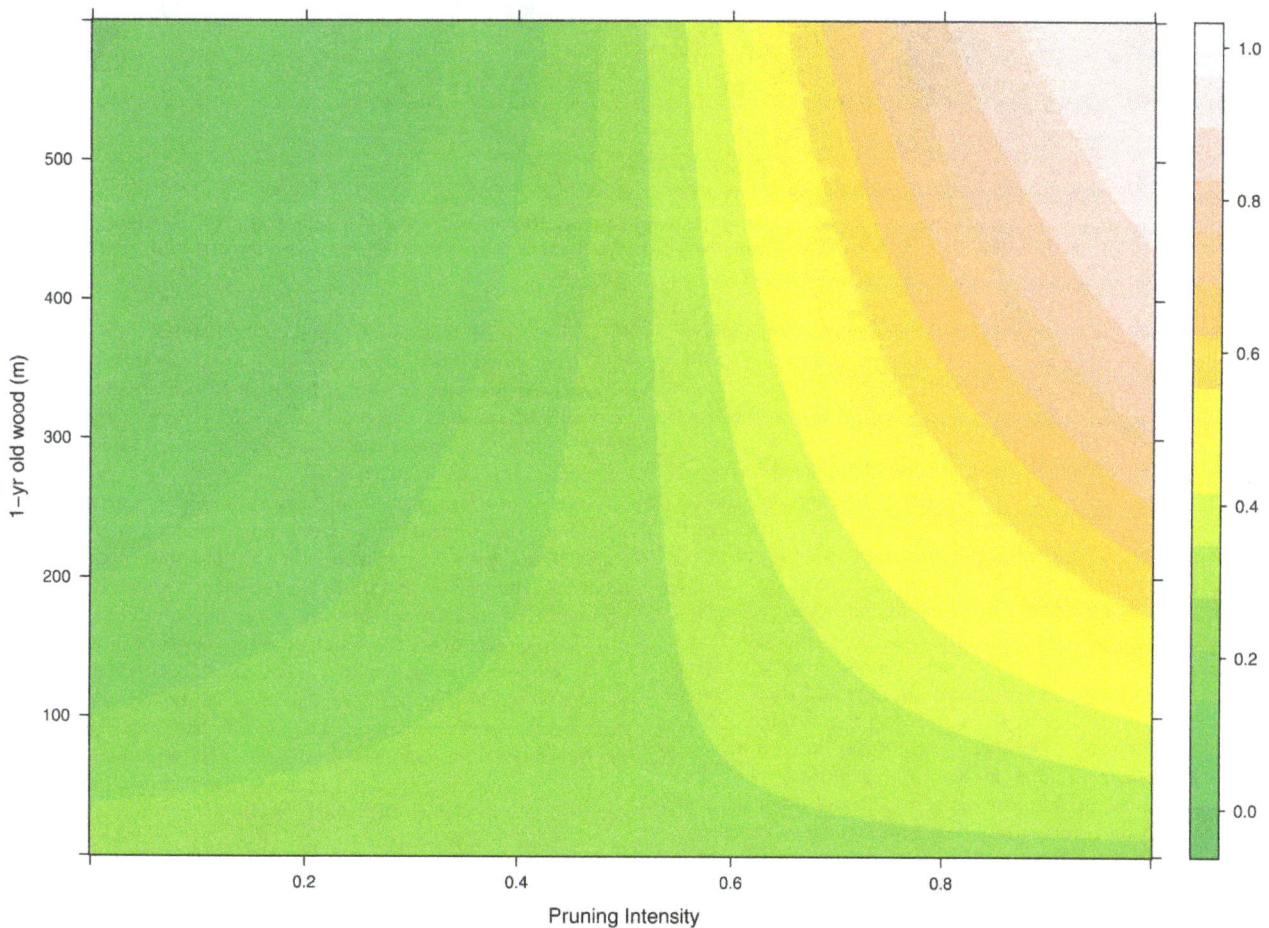

Figure 1. Estimated fraction of long shoots in the growing season (P_{LS}) of a virtual peach *Prunus persica* tree as function of 1-yr old wood before winter pruning (L_{w1}) and pruning intensity (PI, i.e. fraction of 1-yr old shoot removed before bud break).

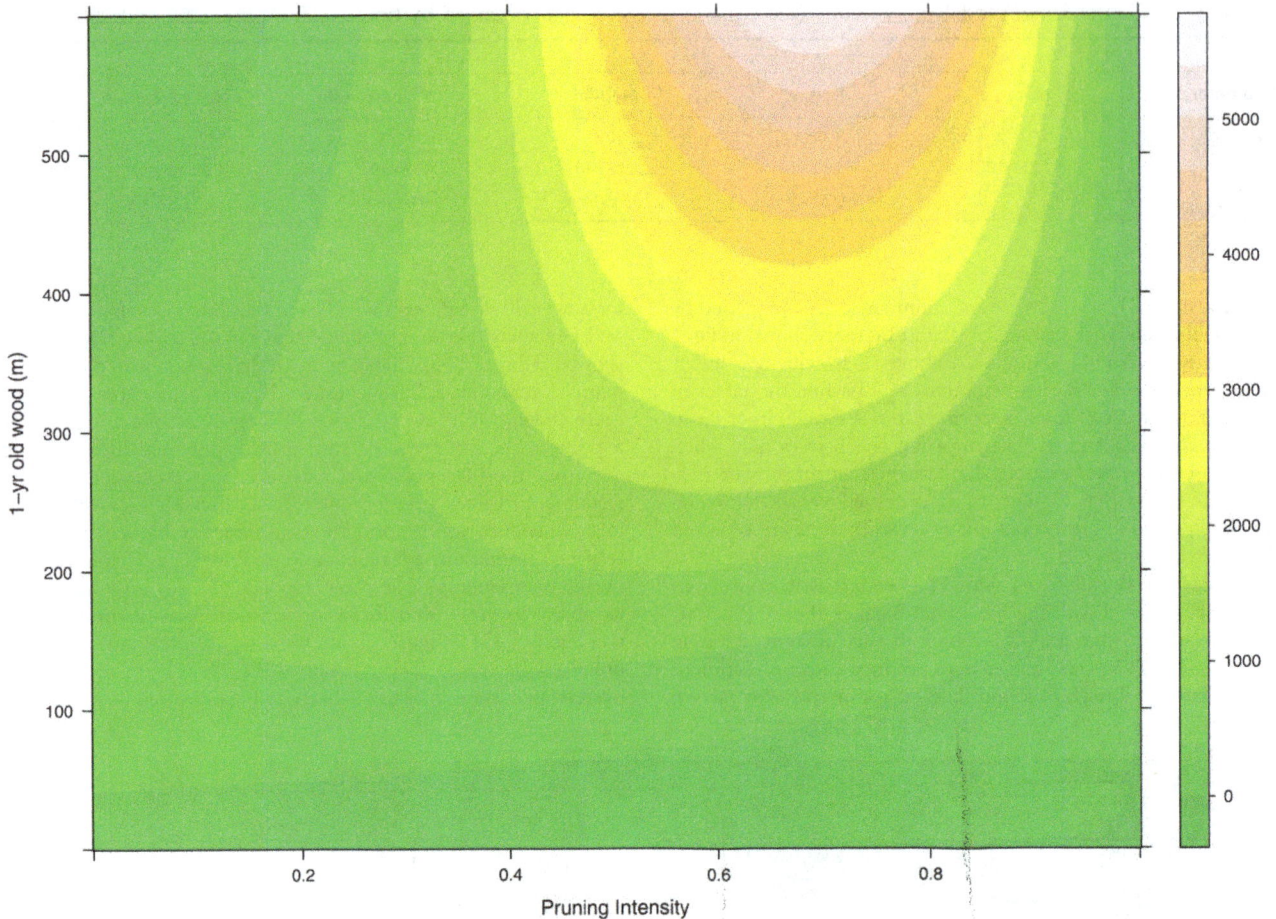

Figure 2. Estimated abundance of long shoots (N_LS), in the growing season of a virtual peach *Prunus persica* tree as function of 1-yr old wood before winter pruning (L_w1) and pruning intensity (PI,i.e. fraction of 1-yr old shoot removed before bud break).

to herbivory [16], it is more likely in environments with high nutrient availability [17], it is well documented in fruit trees ([18]; [19]; [20]) and it is well known by farmers that remove plant biomass with the final aim to increase shoot growth and related fruit production [2].

Farmer behavior in peach commercial orchards is well predicted by our simple model suggesting to exert low or no winter pruning intensity over small trees (i.e. trees with $L_{W1} < 80$ m, usually corresponding to trees <3-yr old) and gradually to increase pruning intensity until removing up to 70% of L_{W1} [21]. In late maturing peach orchards, increasing N_{LS} leads to an increase in fruit production. In fact, N_{LS} in a given year affects the number of fruits of the next year since peach flower buds and hence fruits are produced on 1-yr old long shoots. However fruit distribution on short and long shoots might highly vary between cultivars; other species bear fruits only on SS (e.g. cherry) with winter pruning increasing vegetative growth but decreasing yield in the following season [22] and other such as apple bear fruits on both LSs and SSs. Although farmer objective is likely to differ for different cultivars and species, our model would be useful to determine optimal winter pruning intensity according to different farmer objectives such as minimizing N_{LS} or obtaining an optimal ratio between LSs and SSs.

Grechi et al. [5], in a work focused on consequence of winter pruning on peach tree-aphid interactions, proposed an exponential relationship (i.e. $P_{LS} = \alpha \cdot \exp(\beta \cdot PI)$) linking P_{LS} to the solely PI. Although that relationship highlighted the importance of PI on P_{LS}, it can provide unrealistic biological results with $P_{LS} > 1$ and it neglects the previous growth of the tree. In the present work we overcome these main drawbacks since the image of function (1) ranges between 0–1 and the effect of previous growth on P_{LS} is considered through its proxy L_{W1}.

We are conscious that a better insight into plant partitioning of new shoots in short and long ones will be achieved only by future experiments gathering information not just on plants having different value of L_{W1} and subjected to levels of winter PI, but also on plants having different ages and subjected to different pruning practices (e.g. summer *vs.* winter pruning, centrifugal *vs.* conventional etc.) and environmental stressors. More comprehensive dataset would also permit a validation of the model and possibly increase its predictive power. Yet, despite the above-mentioned limitations, our results are consistent with functional-balance theory and common cultivar practices, and the proposed model can help in modeling the effect of winter pruning above tree growth and fruit production.

Supporting Information

Figure S1 Estimated versus observed fraction of long shoots.

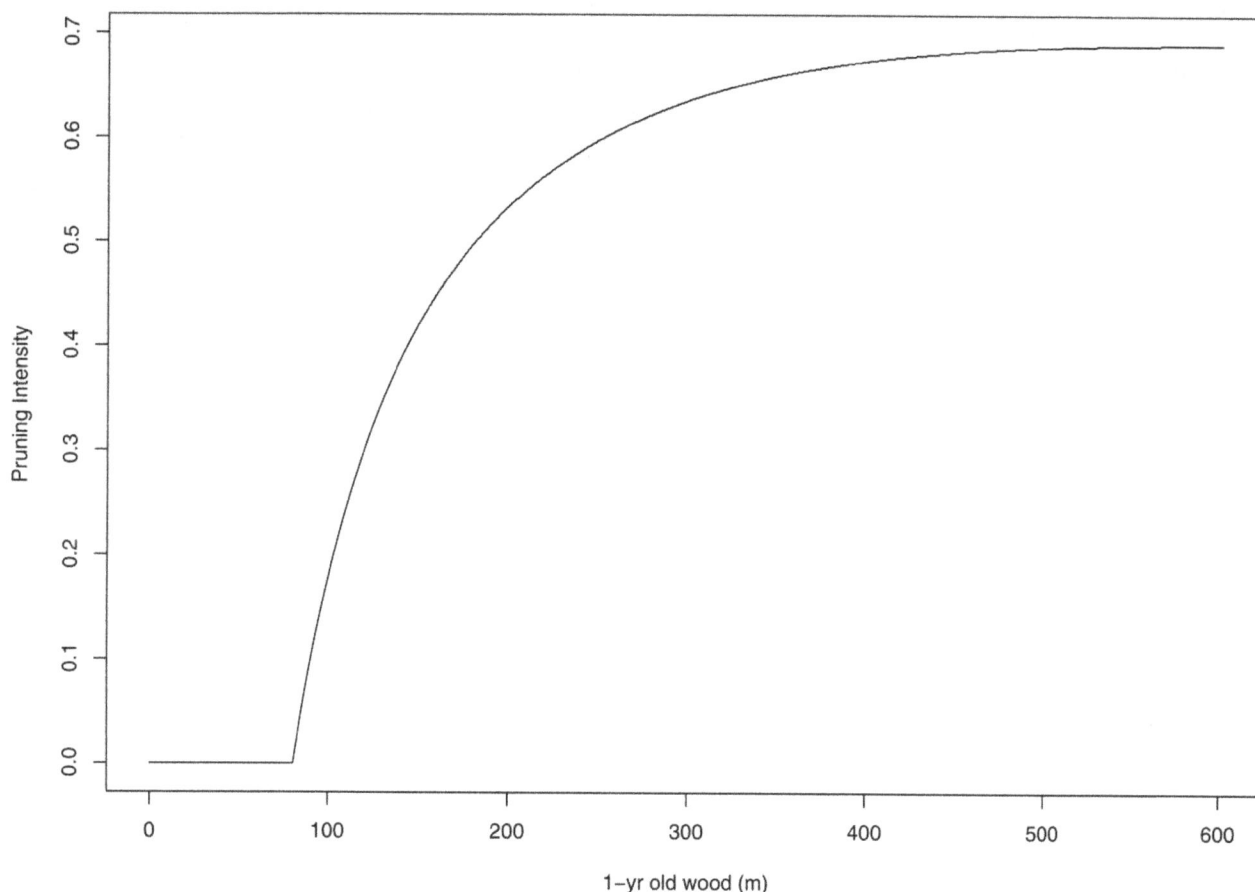

Figure 3. Pruning intensity (PI, i.e. ratio of 1-yr old shoot removed before bud break) determining the highest number of long shoots, as a function of 1-yr old wood before pruning (L_{w1}).

Figure S2   Variability of estimated parameters assessed via bootstrap (1000 iterations): bivariate scatter plots, linear fits and median values below and above the diagonal; histograms on the diagonal. Pearson correlation are equal to -0.89 0.49 and -0.80 respectively between log a–c, a–e, and c–e.

Author Contributions

Conception and design of the work: DB MG FL. Analysis and interpretation of data: DB MG FL. Drafting the article: DB. Critical article revision for important intellectual content: MG FL. Final approval of the version to be published: DB MG FL.

References

1. Zimmermann M, Brown C (1971) Trees: structure and function. Springer-Verlag.
2. Faust M (1989) Physiology of temperate zone fruit trees. John Wiley & Sons, Inc. p.
3. Davidson R (1969) Effect of Root/Leaf Temperature Differentials on Root/Shoot Ratios in Some Pasture Grasses and Clover. Ann Bot 33: 561–569.
4. Thornley J, Johnson I (2000) Plant and Crop Modeling: A Mathematical Approach to Plant and Crop Physiology. The Blackburn Press.
5. Grechi I, Sauge M, Sauphanor B, Hilgert N, Senoussi R, et al. (2008) How does winter pruning affect peach tree- Myzus persicae interactions? Entomologia Experimentalis et Applicata 128: 369–379.
6. Bussi C, Bruchou C, Lescourret F (2011) Response of watersprout growth to fruit load and intensity of dormant pruning in peach tree. Scientia Horticulturae 130: 725–731.
7. Stephan J, Lauri P, Dones N, Haddad N, Talhouk S, et al. (2007) Architecture of the pruned tree: impact of contrasted pruning procedures over 2 years on shoot demography and spatial distribution of leaf area in apple (*Malus domestica*). Annals of botany 99: 1055–1065.
8. Gordon D, Dejong TM (2007) Current-year and subsequent-year effects of crop-load manipulation and epicormic-shoot removal on distribution of long, short and epicormic shoot growth in Prunus persica. Annals of botany 99: 323–332.
9. Fumey D, Lauri P, Guédon Y, Godin C, Costes E (2011) How young trees cope with removal of whole or parts of shoots: An analysis of local and distant responses to pruning in 1-year-old apple (Malus xdomestica; Rosaceae) trees. American journal of botany 98: 1737–1751.
10. Efron B (1979) Bootstrap Methods: Another Look at the Jackknife. The Annals of Statistics 7: 1–26.
11. Grechi I, Hilgert N, Sauphanor B, Senoussi R, Lescourret F (2010) Modelling coupled peach tree–aphid population dynamics and their control by winter pruning and nitrogen fertilization. Ecological Modelling 221: 2363–2373.
12. Fournier D, Costes E, Guedon Y (1998) A comparison of different fruiting shoots of peach tree. Acta Horticulturae: 557–565.
13. Wilson B (1966) Development of the shoot system of Acer Rubrum L. Petersham Mass.: Harvard University. p.
14. Greenwood M, Day M, Schatz J (2010) Separating the effects of tree size and meristem maturation on shoot development of grafted scions of red spruce (Picea rubens Sarg.). Tree physiology 30: 459–468.
15. Belsky A (1986) Does herbivory benefit plants? A review of the evidence. American Naturalist 127: 870–892.
16. McNaughton S (1984) Grazing lawns: animals in herds, plant form, and coevolution. American Naturalist 124: 863–886.

17. Maschinski J, Whitham T (1989) The continuum of plant responses to herbivory: the influence of plant association, nutrient availability, and timing. American Naturalist 134: 1–19.

18. Lakso A (1984) Principles of tree designs for production and mechanical harvest. Great Lakes Fruit Grower News 23: 50–51.

19. Webster A, Sheperd U (1984) The effects of summer shoot tipping and rootstock on the growth, floral bud production, yield and fruit quality of young sweet cherries. Journal of horticultural science 59: 175–182.

20. Rom C, Ferree D (1985) Time and severity of summer pruning influences on young peach tree net photosynthesis, transpi- ration, and dry weight distribution. Journal of the American Society of Horticultural Sciences 110: 455–46.

21. Giauque P (2003) Conduite du Verger de Pêcher: Recherche de la Performance. Paris, France

22. Elfving D, Forshey C (1976) Growth and fruiting response of vigorous apple branches to pruning and branch orientation treatments. Journal of the American Society for Horticultural Science 101: 290–293.

Spatial and Temporal Variations of Ecosystem Service Values in Relation to Land Use Pattern in the Loess Plateau of China at Town Scale

Xuan Fang[1], Guoan Tang[1]*, Bicheng Li[2], Ruiming Han[3]

1 Key Laboratory of Virtual Geographic Environment, Ministry of Education, School of Geography Science, Nanjing Normal University, Nanjing, China, **2** Research Center of Soil and Water Conservation and Ecological Environment, Chinese Academy of Sciences, Yangling, Shaanxi, China, **3** School of Geography Science, Nanjing Normal University, Nanjing, China

Abstract

Understanding the relationship between land use change and ecosystem service values (ESVs) is the key for improving ecosystem health and sustainability. This study estimated the spatial and temporal variations of ESVs at town scale in relation to land use change in the Loess Plateau which is characterized by its environmental vulnerability, then analyzed and discussed the relationship between ESVs and land use pattern. The result showed that ESVs increased with land use change from 1982 to 2008. The total ESVs increased by 16.17% from US$ 6.315 million at 1982 to US$ 7.336 million at 2002 before the start of the Grain to Green project, while increased significantly thereafter by 67.61% to US$ 11.275 million at 2008 along with the project progressed. Areas with high ESVs appeared mainly in the center and the east where largely distributing orchard and forestland, while those with low ESVs occurred mainly in the north and the south where largely distributing cropland. Correlation and regression analysis showed that land use pattern was significantly positively related with ESVs. The proportion of forestland had a positive effect on ESVs, however, that of cropland had a negative effect. Diversification, fragmentation and interspersion of landscape positively affected ESVs, while land use intensity showed a negative effect. It is concluded that continuing the Grain to Green project and encouraging diversified agriculture benefit to improve the ecosystem service.

Editor: Ricardo Bomfim Machado, University of Brasilia, Brazil

Funding: This study was sponsored by the Jiangsu Planned Projects for Postdoctoral Research Funds (No. 1401033C), the National Natural Science Foundation of China (No. 41401441), and the Priority Academic Program Development of Jiangsu Higher Education Institutions (PAPD) (No. 164320H101). The funders had no role in study design, data collection and analysis, decision to publish, or preparation of the manuscript.

Competing Interests: The authors have declared that no competing interests exist.

* Email: tangguoan@njnu.edu.cn

Introduction

Ecosystem contributes to human welfare by providing goods and services directly and indirectly [1–2]. With widely spreading of environmental problems, ecosystem service received increasing attention. Many studies showed human factors, such as urban sprawl [3,4,5], socioeconomic changes [6], agricultural policies [7,8], could affect natural or artificial ecosystems. Land use, an original and foundational human activity and represents the most substantial human alteration to systems on the planet of earth for long-term study [9], plays an important role in providing ecosystem services, including biodiversity, water filtration, retention of soil, etc. [10] Inappropriate land use may lead to significant degradation of local and regional ecological services [11]. Moreover, there were studies showed that ecosystem service trade-offs could successful apply to land use planning [12,13]. Understanding the relationship between ecosystem services and land use change is essential for maintaining a healthy ecosystem and getting sustainable services.

The growing body of literatures focused on how ecosystem service changes in response to land use change of different regions [14,15,16,17,18]. However, these studied focus on the impact of land use type on ecosystem service, while the spatial pattern of land that reflects ecological processed and functions [19] get less attention. Monitoring the characteristic of landscape patterns including area, shape, diversity, etc., is helpful to deeply understand the relationship between ecosystem service and land use change and then to provide complete references for land use planning.

The Loess Plateau is the area suffered from the most severe soil erosion in the world, and it is also a major agricultural production region in China [20]. Long-term poor land use has resulted in vegetation destruction and accelerated soil erosion [21]. To control soil erosion and restore the ecosystem, the Grain for Green project converting slope cropland to grassland or forestland was implemented in 1999 by the Chinese Government [22]. The land use on the plateau under the project has changed significantly. Studying the ecosystem service in relation to land use change before and after the Grain to Green project was crucial for ecosystem protection and agricultural sustainability for the area. Researchers have analyzed ecosystem service at different scales within the Loess Plateau [17,18,23]. However, town is a basic administrative area in China. Exploring the characteristic of ecosystem services change at town scale is of practical significance to provide operable land use planning.

Ecosystem service values (ESVs) is monetary assessment of ecosystem services. This paper examined the characteristics of ESVs at Hechuan town, a typical town in the hilly and gully region of the Loess Plateau. The objectives of this study were: 1) to analyze the changes in land use pattern from 1982 to 2008; 2) to access the spatial and temporal variation in ESVs in response to land use during this period; 3) to quantitively analysis the relationship between ESVs and land use pattern; and 4) to discuss how land use management is favorable for ecosystem service supply and the ecological and economic sustainable development.

Data and Methods

2.1 Ethics statement

No specific permits were required for the described studies, and the work did not involve any endangered or protected species.

2.2 Study area

The study area, Hechuan town (106°18′43″~106°32′16″E, 35°54′59″~36°06′05″N), is located in Guyuan city of the Ningxia Hui Autonomous Region of northwest China (Fig. 1), consisting 12 villages with 16,524 people. The reasons that Hechuan Town was chosen as the study area were, on the one hand, Hechuan town has the typical characteristics of Loess Plateau including the

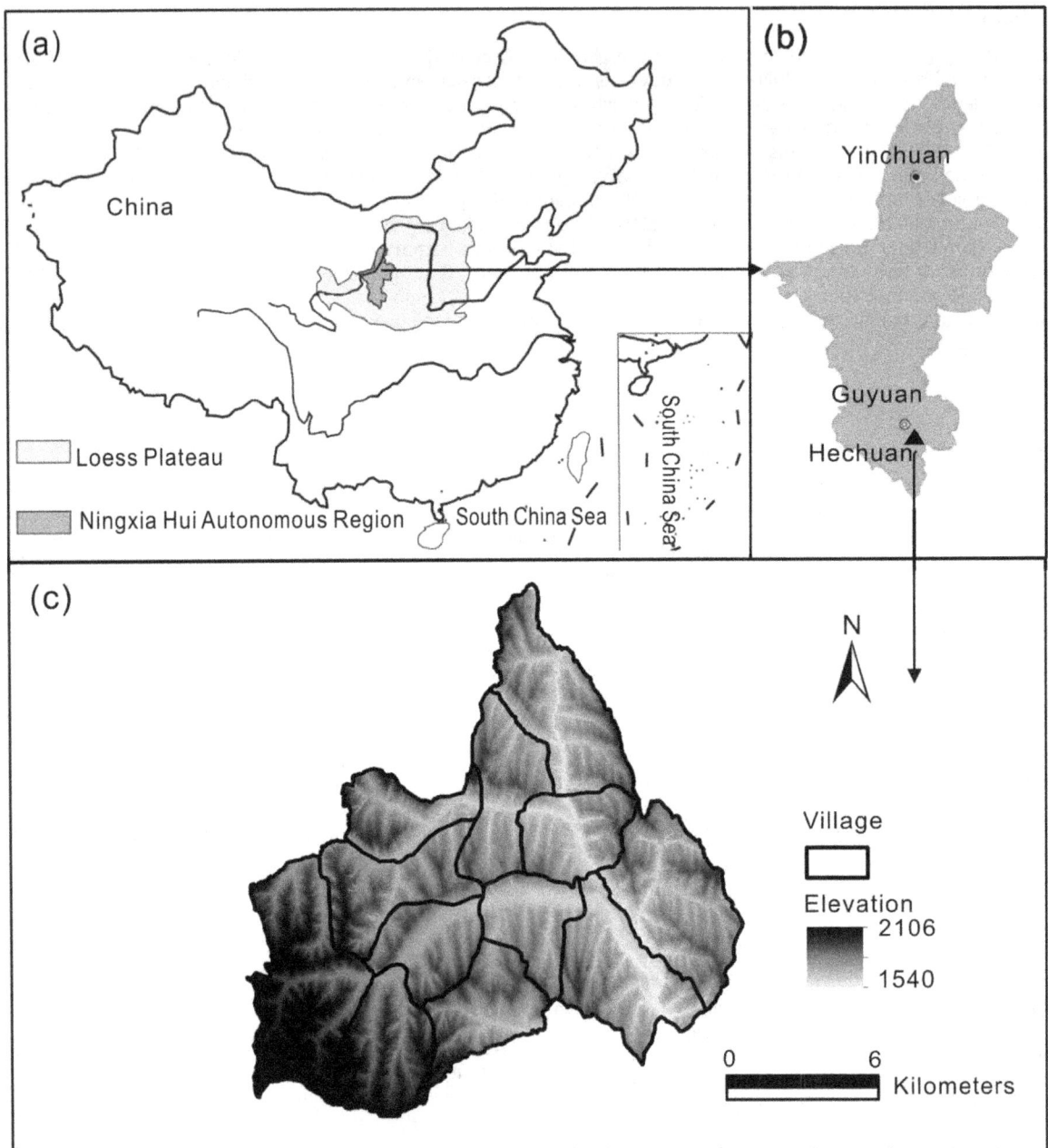

Figure 1. Location of the study area. Ningxia Province and the Loess Plateau, China (a), the location of Hechuan Town in the Loess area of Ningxia Province (b) and, the village boundary and the digital elevation model (DEM) map of Hechuan Town (c).

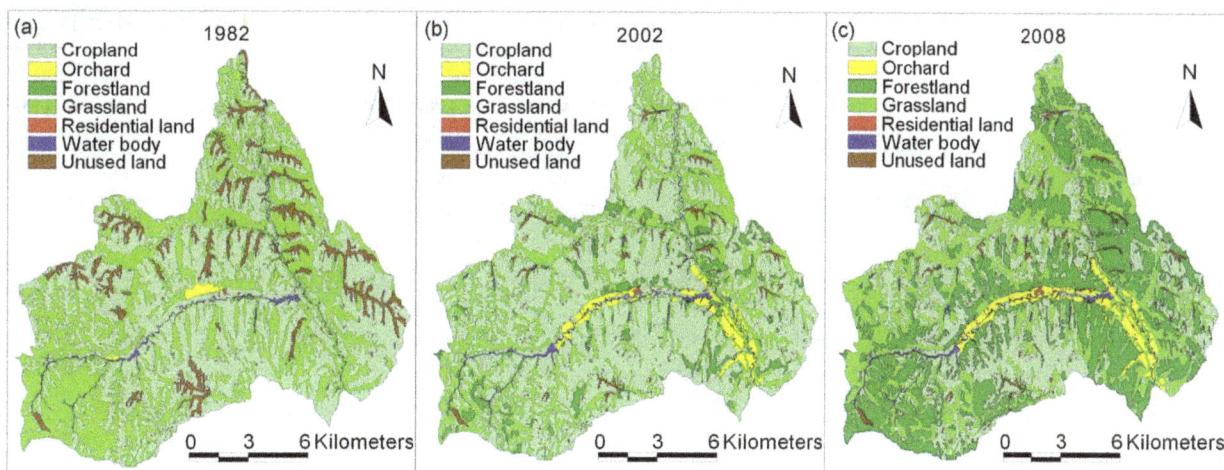

Figure 2. Land use maps of Hechuan town in 1982 (a), 2002 (b) and 2008 (c).

terrain of hill and gull, the fragile ecosystem and the backward economy; on the other hand, there was a long term ecological observation and experiment station in the study area, which facilitated the survey of land use and ecosystems. This town has an altitude ranging from 1540 to 2106 m, covering an area of 215.58 km^2. There exist the topographic differences in the town. The central area with river terrace stretches smoothly with a low elevation. The terrain in the northern area is fragmented while that of southern area is relatively simple. Hechuan town has a semi-arid continental temperate climate with the average annual temperature of 6.9°C and precipitation of 419 mm (1982–2002). Most of the annual precipitation is concentrated between June to September in the form of heavy storms that can cause severe soil erosion. The soil is composed of loessial soil and Dark loessial soils, which is erodible due to its weak cohesion and high infiltrability.

The ecosystem in Hechuan town is fragile with serious soil erosion and frequent natural disasters. Human disturbances of excessive land use, such as deforestation, overgrazing and over-reclamation further destructed the native natural grassland. Therefore, this area has long been in a vicious circle, endless cultivation and poverty. Since the early 1980s, a variety of comprehensive investigation of soil erosion was practiced by Chinese Academy of Sciences. Shanghuang watershed, located in the east of Hechuan town, was taken as a key test area. The

ecological restoration covering the whole town was started from implementing the Grain for Green project after 2002 (launched in 1999 by China government). Since then, abandoned cropland, shrubland (*Caragana korshinskii*, *Hippophae rhamnoides*) and artificial grassland (*Medicago sativa*) was generated, which made a significant change on landscape pattern and ecosystem components providing a variety of ecosystem services. Meanwhile, farming and grazing, the traditional way of living, had to be changed, and raising livestock, orchards, and migrant working diversified their incomes.

2.3 Data acquisition and preprocessing

Land use data was the key data for evaluating landscape pattern and ecosystem service. The land use data of 1982 was obtained by digitizing the land use patches from the 1:10,000 scale topographic maps of 1982, in which the information of land use types and its boundary are clearly shown. The 10 m resolution of remote sensing image could be considered to be corresponding with the scale of 1:50000 [24,25]. The land use data of 1982 acquired from 1:10000 topographic maps was therefore generalized to be at 1:50000 scale [26]. The land use data of 2002 and 2008 were respectively extracted from the 10 m resolution multispectral Spot-5 image of 2002 and 2008 by updating the land use patches of

Table 1. Equivalent weight factor of ecosystem service values (ESVs) per hectare of terrestrial ecosystem in China [30].

	Cropland	Forestland	Grass land	Water body	Barren land
Gas regulation	0.72	4.32	1.5	0.51	0.06
Climate regulation	0.97	4.07	1.56	2.06	0.13
Water supply	0.77	4.09	1.52	18.77	0.07
Soil formation and retention	1.47	4.02	2.24	0.41	0.17
Waste treatment	1.39	1.72	1.32	14.85	0.26
Biodiversity protection	1.02	4.51	1.87	3.43	0.40
Food production	1.00	0.33	0.43	0.53	0.02
Raw material	0.39	2.98	0.36	0.35	0.04
Recreation and culture	0.17	2.08	0.87	4.44	0.24
Total	7.90	28.12	11.67	45.35	1.39

Table 2. The ecosystem service values (ESVs) per hectare of different land use types in Hechuan town (US$·ha-1·yr-1).

	Cropland	Orchard	Forestland	Grass land	Water body	Unused land
Gas regulation	22.570	91.222	135.422	47.022	15.987	1.881
Climate regulation	30.407	88.244	127.585	48.902	64.576	4.075
Water supply	24.138	87.930	128.212	47.649	588.397	2.194
Soil formation and retention	46.081	98.118	126.018	70.219	12.853	5.329
Waste treatment	43.573	47.649	53.918	41.379	465.514	8.150
Biodiversity protection	31.975	99.999	141.378	58.620	107.523	12.539
Food production	31.348	11.912	10.345	13.480	16.614	0.627
Raw material	12.226	52.351	93.416	11.285	10.972	1.254
Recreation and culture	5.329	46.238	65.203	27.272	139.184	7.523
Total	247.647	623.663	881.498	365.828	1421.619	43.573

1982 one by one in visual interpretation method. The interpretation sign was established by understanding the Spot image characteristics and carrying out field surveys in order to further determine the relationship between the true ground and the image. The kappa accuracy index [27] was used to assess the accuracy of the interpretation. The stratified random sampling method was used to generate the reference points on the classified image for the accuracy test. These reference points were located in the field with a GPS with 5-m precision for ground truth. The total kappa indexes are all higher than 0.85, which are higher than the minimum acceptable (0.7) [28]. Considering the characteristic of the land use in study area and the interpretation level of the data and to facilitate the calculation of ESVs, the land use was classified into seven types: cropland, orchard, forestland, grassland, residential area, water area, and unused land (Fig. 2).

To acquire accurate area data of the land use for ESVs estimation and facilitate analyzing the spatial distribution of ESVs, the topographic maps and Spot images were transformed to the same projection and coordinate system (the Albers-Conical-Equal-Area projection system and Krasovsky 1940 coordinate system) before the extraction of land use data, and all acquired land use data were transformed to Arc-grid formats with the same grid size (10 m×10 m). The above data processing was completed using ERDAS and ArcGIS software.

2.4 Analysis on land use pattern

The transfer matrix analysis of land use was produced to understand how land use changed. Landscape metrics analysis was used for spatial pattern analysis of land use. Landscape metrics has been adopted widely; meanwhile, its abilities to indicate ecological process gained increasing attention [29,30,31]. Conceptual flaws in landscape pattern analysis, limitations inherent in landscape metrics and the improper use of pattern analysis may lead to the misuse of landscape metrics [32]. For better explanations and predictions of ecological phenomena from ecological pattern, the landscape metrics in this study was therefore selected by two steps. Firstly, the diversity, the fragmentation and the dominance of landscape were all considered, and then 34 metrics was selected, by understanding the knowledge of the landscape pattern and the ecological services indication of landscape metrics [33,34] and referring to the previous studies on landscape pattern [4,31,35,36]. Secondly, a correlation analysis for the 34 metrics was employed to ensure the low redundancy among landscape metrics. If the coefficient between two metrics was significant at 0.05 level, only one metric of them could be eventually selected.

Landscape-level metrics providing general landscape information and class-level metrics providing more specific information about variations at the local level and spatial patterns of land use classes [37] were used to monitor the characteristics of landscape pattern. The selected landscape-level metrics were patch density (PD), area-weighted mean shape index (SHAPE_AM), Interspersion and Justaposition Index (IJI), and Shannon's diversity index (SHDI). The selected class-level metrics were PD, the percentage of landscape (PLAND), SHAPE_AM and IJI. PD and SHAPE_AM could show the fragmentation of landscape. SHDI and PLAND reflect the dominance of some land use type and the diversity of landscape, respectively. IJI reflects whether the patches or classes are contiguous. Landscape metrics analysis was conducted with above metrics by FRAGSTATS 3.3, in which the eight-neighbor rule was used to derive the patch number. Besides these metrics, the land use intensity index (LUII) was also used to describe the landscape pattern. It was calculated by the following equation [31]:

$$LUII = \sum_{i=1}^{n} A_i \times C_i \tag{1}$$

where $LUII$ is the land use intensity index, A_i is the percentage of for a give land use type i, and C_i is the coefficient value of intensity for a give land use type i, that is assigned 4 for build-ups, 3 for farmland and 2 for forest, orchard, grassland and water bodies, and 1 for unused land.

2.5 Estimation of ESVs

Costanza et al.'s model of ESVs estimation was adopted in this study [1,2]. The model classified ecosystem service into 17 types of service functions and estimated the ESVs by placing an economic value on different biomes [34]. For the defects of this model, such as overestimating the agriculture ESVs and underestimating the wetland ESVs, Xie et al. proposed refined coefficients for ESVs assessment both solving the above problem and making it apply to China [33,34]. Based on this model, the total ESVs in the study area was calculated using the following formulas:

$$ESV_k = \sum_f A_k VC_{kf} \tag{2}$$

Table 3. Land use transition matrix from 1982 to 2002 and from 2002 to 2008 (%).

1982	2002								
	Cropland	Orchard	Forestland	Grassland	Residential land	Water body	Unused land	Total	Loss
Cropland	44.42	2.33	1.18	2.50	0.35	0.04	0.01	50.83	6.41
Orchard	0.17	0.05	0.26	0.09	0.00	0.00	0.00	0.57	0.53
Forestland	0.10	0.01	0.15	0.02	0.00	0.00	0.00	0.28	0.13
Grass land	12.63	0.39	4.74	22.15	0.01	0.08	0.00	40.01	17.86
Residential land	0.05	0.01	0.00	0.01	0.14	0.00	0.00	0.22	0.07
Water body	0.01	0.00	0.01	0.00	0.00	0.80	0.00	0.81	0.02
Unused land	1.39	0.00	0.16	3.98	0.00	0.07	1.66	7.27	5.61
Total	58.76	2.78	6.51	28.77	0.51	0.99	1.68	100.00	
Gain	14.35	2.73	6.36	6.62	0.37	0.19	0.01		

2002	2008								
	Cropland	Orchard	Forestland	Grass land	Residential land	Water body	Unused land	Total	Loss
Cropland	27.09	1.00	20.75	9.84	0.07	0.00	0.00	58.76	31.68
Orchard	0.00	2.75	0.01	0.00	0.01	0.00	0.00	2.78	0.03
Forestland	0.05	0.04	5.84	0.57	0.01	0.01	0.00	6.51	0.67
Grass land	0.12	0.00	7.17	21.42	0.02	0.04	0.00	28.77	7.35
Residential land	0.00	0.00	0.00	0.00	0.51	0.00	0.00	0.51	0.00
Water body	0.02	0.02	0.00	0.01	0.00	0.94	0.00	0.99	0.05
Unused land	0.00	0.00	0.27	0.08	0.00	0.00	1.33	1.68	0.35
Total	27.27	3.82	34.05	31.91	0.62	1.00	1.33	100.00	
Gain	0.18	1.07	28.21	10.50	0.11	0.05	0.00		

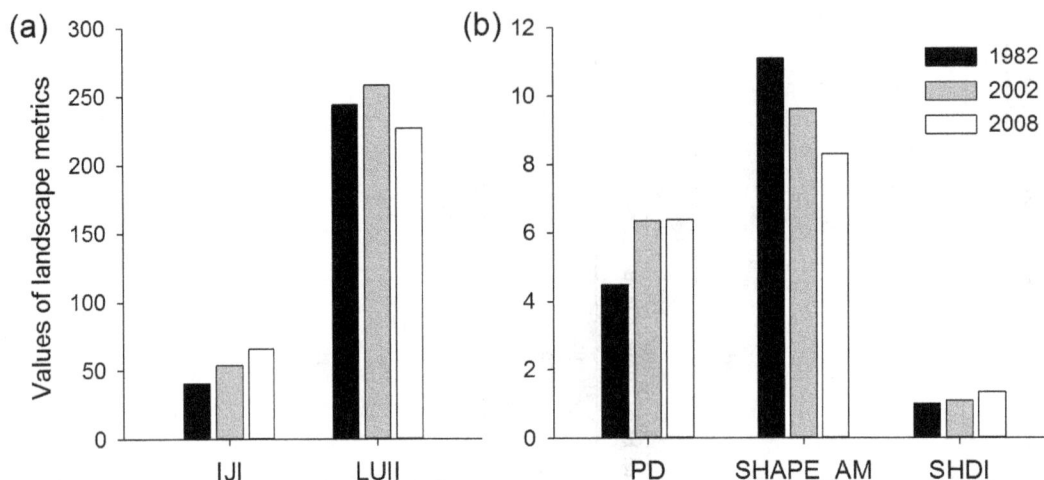

Figure 3. Landscape metrics at the landscape level in Hechuan Town in 1982, 2002 and 2008. IJI: Interspersion and Justaposition Index; LUII: land use intensity index; PD: patch density; SHAPE_AM: area-weighted mean shape index; SHDI: Shannon's diversity index.

$$ESV_f = \sum_k A_k VC_{kf} \qquad (3)$$

$$ESV = \sum_k \sum_f A_k VC_{kf} \qquad (4)$$

where ESV_k, ESV_f, ESV are the ESVs of land use type k, the ESVs of ecosystem service function type f, and the total ESVs respectively. A_k is the area (ha) for land use types. VC_{kf} is the value coefficient (US$·ha-1·yr-1) for land use type k and ecosystem service function type f, which is the key for ESVs estimating. Xie et al.'s model was used to determine VC_{kf}, which can be expressed as follows:

$$VC_{kf} = R_{kf} \times V_f \qquad (5)$$

where R_{kf} is the equivalent weight factor of ecosystem service, V_f is food production values of agriculture land per area per year.

The equivalent weight factor was presented for customizing Chinese terrestrial ecosystem based on Costanza et al.'s model by surveying 500 Chinese ecologists (Table 1) [34]. It is the ratio of the ESVs to the economic value of average natural food production provided by agricultural land per hectare per year. The factors of land use types in our study were basically assigned based on the nearest ecosystems in Xie et al.'s model. However, minor adjustments were made. The equivalent weight factor of orchard which was not put forward clearly in Xie et al.'s model was determined by the mean of grassland and forestland by referring some researches [5,18]. The factor of unused land equates to that of barren land, and that of residential land was determined to zero.

The value of food production service of agriculture land per area per year was considered to be 1/7 of the actual price of food production in Xie et al.'s model. With the average actual food production of cropland in Hechuan town from 1982 to 2008 of 901.77 kg/ha which was get from *Statistic yearbook of the Yuanzhou District, Guyuan City, Ningxia Hui Autonomous Region* and the average grain price of US$ 0.243 per kilogram (i.e. an equivalent of RMB Yuan 1.69 according to the average exchange

rate of 2008) in 2008, the value of food production service of cropland per area per year was calculated to be US$ 31.348 (i.e. an equivalent of RMB Yuan 217.713 according to the average exchange rate of 2008). ESVs of one unit area of each land use types were then assigned as shown in Table 2.

After the ESVs were calculated by above processing, a sensitivity analysis was conducted to test the land use type's representative for ecosystem types and the certainty of the coefficients value for ecosystem service. A coefficient of sensitivity (CS) was used to indicate the degree of sensitivity of ESVs to a coefficients value, calculated by the following formula [5]:

$$CS = \left| \frac{(ESV_j - ESV_i)/ESV_i}{(VC_{jk} - VC_{ik})/VC_{ik}} \right| \qquad (6)$$

where ESV_j an ESV_i are the total ESVs of the initial status j and the adjusted status i, and VC_{jk} and VC_{ik} are the initial and adjusted coefficients. A 50% adjustment in the coefficients was made in the study. The greater the CS responded to the adjustment, the more critical is the use of an accurate coefficient [38]. A CS lower than 1 indicates the ESVs is inelastic to the coefficient and the estimation of ESVS is reliable. Otherwise, a CS greater than 1 indicates the estimation of ESVs is sensitive to the coefficient.

2.6 Correlation and regression analysis

The data of ESVs and landscape metrics was used to analysis the relationship between ecosystem service and land use pattern change. Because the spatial variation of landscape pattern exist among 12 villages in Hechuan town, the land use data of the three years (1982, 2002 and 2008) for the 12 villages can be considered as representing different landscape pattern on a time-for-space perspective [39]. Therefore, there were totally 36 sample data. Correlation and regression was employed for the relationship analysis, in which Multiple stepwise regression was specifically chosen considering the multicollinearity among landscape metrics. The dependents were the nine categories and total ESVs, while the corresponding independents were the landscape-level and class-level landscape metrics.

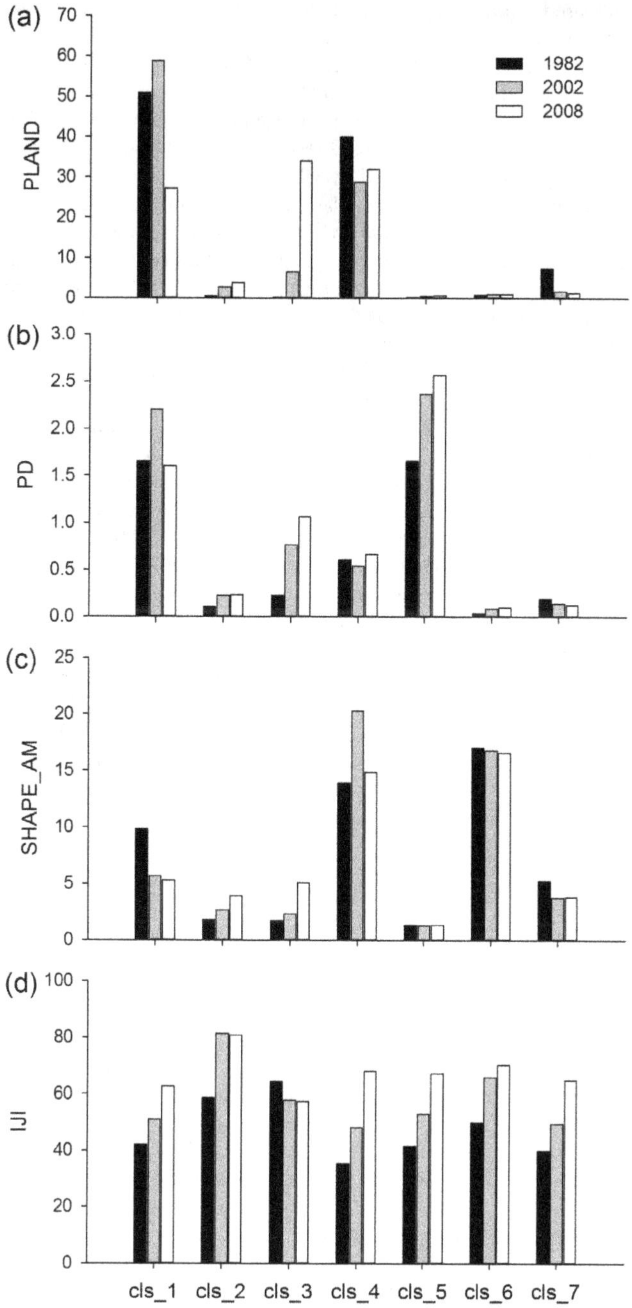

Figure 4. Landscape metrics at the class-level in Hechuan Town in 1982, 2002 and 2008. cls_1, cls_2, cls_3, cls_4, cls_5, cls_6, and cls_7 represent cropland, orchard, forestland, grassland, residential land, water body and unused land. PLAND: the percentage of landscape; PD: patch density; SHAPE_AM: area-weighted mean shape index; IJI: Interspersion and Justaposition Index.

Table 4. The change of ecosystem service values (ESVs) in Hechuan Town from 1982 to 2008.

ESVs		Cropland	Orchard	Forestland	Grass land	Water body	Unused land	Total
ESVs (10⁶ US$ yr⁻¹)	1982	2.714	0.077	0.051	3.155	0.249	0.068	6.315
	2002	3.137	0.374	1.237	2.269	0.304	0.016	7.336
	2008	1.456	0.514	6.470	2.517	0.305	0.013	11.275
Change of ESVs (10⁶ US$ yr⁻¹)	1982–2002	0.423	0.297	1.186	−0.887	0.054	−0.053	1.021
	2002–2008	−1.681	0.140	5.234	0.248	0.002	−0.003	3.939
	1982–2008	−1.258	0.437	6.419	−0.638	0.056	−0.056	4.960
Change of ESVS (%)	1982–2002	2.248	55.387	333.431	−4.045	3.145	−11.081	2.328
	2002–2008	−7.716	5.404	60.934	1.574	0.072	−2.992	7.731
	1982–2008	−6.674	81.578	1805.410	−2.913	3.232	−11.771	11.309
Average annual Change (%yr⁻¹)	1982–2002	0.112	2.769	16.672	−0.202	0.157	−0.554	0.117
	2002–2008	−1.286	0.901	10.155	0.262	0.012	−0.498	1.289
	1982–2008	−0.256	3.137	69.439	−0.112	0.124	−0.452	0.435

Results

3.1 Changes of land use pattern

Table 3 showed the land use transition matrix. From 1982 to 2002, cropland as the dominant land use type increased from 50.83% to 58.76%. Grassland was the land use type with the largest change in area, decreasing from 40.01% to 28.77%. Orchard increased by 6.24% of total area, indicating the economic driving force of fruit trees on land use change. Forestland

Table 5. Values of different ecosystem service functions in 1982, 2002, and 2008.

	1982			2002			2008		
	ESVs (10^6 US$\cdot yr^{-1}$)	%	Rank	ESVs (10^6 US$\cdot yr^{-1}$)	%	Rank	ESVs (10^6 US$\cdot yr^{-1}$)	%	Rank
Gas regulation	0.678	10.73	6	0.826	11.26	6	1.529	13.56	5
Climate regulation	0.791	12.53	5	0.936	12.75	5	1.540	13.65	4
Water supply	0.800	12.67	4	0.960	13.09	4	1.610	14.28	3
Soil formation and retention	1.141	18.06	1	1.260	17.17	1	1.764	15.65	1
Waste treatment	0.938	14.85	2	1.015	13.84	3	1.078	9.56	6
Biodiversity protection	0.915	14.49	3	1.054	14.37	2	1.738	15.42	2
Food production	0.466	7.38	7	0.506	6.90	7	0.367	3.25	9
Raw material	0.247	3.91	9	0.390	5.32	8	0.881	7.81	7
Recreation and culture	0.339	5.37	8	0.388	5.29	9	0.768	6.81	8
Total	6.315	100.00		7.336	100.00		11.275	100.00	

increased from 0.57% to 2.78%, reflecting that ecological restoration began to gain attention. From 2002 to 2008, cropland and forestland changed significantly, decreasing from 58.76% to 27.27% and increasing from 6.51% to 34.05% respectively. Land use structure was transferred from cropland dominated (58.76%) to cultivated land (27.27%), forestland (34.05%) and grassland (31.91%) relatively balanced distributed.

The most notable change of land use from 1982 to 2002 was the conversion from grassland to cropland and forestland with 12.63% and 4.74% of the total area respectively. The conversions from cropland (2.50%) and unused land (3.98%) to grassland were not adequate to compensate for the grass loss. From 2002 to 2008, the notable changes of land use were cropland to forestland, cropland to grassland, and grassland to forestland, with the rates of 20.75%, 9.84%, and 7.17% respectively. It was found that the conversion among land use types was more outstanding and concentrated than that before 2002, reflecting that the Grain for Green project as an ecological policy had great influence on land use change.

The results of landscape-level metric analysis were exhibited in Fig. 3. The significant increased PD from 1982 to 2002 reflected the landscape fragmentation. It was relative to the increase of patches on the land use types with intense human disturbance, such as cropland, residential land and artificial reservoir. Oppositely, the slight change of PD from 2002 to 2008 reflected that human disturbance became stable. The change of human disturbance was also demonstrated by the change of LUII which increased before 2002 and decreased after 2002. SHAPE_AM decreased in the study period, showing the landscape became more regular in shape. The increase of IJI suggested that the landscape became more contiguous and the ecological connectivity among land use types increased. SHDI increase obviously from 2002 to 2008, which related to that the land use structure became even.

Fig. 4 showed the change of class-level metrics. The PLAND of land use types indicated that cropland, forestland, and grassland had significantly influence on land use pattern. PD in orchard, forestland, and residential land increased obviously, attributing to the increasing area of these land use types and the fragmental terrain. SHAPE_AM showed that cropland and unused land became more regular in shape, while orchard and forestland more complicated. IJI increased generally in land use types. Orchard was the most contiguous with high IJI, which was relative to its concentrated distribution across the river terrace.

3.2 ESVs from 1982 to 2008

The ESVs of each land use type and the total ESVs was shown in Table 4. The total ESVs of Hechuan town was US$ 6.315, US$ 7.336 and US$ 11.275 million in 1982, 2002 and 2008, respectively. From 1982 to 2002, the decline of ESVs caused by the decrease of grassland was offset by the increase of forestland, orchard and cropland, resulting that the total ESVs increased by US$ 1.021 million. From 2002 to 2008, the total ESVs increased by US$ 3.939 million, mainly due to the increase of forestland. The average annual change rate of total ESVs before and after 2002 was quite different, that is 0.81% and 8.95% respectively. It indicated the Grain to Green project implemented since 2002 had a significant effect on the ecosystem service. It was also shown from the value of ESVs produced by forestland occupying 57.39% of the total ESVs. Overall, the total ESVs increased US$ 4.960 million during the study period, mainly due to the increase of ESVs by the increase of forestland and orchard far beyond the decrease of ESVs by the decrease of cropland and grassland. It was essentially because of the higher coefficient value of forestland and orchard than that of cropland and grassland.

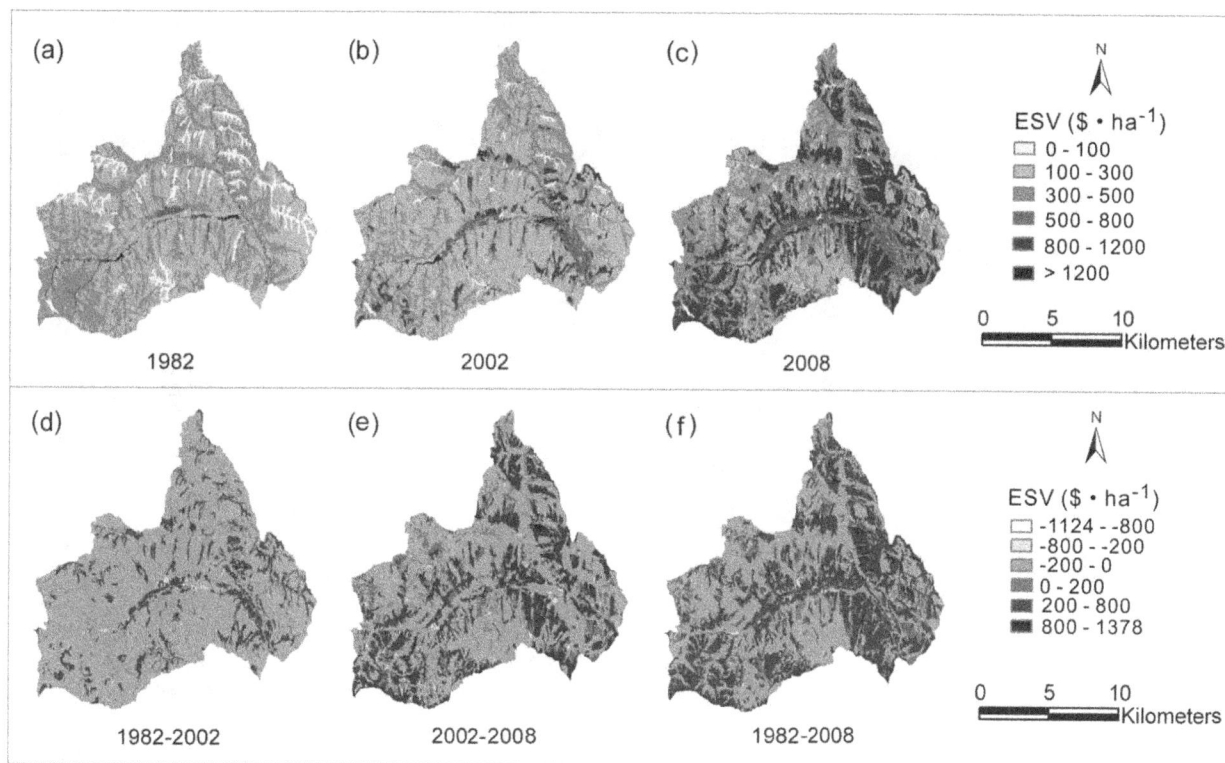

Figure 5. Spatial and temporal distribution of ecosystem service values (ESVs) in Hechuan Town from 1982 to 2008. The spatial distribution of ESVs in 1982 (a), 2002 (b) and 2008 (c), and the spatial-temporal changes of ESVs between time intervals from 1982 to 2002 (d), 2002 to 2008 (e) and 1982 to 2008 (f).

The ESVs of each ecosystem function type was shown in Table 5. Expect for food production, the values of ecosystem service functions increased especially after 2002. The decrease of food production was due to the great decline of cropland in the Grain to Green project. The ESVs proportion of each ecosystem function type to the total ESVs represented the contribution of each ecosystem function to the total ESVs. It was found that the functions of soil formation and retention, waste treatment, and food production were decline during 1982 to 2008, while other functions were improved. The rank of the contribution by each ecosystem service function was also estimated. It was basically stable except for relatively obvious decline in the rank of waste treatment and food production. In 2008, the rank order for each ecosystem service was as follows from high to low, soil formation and retention, biodiversity protection, water supply, climate regulation, gas regulation, waste treatment, raw material, recreation and culture, and food production. Soil formation and retention was the highest during the study period.

3.3 Spatial distribution of ESVs

Maps of ESVs in different periods (Fig. 5) showed the spatial distribution of ESVs of unit area in Hechuan town, directly reflecting the difference of ESVs among land use types. In 1982, the ESVs>4000 mostly appeared in the center of the town where river and river terrace located. It was because water body and orchard which intensely distributed in river terrace for its high water demand both had high ESVs. Therefore, due to the orchard increasing intensely and the forest increasing scatteredly, the increase of ESVs also mainly happened across the river terrace in 2002. Since 2008, the ESVs>4000 spread widely with the increase of forestland transformed from cropland. The lowest ESVs mostly

occurred in the gully where unused land was distributed in 1982. With vegetation recovery in the gully, the low ESVs happened from gully to terraced hillside where cropland with low ESVs was distributed in 2008. Fig. 5d–f showed the temporal change of ESVs spatial distribution. The change characteristic of 2002 to 2008 was adjacent to that during the total study period, reflecting that the change of ESVs mainly occurred after 2002, just after the Grain to Green project.

3.4 Relationship between ESVs and land use pattern

From the above analysis on the change of land use and ESVs in quantity and spatial distribution, we could infer there was some relationship between land use change and ecosystem service. To quantitively understand the relationship, the correlation analysis and regression analysis between ESVs and landscape pattern metrics was conducted.

Table 6 showed there existed significant correlations between ESVs and many landscape metrics ($p<0.01$), which explained that landscape pattern affected ESVs significantly. For example, the correlation coefficients between total ESVs and landscape metrics showed that there existed significantly positive relationship between SHDI (0.433), PLAND_3 (0.677), SHAPE_AM_3 (0.744), IJI_4 (0.513) and ESVs, and negative relationship between LUII (−0.634), PLAND_1 (−0.752) and ESVs. It reflected that the diversity and intensity of land use had important effects on total ESVs. It also reflected that cropland, forestland and grassland were the land use types which had significant effects on total ESVs. On quantity,the less the cropland and the more the forestland, the higher the total ESVs were. As to the landscape shape, the more regular the cropland and the more complex the forestland, the higher the total ESVs were. The higher the IJI of grassland, the

Table 6. Correlation coefficients between ecosystem service values (ESVs) and landscape pattern metrics.

	TESVs	ESVs_1	ESVs_2	ESVs_3	ESVs_4	ESVs_5	ESVs_6	ESVs_7	ESVs_8	ESVs_9
PD	0.035	0.497*	0.509*	0.539*	0.477*	0.516*	0.499*	-0.221	0.547	0.478*
SHAPE_AM	0.326	-0.216	-0.220	-0.188	-0.177	-0.088	-0.205	0.026	-0.290	-0.166
UI	0.292	0.624*	0.635*	0.639*	0.597*	0.534*	0.621*	-0.293	0.687	0.586*
SHDI	0.433*	0.763*	0.764*	0.766*	0.741*	0.507*	0.765*	-0.636*	0.775*	0.765*
LUII	-0.634*	-0.681*	-0.658*	-0.618*	-0.675*	-0.113	-0.684*	0.977*	-0.599*	-0.734*
PLAND_1	-0.752*	-0.810*	-0.795*	-0.772*	-0.811*	-0.334	-0.815*	0.952*	-0.742*	-0.853*
PD_1	0.045	-0.055	-0.063	-0.056	-0.054	-0.113	-0.051	-0.134	-0.091	-0.022
SHAPE_AM_1	-0.476*	-0.369	-0.358	-0.330	-0.368	-0.063	-0.369	0.495	-0.328	-0.390
UJI_1	0.189	0.542*	0.552*	0.530*	0.510*	0.405	0.533*	-0.199	0.619	0.485*
PLAND_2	0.323	0.527*	0.541*	0.590*	0.520*	0.595*	0.534*	-0.246	0.558	0.525*
PD_2	0.420	0.457*	0.471*	0.515*	0.449	0.529*	0.463*	-0.193	0.491	0.452
SHAPE_AM_2	0.159	0.450	0.468*	0.541*	0.439	0.629*	0.460*	-0.149	0.493	0.452
UJI_2	0.207	0.392	0.409	0.483*	0.376	0.583*	0.402	-0.115	0.438	0.396
PLAND_3	0.677*	0.984*	0.983*	0.941*	0.975	0.558*	0.980*	-0.770*	0.988*	0.961*
PD_3	0.276	0.631*	0.637*	0.629	0.625	0.477*	0.629*	-0.383	0.653	0.606*
SHAPE_AM_3	0.744*	0.828*	0.827*	0.780*	0.820	0.449*	0.821*	-0.623*	0.836*	0.799*
UJI_3	0.040	0.231	0.241	0.291	0.208	0.356	0.236	-0.069	0.276	0.231
PLAND_4	0.224	-0.192	-0.212	-0.203	-0.159	-0.298	-0.181	-0.264	-0.311	-0.110
PD_4	-0.294	0.199	0.201	0.190	0.160	0.101	0.192	-0.086	0.257	0.171
SHAPE_AM_4	0.455	-0.061	-0.065	-0.049	-0.028	-0.029	-0.053	-0.086	-0.125	-0.022
UJI_4	0.513*	0.717*	0.719*	0.705*	0.697*	0.457*	0.715*	-0.539*	0.739*	0.701*
PLAND_5	-0.035	0.290	0.313	0.381	0.279	0.578*	0.297	0.081	0.357	0.271
PD_5	-0.047	0.244	0.269	0.322	0.244	0.548*	0.248	0.188	0.314	0.208
SHAPE_AM_5	0.118	0.081	0.082	0.053	0.072	-0.009	0.075	-0.004	0.105	0.055
UJI_5	0.307	0.461*	0.470*	0.525*	0.446*	0.512*	0.471*	-0.312	0.482	0.477*
PLAND_6	0.047	0.139	0.167	0.360	0.148	0.852	0.170	0.064	0.153	0.207
PD_6	-0.160	0.122	0.137	0.198	0.105	0.371	0.128	0.080	0.172	0.118
SHAPE_AM_6	0.378	0.088	0.086	0.140	0.080	0.141	0.099	-0.218	0.067	0.137
UJI_6	0.020	0.201	0.217	0.276	0.197	0.438	0.208	0.028	0.239	0.197
PLAND_7	-0.385	-0.447	-0.474*	-0.525*	-0.507*	-0.769*	-0.456*	-0.041	-0.442	-0.422
PD_7	-0.270	-0.105	-0.129	-0.182	-0.154	-0.494*	-0.114	-0.245	-0.108	-0.089
SHAPE_AM_7	-0.236	-0.313	-0.329	-0.434	-0.348	-0.650*	-0.333	0.150	-0.279	-0.358
UJI_7	0.210	0.416	0.408	0.287	0.402	-0.083	0.394	-0.254	0.443	0.341

TESVs: the total ecosystem service values (ESVs); ESVs_1: the ESVs of gas regulation; ESVs_2 climate regulation; ESVs_3: the ESVs of water supply; ESVs_4: the ESVs of soil formation and retention; ESVs_5: the ESVs of waste treatment, ESVs_6: the ESVs of biodiversity protection; ESVs_7 the ESVs of food production; ESVs_8: the ESVs of raw material; ESVs_9: the ESVs of recreation and culture. PD: patch density; SHAPE_AM: area-weighted mean shape index; IJI: Interspersion and Justaposition Index; SHDI: Shannon's diversity index; LUII: land use intensity index; PLAND: percentage of landscape. The 1, 2, 3, 4, 5, 6, 7 after the above landscape metrics respects different landscape, that is cropland, orchard, forestland, grassland, residential land, water body and unused land, respectively.
*significant at 0.01 level.

Table 7. Regression analysis between ecosystem service values (ESVs) and landscape patterns (n = 36).

Dependent	Standardized coefficients regression	R^2	Sig.
Gas regulation	0.878×PLAND_3+0.166×PLAND_2-0.099×PLAND_1-0.068×IJI_1	0.990	*
Climate regulation	0.790×PLAND_3-0.197×PLAND_7-0.190×LUII+0.081×PLAND_2	0.998	*
Water supply	0.665×PLAND_3-0.317×PLAND_7-0.254×LUII+0.106×PLAND_2	0.955	*
Soil formation and retention	0.684×PLAND_3-0.301×PLAND_7-0.284×LUII+0.066×PLAND_2	0.998	*
Waste treatment	0.672×PLAND_6+0.365×PLAND_3-0.352×PLAND_7+0.051×PLAND_2+0.049×PLAND_5	0.993	*
Biodiversity protection	0.059×SHDI +0.861×PLAND_3-0.033×SHAPE_AM_3+0.133×PLAND_1	0.967	*
Food production	0.742×LUII-0.173×PLAND_3-0.052×SHDI+0.106×PLAND_1	0.991	*
Raw material	0.964×PLAND_3+0.091×SHDI+0.068×LUII	0.981	*
Recreation and culture	−0.747×PLAND_1+0.380×IJI_1	0.853	*
Total	-0.588×PLAND_1+0.569× SHAPE_AM_3-0.303×SHDI	0.709	*

*significant at 0.01 level.
PLAND_1: the percentage of cropland; PLAND_2: the percentage of orchard; PLAND_3: the percentage of forestland; PLAND_5: the percentage of residential land; PLAND_6: the percentage of water body; PLAND_7: the percentage of unused land; SHAPE_AM_3: the area-weighted mean shape index of forestland; IJI_1: the Interspersion and Justaposition Index of cropland; LUII: land use intensity index; SHDI: Shannon's diversity index.

higher the total ESVs were. This indicated that the connectivity of grassland was important for ecosystem service.

Correlation also occurred between ESVs of all the functions and landscape metrics (Table 6). However, the relationships between ESVs of different functions and landscape pattern were different. For example, the correlation between food production and landscape pattern was almost opposite from that between other ecosystem functions and landscape pattern. For example, PLAND_1 had a positive effect on food production; SHDI, PLAND_3, SHAPE_3, and IJI_4 had a negative effect on food production. It could infer that there were contradictions between food production and other ecosystem functions.

As shown in Table 7, the result of regression analysis further explained that the ESVs was correlated significantly with landscape pattern. The total ESVs could be predicted by PLAND on cropland, SHAPE_AM on grassland, and SHDI. ESVs of all kinds of ecosystem functions also could be explained by landscape metrics. These regression equations indicated that landscape-level metrics (such as SHDI and LUII) and class_level metrics (such as PLAND of forestland, orchard, and cropland, unused land, SHAPE of forestland, IJI of cropland) acted as predictors for categories of ecosystem services. Specifically, the proportion of forest (PLAND_3) accounted for almost all of the categories of ecosystem services.

Discussion

4.1 Reliability of ESVs

This study estimated ESVs by multiplying the area for each land use types by the corresponding value coefficients. As discussed in the previous researches, estimations using this method was coarse

with high variation and uncertainty for the following reasons, limitations on the economic evaluation [1], problems of double counting and scales [40,41,42], the complex, dynamic and nonlinear ecosystems [43], the imperfect matches of land use categories as proxies [38] and the accuracy of the ecosystem value coefficients [5]. This study also existed such uncertainty on ESVs estimation. For example, the value coefficient of orchard, determined by the average of forest and grassland, was an approximate estimation and need a further exploration. However, the estimation of temporal variation on ESVs was considered to be more reliable than that of cross-sectional analysis [5]. In addition, the sensitivity analysis of the estimated ESVs with 50% adjustment in the value coefficients was conducted. The result showed that the sensitivity coefficients of all land use categories were lower than 1 (Table 8), which suggested that despite of the above limitations, the estimated ESVs are reliable and useful for subsequent study.

4.2 Relationship between ESVs and landscape pattern

It is usually assumed that land use can affect the ecosystem service. Moreover, a few studies showed that there was a correlation between landscape pattern and ESVs [41,44]. This study signified this statement at town scale on the Loess Plateau. Land use configuration, land use intensity, landscape diversity, fragmentation and connectivity all affected ecosystem service.

The correlation analysis between ESVs and PLAND implied land use structure had significant impact on ecosystem service. Especially, the increase of forestland and the decrease of cropland played an important part in improving the ESVs in the past twenty years. It is closely related to the Grain to Green project comprehensively started in study area since 2002. In the project,

Table 8. The coefficient of sensitivity (CS) resulting from adjustment of ecosystem valuation coefficients.

	Cropland	Orchard	Forestland	Grass land	Water body	Unused land
1982	−0.430	−0.012	−0.008	−0.500	−0.039	−0.011
2002	−0.428	−0.052	−0.169	−0.309	−0.041	−0.002
2008	−0.129	−0.048	−0.574	−0.223	−0.027	−0.001

measures for optimizing land use structure were implemented, including restoring slope cropland into forest and grassland, banning grazing, transforming slopes into terraces, and building reservoirs, etc. Forestland and grassland increased by 423.19% (27.54% of the study area) and 10.93% (3.15% of the study area), and cropland decreased by 53.59%(31.49% of the study area) (Table 3). The increase of ESVs due to the increase of forestland occupied 46.28% of the total ESVs in 2008 (Table 4). The result of the correlation analysis between ESVs and PLAND reflected that vegetation recovery could strongly enhance ecosystem service, and it was coincident with many other studies on the Loess Plateau [17,18,23,45]. LUII, which also related to the proportion of land use types, implied the intensity of human activities. This study showed land use intensity had a negative effect on ecosystem service with negative correlation coefficients (−0.634) (Table 6). It was coincident with some studies on ESVs change under urbanization [5,31]. These studies showed that urbanization which means the increase of land use intensity led to considerable declines in ESVs.

Landscape diversity always presents high positive relevance with biodiversity [46]. Our results were coincident to previous statements given the positive relationships between SHDI and biodiversity conservation. However, there were studies reporting the negative relationships between them, in which the increase of SHDI was the result of rapid urban sprawl [31]. In our study, the increase of SHDI was because land use structure became more balanced, which was the result of the increase of forestland. In addition, landscape diversity could also promote agricultural production [47]. Our study disagreed with this statement, and showed that food production was weakened with landscape diversification. It was because that the increase of SHDI was the result of a larger number of conversion from cropland to forestland. Therefore, the relationship between landscape diversity and biodiversity conservation as well as food production should not be treat as the same but be understood considering the driving force of SHDI change.

Fragmentation could lead to declining habitat quality, lower wildlife survival, and limited movement of soil microorganisms [48], and subsequently cause the decrease of ecosystem service [30]. Our study disagreed with this statement. For example, PD of the total landscape, PD_Forest, PD_orchard and shape_ Forest revealed significantly positive impacts on most categories of ESVs (Table 6–7). The increase of PD and the decrease of connectivity of landscape were usually simultaneous, which is disagreed in our study (Fig. 3 and Fig. 4). The landscape became more contiguous as IJI shown. Table 6 showed the IJI had significantly positive impacts on ESVs. Especially, the increase of IJI of grassland promoted the total ESVs and all categories of ESVs. This maybe because the connectivity of landscape has contribution to habitat corridors [49] and forest production [50].

Based on the relationship between ESVs and landscape pattern, we could improve the ecosystem service by the adjustment of land use policy. On the one hand, continuing to implement the Grain to Green project is helpful for improving ESVS, because it could increase the vegetation coverage, decline the intensity of land use, and make cropland become regular by canceling the slope cropland. On the other hand, diversified agriculture gathering planing fruit trees, planting crops and breeding, which could promote the diversification of land use, should be encouraged to increase both ESVs and farmer's incomes.

Conclusion

ESVs at town scale in the Loess Plateau were estimated in Hechuan town of Ningxia Hui Autonomous Region from 1982 to 2008. It was concluded that ESVs varied with land use change. ESVs in 1982, 2002, and 2008 were US$ 6.315, US$ 7.336 and US$ 11.275 million respectively. Among all the land use types, forestland, grassland and cropland had important contribution (> 90%) on ESVs. The total ESVs increased slowly by 16.17% due to the decrease of grassland from 1982 to 2002, while the total ESVS increased significantly by 67.61% due to the increase of forestland from 2002 to 2008. Areas with high services level were mainly located in the center due to orchard and east due to forestland, while areas with low services level mainly located in the north and south sides due to cropland.

Land use pattern had a significant effect on ecosystem service in our study by analyzing and discussing the relationship between landscape pattern and ESVs. The proportion of forestland had a positive effect on ecosystem service while that of cropland had a negative effect on ESVs. The diversity and interspersion of landscape both had a positive effect on ESVs. Land use intensity which reflects the intensity of human activities had a negative effect on ESVs. Fragmentation had positive effect on ESVs, which was disagreed with the previous studies because the fragmentation in study area was related to the increased patch of such land use types as forestland, water body, orchard.

Based on the results of this study, it was conclude that land use pattern was important for ecosystem service. Therefore, we could improve the ecosystem service by the adjustment of land use policy. Continuing the Grain to Green project is reasonable and significant because it could increase the vegetation coverage and decline land use intensity. Diversified agriculture collecting planing fruit trees, growing food and breeding should be encouraged, because it could not only promote ecosystem service by increasing landscape diversification but also improve people's incomes.

Author Contributions

Conceived and designed the experiments: XF. Analyzed the data: XF. Contributed reagents/materials/analysis tools: GAT BCL. Contributed to the writing of the manuscript: XF GAT RMH.

References

1. Costanza R, Arge DR, Groot DR, Farber S, Grasso M, et al. (1997) The value of the world's ecosystem services and natural capital. Nature 387: 253−260.

2. Costanza R, Cumberland J, Daly H, Goodland R, Norgaard R (1997) An Introduction to ecological economics. Delray Beach Fla USA: St Lucie Press.

3. Kreuter UP, Harris HG, Matlock MD, Lacey RE (2001) Change in ESVs in the San Antonio area, Texas. Ecological Economics 39: 333−346.

4. Ronald CE, Yuji M (2013) Landscape pattern and ESV changes: Implications for environmental sustainability planning for the rapidly urbanizing summer capital of the Philippines. Landscape and Urban Planning 116: 60−72.

5. Li TH, Li WK, Qian ZH (2010) Variations in ESV in response to land use changes in Shenzhen. Ecological Economics 69: 1427−1435.

6. Cai YB, Zhang H, Pan WB, Chen YH, Wang XR (2013) Land use pattern, socio-economic development, and assessment of their impacts on ESV: study on

natural wetlands distribution area (NWDA) in Fuzhou city, southeastern China. Environ Monit Assess 185: 5111−5123.

7. Zaehle S, Bondeau A, Carter RT, Cramer W, Erhard M, et al. (2007) Projected changes in terrestrial carbon storage in europe under climate and land-use change, 1990–2100. Ecosystems 10: 380−401.

8. Eliska L, Jana F, Edward N, David V (2013) Past and future impacts of land use and climate change on agricultural ecosystem services in the Czech Republic. Land Use Policy, 33: 183−194.

9. Vitousek PM, Mooney HA, Lubchenco J, Melillo JM (1997) Human domination of earth's ecosystems. Science 277: 494−499.

10. Nasiri F, Huang GH (2007) Ecological viability assessment: A fuzzy multi-pleattribute analysis with respect to three classes of ordering techniques. Ecol Inform 2: 128−137.

11. Collin ML, Melloul AJ (2001) Combined land-use and environmental factors for sustainable groundwater management. Urban Water 3: 229–237.

12. Schmidta JP, Mooreb R, Alber M (2014) Integrating ecosystem services and local government finances into land use planning: A case study from coastal Georgia. Landscape and Urban Planning 122: 56–67.

13. Ernesto FV, Federico CF (2006) Land-use options for Del Plata Basin in South America: Tradeoffs analysis based on ecosystem service provision. Ecological Economics 57: 140–151.

14. Christine F, Susanne F, Anke W, Lars K, Franz M (2013) Assessment of the effects of forestland use strategies on the provision of ecosystem services at regional scale. Journal of Environmental Management 127: 96–116.

15. Ignacio P, Berta M, Pedro Z, David GDA, Carlos M (2014) Deliberative mapping of ecosystem services within and around Donana National Park (SW Spain) in relation to land use change. Reg Environ Change14: 237–251.

16. Mendoza-Gonzalez G, Martinez ML, Lithgow D, Perez-Maqueo O, Simonin P (2012) Land use change and its effects on the value of ecosystem services along the coast of the Gulf of Mexico. Ecological Economics 82: 23–32.

17. Su CH, Fu BJ (2013) Evolution of ecosystem services in the Chinese Loess Plateau under climatic and land use changes. Global and Planetary Change 101: 119–128.

18. Si J, Nasiri FZ, Han P, Li TH (2014) Variation in ESVs in response to land use changes in Zhifanggou watershed of Loess plateau: a comparative study. Environmental Systems Research 3: 2.

19. Turner MG, Gardner RH, O'Neill RV (2001) Landscape Ecology in theory and practice. New York: Springer-Verlag.

20. Ritsema CJ (2003) Introduction: soil erosion and participatory land use planning on the Loess Plateau in China. Catena 54: 1–5.

21. Fu BJ, Wang YF, Lu YH, He CS, Chen LD, et al. (2009) The effects of land-use combinations on soil erosion: a case study in the Loess Plateau of China. Progress in Physical Geography 33: 793–804.

22. Fu BJ, Chen DX, Qiu Y, Wang J, Meng QH (2002) Land Use Structure and Ecological Processes in the LoessHilly Area, China. Beijing: Commercial Press. (in Chinese).

23. Jing L, Zhiyuan R (2011) Variations in ESV in Response to Land use Changes in the Loess Plateau in Northern Shaanxi Province, China. Int. J. Environ. Res 5: 109–118.

24. Zhang TB, Tang JX, Liu DZ (2006) Feasibility of Satellite Remote Sensing Image About Spatial Resolution. Journal of Earth Sciences and Environment 28: 79–83.

25. Chu YF, Li ES, Lu J, Zhang KK (2007) The Adaptability Analysis to the Satellite Image Spatial Resolution and Mapping Scale. Hydrographic Surveying and Charting 27: 47–50.

26. Li Q, Liu C, Xi CY, Liu ML (2002) Cartographic Generalization of Digital Land Use Current Situation Map. Bulletin of Surveying and Mapping 9: 59–63.

27. Congalton RG (1991) A review of assessing the accuracy of classifications of remotely sensed data. Remote Sensing of Environment 37: 35–46.

28. Wang Y, Gao JX, Wang JS, Qiu J (2014) Value Assessment of Ecosystem Services in Nature Reserves in Ningxia, China: A Response to Ecological Restoration. PloS One 9: e89174. doi:10.1371/journal.Pone.0089174.

29. Ribeiro SC, Lovett A (2009) Associations between forest characteristics and socio-economic development: a case study from Portugal. Journal of Environmental Management 90: 2873–2881.

30. Su S, Jiang Z, Zhang Q, Zhang Y (2011) Transformation of agricultural landscapes under rapid urbanization: a threat to sustainability in Hang-Jia-Hu region, China. Applied Geography 31: 439–449.

31. Su SL, Xiao R, Jiang ZL, Zhang Y (2012) Characterizing landscape pattern and ESV changes for urbanization impacts at an eco-regional scale. Applied Geography, 34: 295–305.

32. Li H, Wu J (2004) Use and misuse of landscape indices. Landscape Ecology 19: 389–399.

33. Xie GD, Lu CX, Xiao Y, Zheng D (2003) The Economic Evaluation of Grassland Ecosystem Services in Qinghai Tibet Plateau, Journal of Mountain Science 21: 50–55. (in Chinese).

34. Xie GD, LU CX, Leng YF, Zheng D, Li SC (2008) Ecological assets valuation of Tibetan Plateau. Journal of Natural Resources 18: 190–196. (in Chinese).

35. Liu DL, Li BC, Liu Xianzhao Z, Warrington DN (2011) Monitoring land use change at a small watershed scale on the Loess Plateau, China: applications of landscape metrics, remote sensing and GIS. Environmental Earth Sciences 64: 2229–2239.

36. Pan WKY, Walsh SJ, Bilsborrow RE, Frizzelle BG, Erlien CM, et al. (2004) Farm-level models of spatial patterns of land use and land cover dynamics in the Ecuadorian Amazon. Agriculture, Ecosystems and Environment 101: 117–134.

37. de Groot RS, Wilson MA, Boumans RMJ (2002) A typology for the classification, description and valuation of ecosystem functions, goods and services. Ecological Economics 41: 393–408.

38. Kreuter UP, Harris HG, Matlock MD, Lacey RE (2001) Change in ESVs in the San Antonio area, Texas. Ecological Economics 39: 333–346.

39. Wu J, Jenerette GD, Buyantuyev A, Redman CL (2011) Quantifying spatiotemporal patterns of urbanization: the case of the two fastest growing metropolitan regions in the United States. Ecological Complexity 8: 1–8.

40. Turner RK, Paavola J, Coopera P, Farber S, Jessamya V, et al. (2003) Valuing nature: lessons learned and future research directions. Ecological Economics 46: 493–510.

41. Hein L, Koppen VK, de Groot RS, van Ierland EC (2006) Spatial scales, stakeholders and the valuation of ecosystem services. Ecological Economics 57: 209–228.

42. Konarska KM, Sutton PC, Castellon M (2002) Evaluating scale dependence of ecosystem service valuation: a comparison of NOAA-AVHRR and Landsat TM datasets. Ecological Economics 41: 491–507.

43. Limburg KE, O' Neill RV, Costanza R, Farber S (2002) Complex systems and valuation. Ecological Economics 41: 409–420.

44. Zhang MY, Wang KL, Liu HY, Zhang CH (2011) Responses of Spatial-temporal Variation of Karst Ecosystem Service Values to Landscape Pattern in Northwest of Guangxi, China. Chin. Geogra. Sci. 21: 446–453.

45. Deng L, Shangguan ZP, Li R (2012) Effects of the grain-for-green program on soil erosion in China. International Journal of Sediment Research 27: 120–127.

46. Nagendra H (2002) Opposite trends in response for the Shannon and Simpson indices of landscape diversity. Applied Geography, 22: 175–186.

47. Shrestha RP, Schmidt-Vogt D, Gnanavelrajah N (2010) Relating plant diversity to biomass and soil erosion in a cultivated landscape of the eastern seaboard region of Thailand. Applied Geography 30: 606–617.

48. Sherrouse BC, Clement JM, Semmens DJ (2011) A GIS application for assessing, mapping, and quantifying the social values of ecosystem services. Applied Geography 31: 748–760.

49. Li M, Zhu Z, Vogelmann JE, Xu D, Wen W, et al. (2011) Characterizing fragmentation of the collective forests in southern China from multitemporal Landsat imagery: a case study from Kecheng district of Zhejiang province.Applied Geography 31: 1026–1035.

50. Long JA, Nelson TA, Wulder MA (2010) Characterizing forest fragmentation: distinguishing change in composition from configuration. Applied Geography 30: 426–435.

Development and Validation of a Weather-Based Model for Predicting Infection of Loquat Fruit by *Fusicladium eriobotryae*

Elisa González-Domínguez[1]*, Josep Armengol[1], Vittorio Rossi[2]

1 Instituto Agroforestal Mediterráneo, Universidad Politécnica de Valencia, Valencia, Spain, 2 Istituto di Entomologia e Patologia vegetale, Università Cattolica del Sacro Cuore, Piacenza, Italy

Abstract

A mechanistic, dynamic model was developed to predict infection of loquat fruit by conidia of *Fusicladium eriobotryae*, the causal agent of loquat scab. The model simulates scab infection periods and their severity through the sub-processes of spore dispersal, infection, and latency (i.e., the state variables); change from one state to the following one depends on environmental conditions and on processes described by mathematical equations. Equations were developed using published data on *F. eriobotryae* mycelium growth, conidial germination, infection, and conidial dispersion pattern. The model was then validated by comparing model output with three independent data sets. The model accurately predicts the occurrence and severity of infection periods as well as the progress of loquat scab incidence on fruit (with concordance correlation coefficients >0.95). Model output agreed with expert assessment of the disease severity in seven loquat-growing seasons. Use of the model for scheduling fungicide applications in loquat orchards may help optimise scab management and reduce fungicide applications.

Editor: Erjun Ling, Institute of Plant Physiology and Ecology, China

Funding: This work was funded by Cooperativa Agrícola de Callosa d'En Sarrià (Alicante, Spain). Three months' stay of E. González-Domínguez at the Università Cattolica del Sacro Cuore (Piacenza, Italy) was supported by the Programa de Apoyo a la Investigación y Desarrollo (PAID-00-12) de la Universidad Politécnica de Valencia. The funders had no role in study design, data collection and analysis, decision to publish, or preparation of the manuscript.

Competing Interests: The authors have declared that no competing interests exist.

* Email: elgondo2@gmail.com

Introduction

Scab, caused by the plant-pathogenic fungus *Fusicladium eriobotryae* (Cavara) Sacc., is the main disease affecting loquat in Spain and in the whole Mediterranean basin [1,2]. The fungus affects young twigs, leaves and fruits, causing circular olive-colored spots that, on fruits, reduce their commercial value [1]. *Fusicladium* spp. are the anamorphic stages of the ascomycete genus *Venturia* but the sexual stage of *F. eriobotryae* has never been found in nature [2].

Although loquat scab is a well-known problem in the areas where loquat trees are cultivated, the biology of *F. eriobotryae* and the epidemiology of the disease have been seldom studied [1,3–8]. These studies have depicted *F. eriobotryae* as a highly rain-dependent pathogen that requires mild temperatures and long wet periods to infect loquat trees.

Environmental requirements for infection and the dispersion patterns have been studied in detail for other *Venturia* spp., such as *Venturia inaequalis* [9–16], *V. nashicola* [17–19], *V. pyrina* [20–24], *F. carpophilum* [25,26], *F. effusum* [27–29], and *F. oleagineum* [30–35]. These studies have been used to elaborate epidemiological models for some of these pathogens including *V. pyrina* [36], *V. nashicola*[37], *V. inaequalis* [38,39], *F. oleagineum* [40], and *F. effusum* [41]. For *V. inaequalis*, the use of epidemiological models to schedule fungicide applications has

reduced the number of treatments [42–45]. To date, no epidemiological model has been developed for *F. eriobotryae*.

Disease modelling is an important step towards the implementation of sustainable agriculture [46,47]. Since the 1990 s, modern crop production has focused on the implementation of less intensive systems with reduced inputs of fertilizers and pesticides, and reduced use of natural resources [46]. Sustainable agriculture has its roots in Integrated Pest Management (IPM) [48]. IPM concepts originated as a reaction to the disruption of agro-ecosystems caused by massive applications of broad-spectrum pesticides in the middle of the last century [46] and also because of concern about the effects of excessive pesticide use on human health [49].

In Europe, the implementation of IPM has been legislatively mandated in recent years because of Directive 2009/128/CE regarding sustainable use of pesticides. Among other actions, the Directive encourages EU Member States to promote low pesticide-input pest control and the implementation of tools for pest monitoring and decision making, as well as advisory services (Art. 14 of the Directive). De facto, the "sustainable use" directive has made IPM mandatory in European agriculture as of 2014. As a consequence, there is an increased interest in the development and use of plant disease models to improve the timing of pesticide applications and to thus limit unnecessary treatments [46,50,51].

Our aims in this paper were (i) to develop a mechanistic, dynamic model to predict infection of loquat fruit by the scab fungus *F. eriobotryae*, and (ii) to evaluate the model against three independent data sets. The model was elaborated based on the principles of "systems analysis" [52,53] and by using recent data on the biology and epidemiology of *F. eriobotryae* obtained under environmentally controlled and field conditions [1,6,7].

Model Development

Based on the available information [1,3–8], the life cycle of *F. eriobotryae* under the Mediterranean climate is described in Figure 1. The fungus oversummers in lesions on branches and leaves and on mummified fruits that remain in the tree after harvest; during summer, high temperatures and low humidity may prevent sporulation on these lesions. Under favorable conditions in the fall, the conidia produced by the oversummering lesions serve as the primary inoculum and infect young leaves or loquat fruits. Conidia are dispersed by splashing rain to nearby fruits and leaves; with suitable temperature and wetness, conidia germinate and penetrate the tissue, probably directly through the cuticle or through stomata. Once infection has occurred and if the temperature is favorable, the fungus grows under the cuticle; conidiophores then erupt through the cuticle and produce new conidia. These conidia cause secondary infections during the entire fruiting season as long as rains disperse them and as long as temperature and wetness duration permit conidial germination, infection, and lesion growth.

Model description

The relational diagram of the model for loquat fruit infection by *F. eriobotryae* is shown in Figure 2, and the acronyms are explained in Table 1. The time step of the model is 1 hour.

The model starts at fruit set and ends at harvest because fruits are assumed to be always susceptible to infection. The model considers the lesions from the previous season on branches, old leaves, and mummified fruits as the sources of primary inoculum. Because the abundance of these lesions in an orchard may vary depending on several conditions–on, for instance, the level of disease or the fungicide treatments in the previous season–and because it is difficult to quantify these lesions, the model assumes that oversummered forms are present in the orchard and that they hold conidia at fruit set and onwards.

The model considers that any measurable rain (i.e., R≥0.2 mm in 1 hour) causes dispersal and deposition of conidia on loquat fruit [7] and triggers an infection process that potentially ends with the appearance of scab symptoms. Each site on the fruit that is occupied by a conidium or conidia is considered a potential infection site and is referred to as a lesion unit (LU). During the infection process, infection on any LU can fail because conidia may fail to germinate or may germinate but then die because of unfavorable conditions. Therefore, the proportion of LUs that become scabbed at the end of the infection process may be less than that occupied by splashing conidia at the beginning of the process.

The model predicts the progress of infection on single LUs, which are the surface unit of the fruit which can become occupied by a scab lesion. This approach is related to the concept of "carrying capacity". In ecology, the carrying capacity is interpreted broadly as the maximum population size that any area of land or water can sustain [54,55]. In plant pathology, the host's carrying capacity for disease is the maximum possible number of lesions that a plant (or an organ) can hold [56]. The carrying capacity is a common concept in plant disease modeling [57–60]. In the model described here, a LU is initially healthy (LUH) but then becomes occupied by: ungerminated conidia

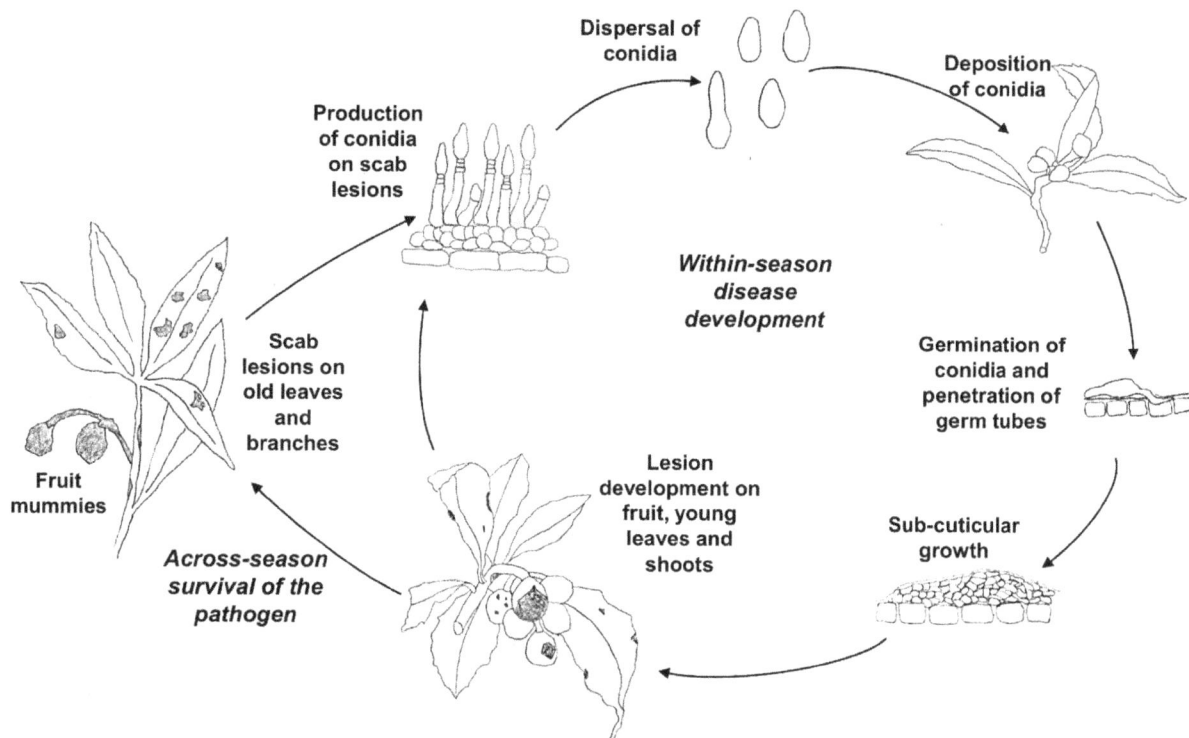

Figure 1. Disease cycle of loquat scab caused by *Fusicladium eriobotryae.*

Table 1. List of variables used in the model.

Acronym	Description	Unit
T	Air temperature	°C
RH	Relative humidity	%
R	Rainfall	mm
VPD	Vapour pressure deficit	hPa
WD	Wetness duration	hours
Teq	Temperature equivalent	°C
LUH	Unit of loquat fruit surface without conidia of *F. eriobotryae*	Number (0–1)
LUUC	Unit of loquat fruit surface with ungerminated conidia of *F. eriobotryae*	Number (0–1)
LUGC	Unit of loquat fruit surface with germinated conidia of *F. eriobotryae*	Number (0–1)
LULI	Unit of loquat fruit surface with latent infection by *F. eriobotryae*	Number (0–1)
LUVI	Unit of loquat fruit surface with visible scab lesions	Number (0–1)
GER	Cumulated conidial germination	Number (0–1)
INF	Cumulated infection	Number (0–1)
SUR	Cumulated conidial survival	Number (0–1)
GER'	Germination rate (first derivative of GERM)	Number (0–1)
INF'	Infection rate (first derivative of INF)	Number (0–1)
SUR'	Survival rate (first derivative of SUR)	Number (0–1)
C	Correction factor	Number (0–1)
DD	Degree days	Number

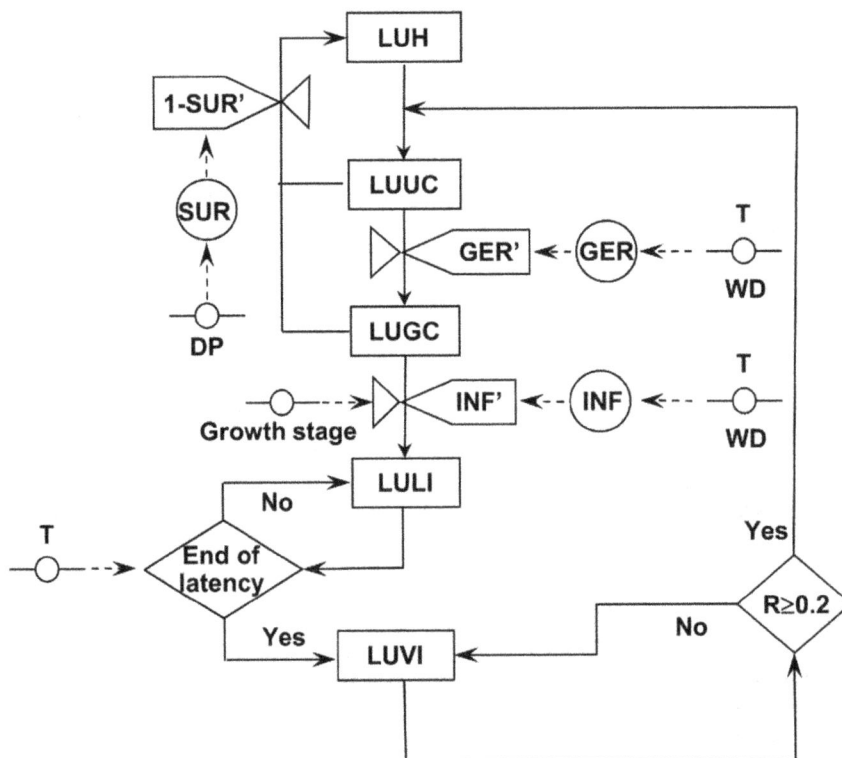

Figure 2. Relational diagram showing how the model simulates infection by *Fusicladium eriobotryae*. Legend: boxes are state variables; line arrows show fluxes and direction of changes from a state variable to the next one; valves define rates regulating these fluxes; diamonds show switches (i.e., conditions that open or close a flux); circles crossed by a line show parameters and external variables; dotted arrows show fluxes and direction of information from external variables or parameters to rates or intermediate variables; circles are intermediate variables. See Table 1 for acronym explanation.

(LUUC) at the time of conidial dispersal; germinated conidia (LUGC) at the time of conidial germination; latent infection after penetration (i.e., hyphae are invading the fruit cuticle; LULI); and visible and sporulating scab lesions at the end of latency (LUVI). Both LUUC and LUGC can fail to progress if ungerminated or germinated conidia die; these LUs then return to being LUHs because they can start a new infection process whenever new conidia are splashed on them.

At any dispersal event on hour h, the model considers that $LUUC_h = 1$. The rate at which $LUUC_h$ advances to $LUGC_h$ depends on a germination rate (GER'), and the rate at which $LUGC_h$ advances to $LULI_h$ depends on an infection rate (INF') (Figure 2). Both GER' and INF' are influenced by temperature (T in °C) and wetness duration (WD, in hours) (i.e., free water on the surface of the loquat fruit) caused by either rain or dew. Fruit surfaces are assumed to be wet on any hour when $R_h > 0$ mm, or $RH_h > 89\%$, or $VPD_h < 1$, where VPD is the vapour pressure deficit (in hPa) calculated using T_h and RH_h, following Buck [61]. The rate at which $LUUC_h$ and $LUGC_h$ returns to LUH_h depends on a survival rate (SUR'), which depends in turn on the length of the dry period (DP), i.e., the number of hours with no wetness on the fruit surface (Figure 2).

GER', INF', and SUR' are calculated at hourly intervals by using the first derivative of the equations described in González-Domínguez et al. [6] in the form:

$$GER' =$$
$$116.249 \times Teq^{4.347} \times \left(1 - Teq^{2.882}\right) \times \qquad (1)$$
$$3.215 \times e^{(-0.376 \times WD)} \times e^{\left[-8.551 \times e^{(-0.376 \times WD)}\right]}$$

$$INF' =$$
$$4.961 \times Teq^{1.700} \times \left(1 - Teq\right)^{0.771} \times \qquad (2)$$
$$0.409 \times e^{-0.087 \times WD} \times e^{\left[4.704 \times e^{(-0.087 \times WD)}\right]}$$

$$SUR' = \frac{0.165}{DP} \qquad (3)$$

where: Teq is the temperature equivalent in the form $Teq = (T - T_{\min})/(T_{\max} - T_{\min})$ where: T is the temperature regime, $T_{min} = 0°C$ and $T_{max} = 35°C$ in equation (1), and $T_{min} = 0°C$ and $T_{max} = 25°C$ in equation (2); WD = number of consecutive hours with wetness; DP = number of consecutive hours with no wetness. When DP = 0, SUR' = 1

At any time of the infection progress (i):

$$LUGC_h = \sum_{i=1}^{i=t} GER'_i \times \left(1 - \sum_{i=1}^{i=t} SUR'_i\right) \times C_i$$

$$LUUC_h = \sum_{i=1}^{i=t} (1 - LUGC_h)_i \times \left(1 - \sum_{i=1}^{i=t} SUR'_i\right) \times C_i$$

$$LULI_h = \sum_{i=1}^{i=t} (INF'_i)$$

$$LUGC_h + LUUC_h + LULI_h + LUH_h = 1$$

where C is a correction factor $(C = 1 - LULI_h)$.

Any infection period triggered by a conidial dispersal event ends when no viable conidia are present on any LUs, exactly when $LUUC \leq 0.01$. An example of model output for a single infection period is shown in Figure 3.

The model considers that any further rain event causes further dispersal and deposition of conidia if >5 hours have passed after the previous dispersal event. This is the time required by a lesion to produce new conidia.

Model output

The model output consists of: (i) the available inoculum on fruits (i.e., the frequency of LUs with ungerminated conidia on each day) as a measure of the potential for infection to occur; (ii) the dynamics of LULI for each infection process; and (iii) the seasonal dynamics of the accumulated values of LULI (ΣLULI) as an estimate of the disease in the orchard.

Examples of model output for the 2011 and 2012 loquat growing seasons are shown in Figures 4 and 5, respectively. The output is based on the weather data registered by a weather station of the Regional Agrometerological Service (http://riegos.ivia.es/) located in Callosa d'En Sarrià, Alicante Province, southeastern Spain.

Model Validation

Three data sets were used to validate the model: (i) incidence of affected fruits in a loquat orchard during growing seasons 2011 and 2012; (ii) disease occurrence on loquat fruits in single-exposure experiments in 2013; and (iii) expert assessment of the disease severity in seven loquat growing seasons.

To operate the model, hourly values of air temperature (T, °C), relative humidity (RH, %), and total rainfall (R, mm) were registered by the weather station of Callosa d'En Sarrià, which is ≤3.5 km from the orchards considered for validation.

Predicted vs. observed disease incidence in orchards

In data set (i), observations were carried out in a loquat orchard in Callosa d'En Sarrià, Alicante Province, southeastern Spain. Details on these data have been previously published [7]. Briefly, fruits from four shoots of each of 46 loquat trees were assessed weekly, and disease incidence was expressed as the percentage of fruits with scab symptoms. The disease incidence was lower in 2011 than 2012, with 27.3% and 97.6% of fruits affected by loquat scab at harvest, respectively [7]. This difference in disease incidence may be related to the fact that the orchard was treated with fungicides for scab control in 2010 but not in 2011 or 2012. Given that the inoculum sources for fruit infection in 2011 was very low because of effective disease control in 2010, a correction factor for LUUC was applied for the infection processes initiated in January 2011, i.e., LUUC = 0.1 instead of = 1 in January 2011.

Model validation was performed by comparing ΣLULI with observed data of disease incidence. Because there is a time lag (i.e., a latency period) between the predicted disease (as ΣLULI) and the disease incidence estimated in the orchard (DI), DI was shifted back by one latency period for comparison between predicted and observed disease. Sanchez-Torres et al. [1] observed a latency period of 21 days at a constant temperature of 20°C, which is a degree-day accumulation (DD base 0°C) of 420. Therefore, DI was shifted back by either 21 days or 420 DD. To calculate the

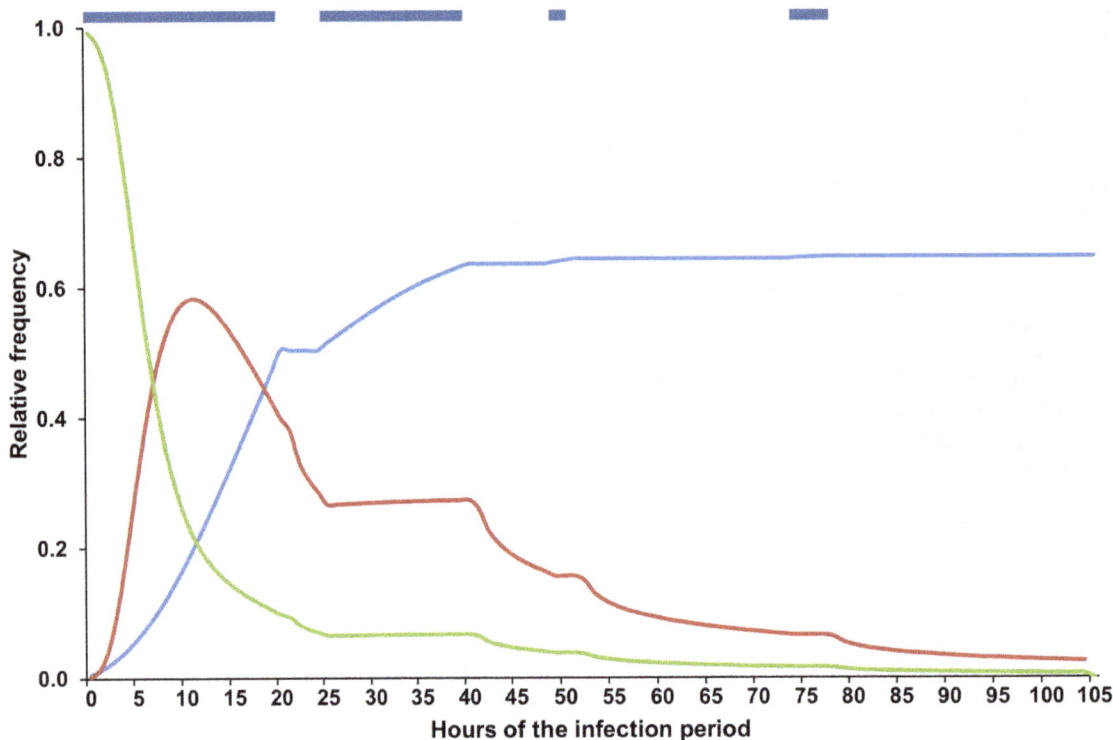

Figure 3. Dynamics of lesion units (LUs) during an infection period of *Fusicladium eriobotryae.* The graph shows the relative frequency of LUs occupied by ungerminated conidia (LUUC, in green), germinated conidia (LUGC, in red), and latent infections (LULI in blue). Blue bars at the top indicate hours with free water on the fruit surface. An infection period starts when a rain event splashes conidia on LUs and ends when no viable conidia are present on any LUs, i.e., when LUUC≤0.01.

DD, the average temperature of each day was considered with base temperature of 0°C.

Predicted vs. observed disease incidence in single-exposure experiments

In data set (ii), data were collected in an abandoned loquat orchard in Callosa d'En Sarrià from 4 February to 15 April 2013. On 25 January, 200 random shoots bearing fruits were covered with water-resistant paper bags (one shoot per bag) to prevent deposition of rain-splashed conidia. On 4 February, 10 random bags were opened to receive splashed inoculum; after seven additional days, the bags were closed again. Ten other randomly selected bags were opened on 11 February and closed again 7 days later. This operation was repeated until nine groups of shoots had been sequentially exposed to rain. At the end of the experiment (15 April 2013), disease incidence (percentage of fruits affected by loquat scab) and severity were assessed in each group of shoots. Disease severity refers to the percentage of fruit area covered by scab lesions and was measured as described by González-Domínguez et al. [8].

Model validation was performed by comparing the model output in the week when a group of shoots was exposed to splashing rain with final disease severity in that group.

Expert assessment

For data set (iii), Esteve Soler (technical advisor of the 'Cooperativa Agricola de Callosa d'En Sarrià') was asked to provide a subjective estimate of the severity (low, medium, or high) of loquat scab in the area for eight growing seasons (from 2005/2006 to 2012/2013). Mr. Soler's estimates were based on his

extensive experience in managing loquat orchards, on his scouting activities in the orchards of the cooperative, and on the number of fungicide treatments that were required to control the disease in the area.

For each season, the model was operated from 1 November to 31 March, and the numbers of disease outbreaks predicted by the model were counted. A disease outbreak was defined as $\Sigma LULI > 0.1$ in 1 day, when no outbreaks were predicted in the previous 5 days. Average and standard error of the number of predicted outbreaks were calculated for each category (low, medium, or high) of scab severity derived from the expert assessment.

Data analysis

Linear regression was used to compare the predicted and observed data of data sets (i) and (ii). To make data homogeneous, $\Sigma LULI$ values at the time of each disease assessment in the orchards were rescaled to the $\Sigma LULI$ at the end of the season; disease incidence was also rescaled to the final disease incidence. A t-test was used to test the null hypotheses that "a" (intercept of regression line) was equal to 0 and that "b" (slope of regression line) was equal to 1 [62]. The distribution of residuals of predicted versus observed values was examined to evaluate the goodness-of-fit. The concordance correlation coefficient (CCC) was calculated as a measure of model accuracy [63]; CCC is the product of two terms: the Pearson product-moment correlation coefficient between observed and predicted values and the coefficient Cb (bias estimation factor), which is an indication of the difference between the best fitting line and the perfect agreement line (CCC = 1 indicates perfect agreement). The following indexes of goodness-of-fit were also calculated [64]: NS model-efficacy

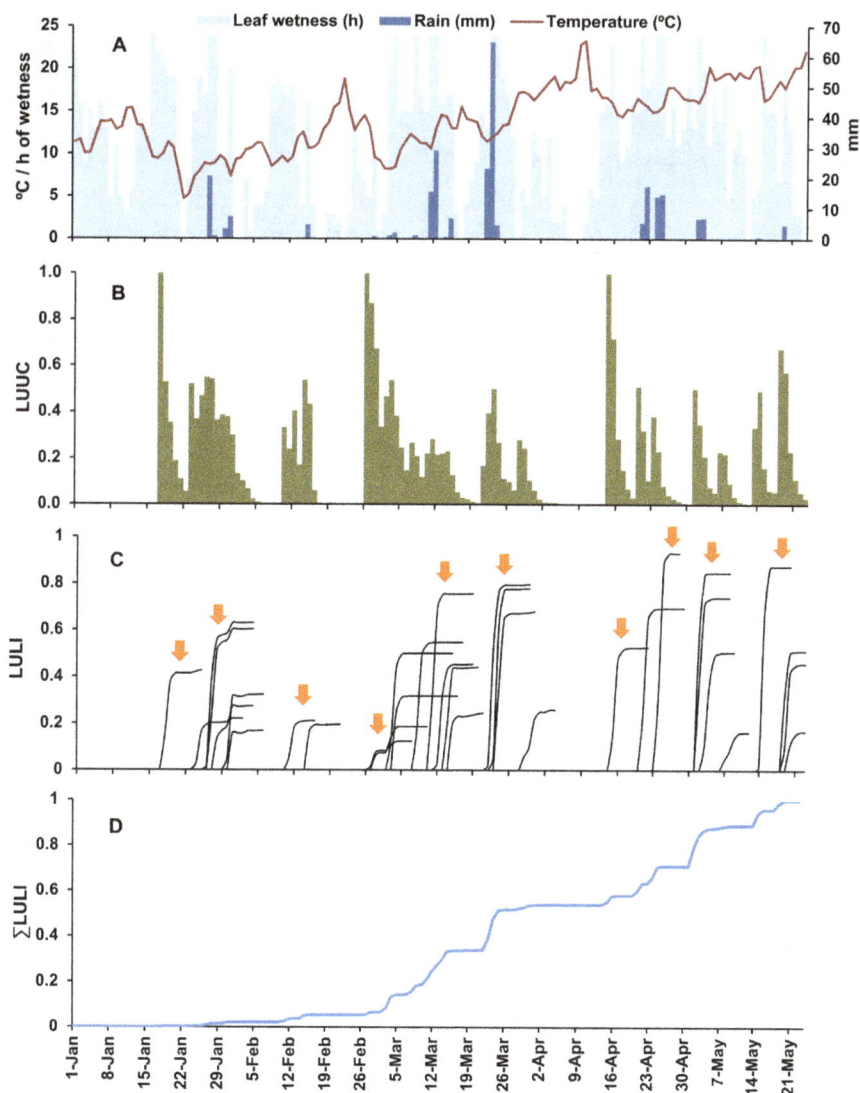

Figure 4. Weather data and model output in 2011. A: daily weather data; B: predicted frequency (%) of lesion units (LUs) with ungerminated conidia; C: predicted increase of LUs with latent infections (LULIs) for each infection period (arrows represent clusters of infection periods, clustering is based on an interval of at least 5 days between the beginning of two consecutive clusters); D: predicted seasonal dynamics of the cumulative values of LULI (ΣLULI).

coefficient, which is the ratio of the mean square error to the variance in the observed data, subtracted from unity (when the error is zero, NS = 1, and the provides a perfect fit); the W index of agreement which is the ratio between mean square error and total potential error (W = 1 represents a perfect fit); model efficiency (EF) which is a dimensionless coefficient that takes into account both the index of disagreement and the variance of the observed values (when EF increases toward 1, the fit increases); and the coefficient of residual mass (CRM) which is a measure of the tendency of the to overestimate or underestimate the observed values (a negative CRM indicates a tendency of the model toward overestimation).

For data set (iii), a one-way analysis of variance (ANOVA) was performed to determine whether the numbers of outbreaks predicted by the model in each category of loquat scab severity defined by the expert (i.e., low, medium, or high) were significantly different from one another.

Results of Model Validation

Predicted vs. observed disease incidence in orchards

In 2011 between 1 January (fruit set) and 23 May (harvest), 257.6 mm of rain fell, distributed in three main periods: the last week of January, the second week of March (with 64.8 mm of rain in 1 day), and the last 2 weeks of April (with daily temperature > 15°C) (Fig. 4A). According to the model, a total of 33 infection periods were triggered by these rain events, and the first was on 17 January (Fig. 4B and 4C). In the analysis of this model output, infection periods were clustered in "infection clusters" based on an interval of a minimum of 5 days elapsed between the beginning of two consecutive infection clusters (i.e., the protection provided by a copper-based fungicide application as described in [65]); therefore, there were 10 infection clusters in the considered period. ΣLULI began to increase from mid-January to mid-February (with three infection clusters), but three infection clusters in March resulted in a substantial increase in ΣLULI to 0.5;

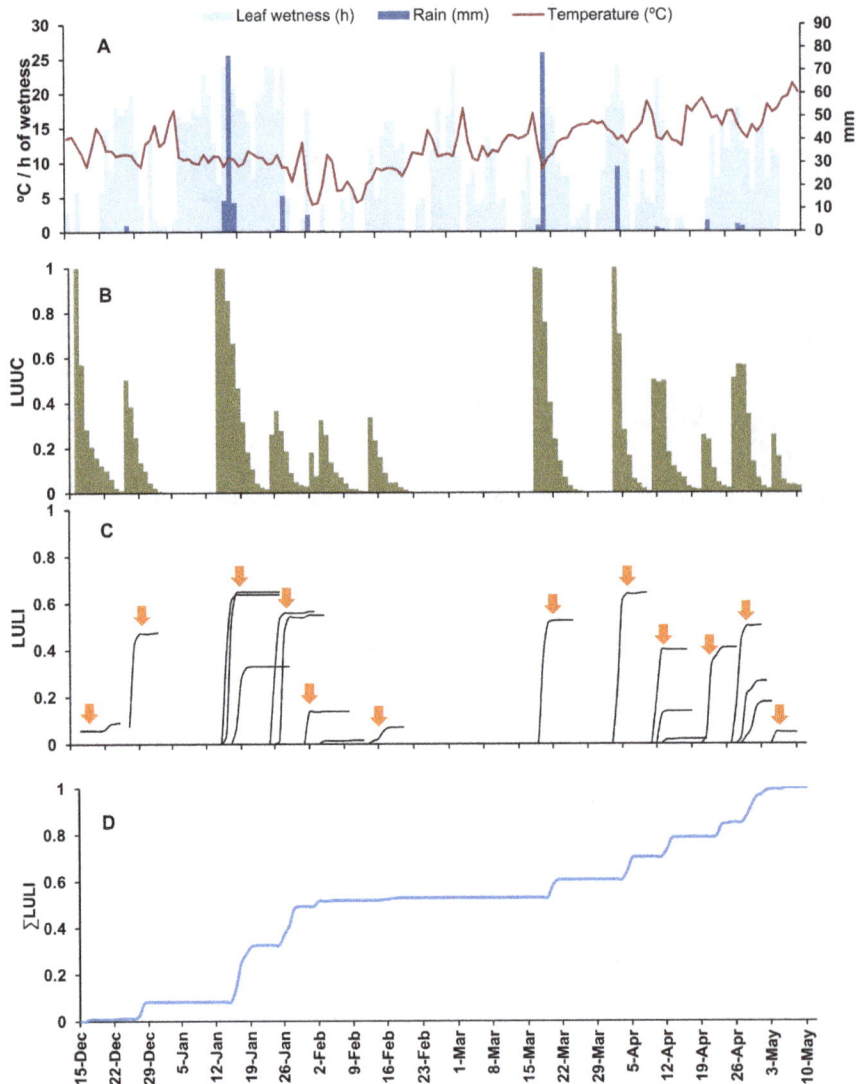

Figure 5. Weather data and model output in 2012. A: daily weather data; B: predicted frequency (%) of lesion units (LUs) with ungerminated conidia; C: predicted increase of LUs with latent infections (LULIs) for each infection period (arrows represent clusters of infection periods, clustering is based on an interval of at least 5 days between the beginning of two consecutive clusters); D: predicted seasonal dynamics of the cumulative values of LULI (ΣLULI).

March had 12 infections periods, and the repeated and abundant rain events provided >18 h of wetness on most days (Fig. 4). From mid-April to the end of the considered period, a constant increase in ΣLULI was associated with abundant rain events and increasing temperature, which triggered four infection clusters (Fig. 4).

In 2012, although the total volume of rain that fell from 15 December to 10 May was similar (255.8 mm) to that in 2011, there were fewer rain events. The model predicted 20 infection periods that were grouped into 12 infection clusters (Fig. 5B and 5C). In 2012, rainy periods were separated by dry periods; from the end of January to mid-March, dry periods caused no substantial infection to develop (Fig. 5C). Therefore, there were two main periods of ΣLULI increase: the last half of January and from the end of March to May (Fig. 5D).

Goodness-of-fit of predicted (ΣLULI) versus observed data (loquat scab incidence) was greater when a fixed period of 21 days was considered for the latency. In this case, values of R^2, CCC, r,

Cb, NS, W, and EF were >0.95 (Table 2). However, when a latency period of 420 DD was considered the values of R^2 and CCC were <0.88, and values of model efficacy (NS) and model efficiency (EF) were 0.75 (Table 2) as a consequence of the high dispersion of residues in 2012 (Figure 6). The model slightly overestimated scab incidence when a latency of 21 days was used (CRM = −0.009) and underestimated scab incidence when DD were used (CRM = 0.182) (Table 2; Figure 6). For both latency options, the regression equations of predicted versus observed data had slopes and intercepts that were not significantly different from 1 and 0, respectively.

Predicted vs. observed disease incidence in single-exposure experiments

From 4 February to 15 April 2013, the model predicted 15 loquat scab infection periods but disease outbreaks were substantial (i.e., they resulted in a >10% increase in severity) in only two exposure periods. In these two cases, LULI values were >0.1;

Table 2. Statistics and indices used for evaluating the goodness-of-fit of loquat scab infection predicted by the model versus disease observed in field.

Data set[a]	A[b]	b	P(a=0)	P(b=1)	R²	CCC	r	Cb	NS	W	EF	CRM
Data set 1 (latency = 21 days)	0.038	0.939	0.190	0.070	0.952	0.974(0.946–0.988)	0.975	0.999	0.951	0.987	0.951	−0.009
Data set 1 (latency = 420 DD)	−0.066	0.928	0.274	0.110	0.841	0.882(0.758–0.944)	0.921	0.960	0.753	0.939	0.752	0.182
Data set 2	0.043	0.965	0.02	0.1	0.984	0.986(0.942–0.996)	0.993	0.993	0.971	0.992	0.971	−0.247

[a]Data set 1 corresponds to comparison of daily accumulated LUVI predicted by the model versus observed data of loquat scab incidence in an orchard in southeastern Spain during 2 years (2011 and 2012). The model used a latency period of 21 days (first row) or 420 DD (second row). Data set 2 compares the increase of model output in weeks in which loquat shoots were exposed to splashing rain (triggering infection) with final disease severity in those shoots.

[b]a and b, parameters of the regression line of the predicted against observed values; P, probability level for the null hypotheses that a = 0 and b = 1; R², coefficient of determination of the regression line; CCC, concordance correlation coefficient; r, Pearson product-moment correlation coefficient; Cb, bias estimation factor; NS, model efficacy; W, index of agreement; EF, model efficiency; CRM, coefficient of residual mass.

when there were no or light outbreaks, *LULI* values were <0.06 (Figure 7). The goodness-of-fit of predicted versus observed for data set (ii) (Table 2) provided values >0.97 for R^2, CCC, r, Cb, NS, W, and EF. Although the slope was not significantly different from 1, the intercept was different from 0 at $P = 0.02$ (Table 2). The negative value of CRM indicated that the model somewhat overestimated disease, mainly when observed disease severity was low (Figure 7).

Expert assessment

The loquat scab epidemics that occurred in the eight seasons of data set (iii) were considered by the expert to be of low, medium, or high severity in two, three, and three seasons, respectively. The number of outbreaks predicted by the model ranged from 4 to 17 among the eight seasons; in average, 8.5 ± 0.5 outbreaks were predicted for years with low value of loquat scab severity, 10 ± 3 for year with medium value and 12 ± 2.9 for years with high value. Although the average number of outbreaks predicted by the model increased as the expert assessment of disease severity increased, the number of predicted epidemics did not significantly differ among the severity categories ($P = 0.71$).

Discussion

In this work, a dynamic model was developed to predict infection of loquat fruits by conidia of *F. eriobotryae*. The model uses a mechanistic approach to describe the infection process [53,66,67]: the model splits the disease cycle of *F. eriobotryae* into different state variables, which change from one state to the following state based on rate variables or switches that depend on environmental conditions by means of mathematical equations. The mathematical equations were developed using published data on *F. eriobotryae* conidial dispersion patterns [7] and on *F. eriobotryae* growth, conidial germination, and infection under different environmental conditions [1,6]. In the absence of precise information, assumptions were made based on available knowledge.

Model validation showed that the model correctly predicted the occurrence of infection periods and the severity of any infection period, as demonstrated by the goodness-of-fit for the data collected on fruits exposed to single rainy periods. Because the purpose of the model is to be part of a warning system for loquat scab management, the ability to correctly predict infection periods is crucial. Accuracy of the model was also confirmed by the comparison of model output with expert assessment. Even though the numbers of predicted outbreaks did not differ among seasons that the expert had categorized as having low, medium, or high disease severity, the number of predicted outbreaks increased with increases in assessed disease severity.

For model validation, the latency period required for the appearance of scab was expressed as a fixed number of days or of degree-days (DD) based on results from Sánchez-Torres et al. [1]. Goodness-of-fit of model prediction was overall better using a fixed period of 21 days instead of 420 DD. In particular, the model underestimated the disease in the early season of 2012 when DD were used. The underestimation was probably caused by low temperatures in that period, which delayed DD accumulation. This result is questionable, because the physiological development of fungi is usually more closely related to DD than to calendar days [68,69]. In this work, DD was fixed based on the latency period observed in loquat plants kept at the optimal temperature for *F. eriobotryae* development, i.e., 21 days at 20°C [1]. Therefore, the DD value used in this study did not account for the non-linear response of *F. eriobotryae* growth to temperatures between 5 and

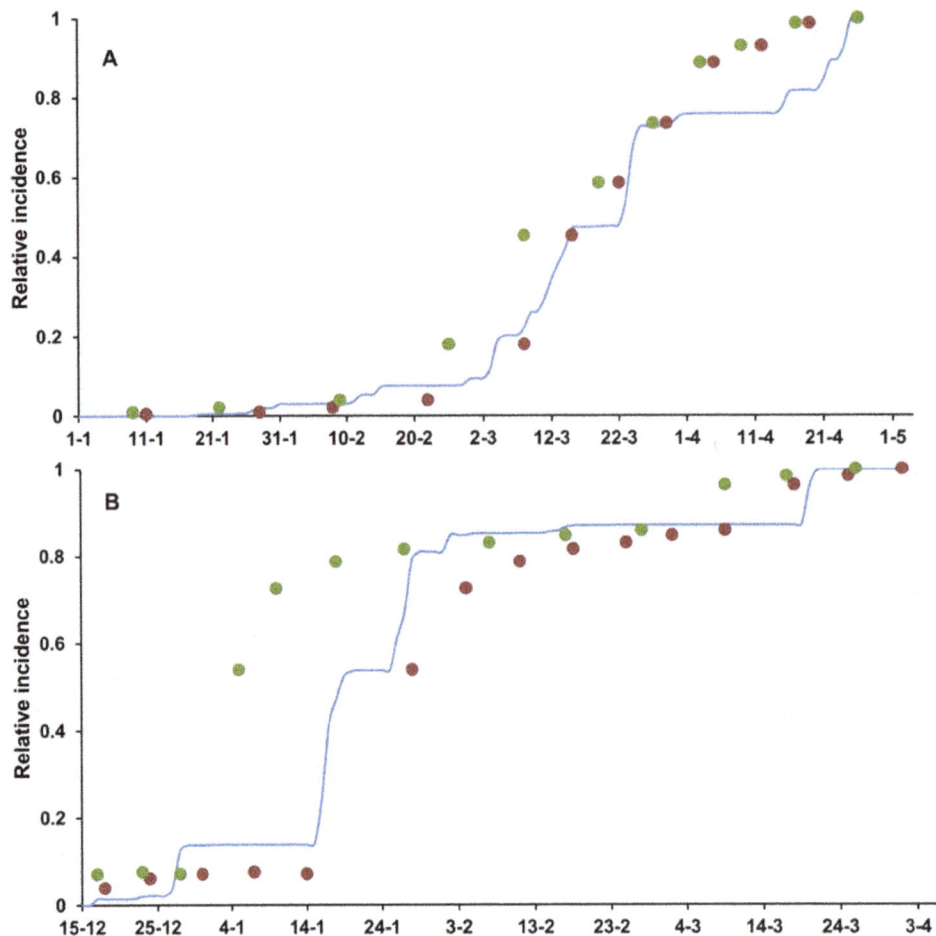

Figure 6. Comparison between model output and scab observed on loquat fruit in southeastern Spain. (A) data from 2011 and (B) data from 2012. Blue lines represent the rescaled infection predicted by the model as the seasonal summation of the lesion units with latent infections (ΣLULI). Points represent rescaled incidence of loquat fruit with scab observed in the orchards; rescaled incidence is shifted back by 21 days (red points) or 420 DD (base 0°C, green points) to account for the latency period, i.e., the time elapsed between infection and visible symptoms in the form of sporulating scab lesions.

30°C [6]. If a function for predicting the appearance of scab symptoms is needed in the model, such a function should be temperature dependent, as it is in models for *V. nashicola* [37] and *F. oleagineum* [40]. Salerno et al. [4] repeatedly exposed potted loquat plants under the canopies of affected trees for 3 days and then incubated these plants under a roof until the appearance of symptoms. Scab appeared in 11 to 26 days at temperatures ranging from 11.4 to 17°C (with a DD range of 157 to 340) and after >220 days at temperatures >20°C. Ptskialadze [5] found scab symptoms on both leaves and fruits 34 and 16 days after infection at 1–4°C and 21–25°C, respectively. Even though the calculation of latency can be improved, the model error in predicting disease onset due to a fix latency period may not reduce the ability of the model to correctly predict infection periods or reduce the value of the model for timing fungicides applications.

The model capitalized on recent research concerning loquat scab [1,6,7]. These studies have considered most of the components of the disease cycle, including dispersion of conidia, infection, incubation, and latency. Nevertheless, other components should be elucidated to improve our knowledge and thus to improve the model [66]. Currently, the model assumes that inoculum sources are always present in scab-affected loquat orchards and that viable *F. eriobotryae* inoculum is always present

at fruit set (i.e., when the model begins operating) and beyond. Salerno et al. [4] found that lesions appear in autumn on leaves that were infected the previous spring, and Prota [3] found that the lesions appearing in autumn produce conidia for 5 to 6 months and that those viable conidia are present all year long. These observations were carried out in Sicily and Sardinia, respectively (i.e., under a Mediterranean climate); therefore, the model assumptions seem plausible. The assumptions that inoculum sources and viable conidia are always present in scab-affected loquat orchards are both precautionary because they can lead to over prediction of infection (which would occur if weather conditions were suitable for infection but no viable conidia were available) and thus to unnecessary applications of fungicides or other disease management measures. Because unnecessary fungicide applications entail costs for growers, consumers, and the environment [51], the model should be expanded to include the oversummering and availability of conidia.

With respect to oversummering, modeling the dormant stage of fungal pathogens is challenging [66], and the dormant stage has therefore been included in only a few models [70–74]. For this purpose, two key aspects must be addressed: (i) the inoculum dose (i.e., the quantity of inoculum that oversummers), which depends on the severity of the disease in each orchard at the end of the

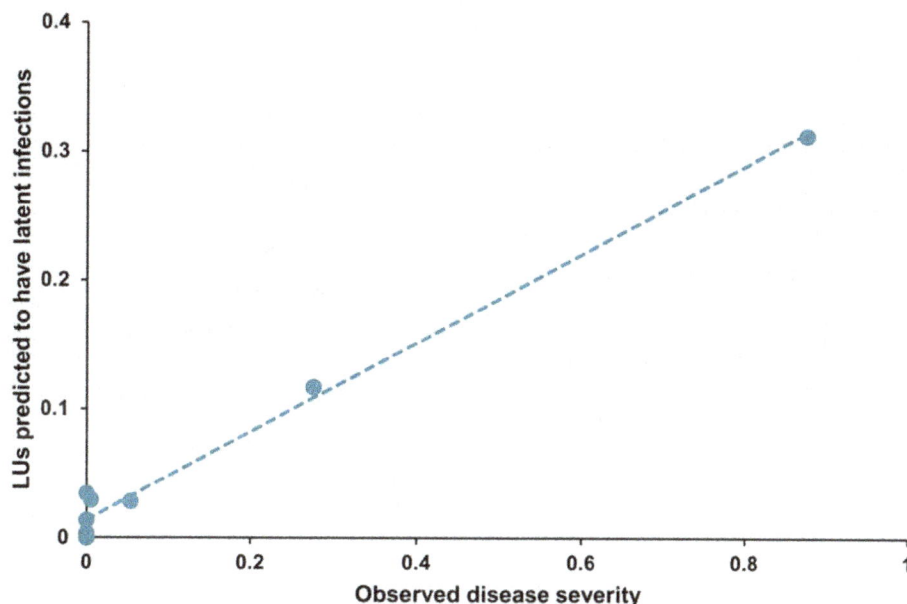

Figure 7. Comparison between model output and scab on loquat fruit in single-exposure experiments. Experiments were carried out in a loquat orchard in southeastern Spain in 2013. Observed data (X axis) are expressed as the rescaled disease severity in 11 groups of fruits that were exposed (for 7-day-long moving periods) to splashing rain in a severely affected orchard; model output (Y axis) is expressed as the summation of the lesion units with latent infections (ΣLULI) in the exposure period.

previous season; and (ii) the time when the primary inoculum begins to be available for infection. In other models, the inoculum dose was directly measured in the field [70,74] or broadly estimated as low/high disease pressure [72]. In our case, incorporation into the model of the specific farmer's assessment of the disease severity in the previous season may represent useful information regarding the potential primary inoculum dose. Modeling the sporulation patterns of *F. eriobotryae* may make it possible to estimate the available inoculum at each infection period. This estimation may consequently improve the ability of the model to predict the severity of each infection period. To account for the presence of inoculum in a model for *V. inaequalis*, Xu et al. [39] assumed a minimum interval of 7 hours between two successive infection processes to allow lesions to recover and sporulate, even though this approximation could introduce errors, because sporulation is highly dependent on temperature and RH [14].

Even without the above possible improvements, the present model can contribute to the practical control of loquat scab. The underutilization of disease predictive systems by farmers have been broadly discussed [46,47,51,75,76]. Rossi et al. [53] summarized the steps necessary for the practical implementation of a model as: (i) develop a computerized version of the model; (ii) create a network of agro-meteorological stations for collecting weather data; (iii) design a strategy for decision-making based on the model output; (iv) develop tools for supporting decision-making (e.g., decision support systems or disease warning systems); and (v) build user's confidence in the model by demonstrating the advantages of its use in comparison with the current options. Efforts devoted to the last three steps are crucial for the future applicability of the model [53] and requires a deep knowledge of the cultural context in which the model will be delivered, the farmers' perception of risk, and the current management of the disease [47].

In the main loquat cultivation areas of Spain, the regional plant protection services use the Mills-Laplante tables [77], which were developed to control apple scab, to estimate the risk of infection by *F. eriobotryae* [78]. Researchers have indicated that the Mills-Laplante tables over-predict the number of infections for apple scab [37,79]. That the tables could over-predict the number of loquat scab infections has also been discussed, because the conidia of *F. eriobotryae* require longer times for leaf infection than those described by the Mills-Laplante tables for *V. inaequalis* and because the temperature range in which *F. eriobotryae* infection occurs is quite different [6,7]. Thus, the present model represents an improvement in loquat scab management, i.e., it should optimise scab management by helping loquat growers to schedule and probably to reduce fungicide applications.

The long-term existence of a warning system for loquat scab monitoring in Spain [80] may facilitate the implementation of the model developed in this area because i) extension agents and advisors are familiar with the use and interpretation of epidemiological models, and ii) loquat farmers are accustomed to considering the concept of "infection risk" when scheduling fungicide applications.

Because model building is "a never-ending story" [47,81], researchers will likely continue to improve the loquat scab model described here. As discussed in this manuscript, it will be necessary to define a relationship between model output and infection severity so as to identify appropiate thresholds for deciding when the treatments are needed [53,63].

Acknowledgments

We would like to thank E. Soler his contribution for model validation.

Author Contributions

Conceived and designed the experiments: EGD VR JA. Performed the experiments: EGD VR. Analyzed the data: EGD VR. Contributed reagents/materials/analysis tools: JA VR. Contributed to the writing of the manuscript: EGD JA VR.

References

1. Sánchez-Torres P, Hinarejos R, Tuset JJ (2009) Characterization and pathogenicity of *Fusicladium eriobotryae*, the fungal pathogen responsible for loquat scab. Plant Dis 93: 1151–1157.

2. Gladieux P, Caffier V, Devaux M, Le Cam B (2010) Host-specific differentiation among populations of *Venturia inaequalis* causing scab on apple, pyracantha and loquat. Fungal Genet Biol 47: 511–521.

3. Prota U (1960) Ricerche sulla «ticchiolatura del Nespolo del Giappone e sul suo agente (*Fusicladium eriobotryae* Cav.). I. Osservazioni sull'epidemiologia della malattia e sui caratteri morfo-biologici del parassita in Sardegna. Stud di Sassari 8: 175–196.

4. Salerno M, Somma V, Rosciglione B (1971) Ricerche sull'epidemiologia della ticchiolatura del nespolo del Giappone. Tec Agric 23: 3–15.

5. Ptskialadze L (1968) The causal agent of loquat scab and its biological characteristics. Rev Appl Mycol 47: 268.

6. González-Domínguez E, Rossi V, Armengol J, García-Jiménez J (2013) Effect of environmental factors on mycelial growth and conidial germination of *Fusicladium eriobotryae*, and the infection of loquat leaves. Plant Dis 97: 1331–1338.

7. González-Domínguez E, Rossi V, Michereff SJ, García-Jiménez J, Armengol J (2014) Dispersal of conidia of *Fusicladium eriobotryae* and spatial patterns of scab in loquat orchards in Spain. Eur J Plant Pathol 139: 849–861.

8. González-Domínguez E, Martins RB, Del Ponte EM, Michereff SJ, García-Jiménez J, et al. (2014) Development and validation of a standard area diagram set to aid assessment of severity of loquat scab on fruit. Eur J Plant Pathol 139: 413–422.

9. Becker CM, Burr TJ (1994) Discontinuous wetting and survival of conidia of *Venturia inaequalis* on apple leaves. Phytopathology 84: 372–378.

10. Hartman JRR, Parisi L, Bautrais P (1999) Effect of leaf wetness duration, temperature, and conidial inoculum dose on apple scab infections. Plant Dis 83: 531–534.

11. Holb IJ, Heijne B, Withagen JCM, Jeger MJ (2004) Dispersal of *Venturia inaequalis* ascospores and disease gradients from a defined inoculum source. J Phytopathol 152: 639–646.

12. Rossi V, Giosue S, Bugiani R (2003) Influence of air temperature on the release of ascospores of *Venturia inaequalis*. J Phytopathol 151: 50–58.

13. Stensvand A, Gadoury DM, Amundsen T, Semb L, Seem RC (1997) Ascospore release and infection of apple leaves by conidia and ascospores of *Venturia inaequalis* at low temperatures. Phytopathology 87: 1046–1053.

14. Machardy WE (1996) Apple scab. Biology, epidemiology and management. St. Paul: APS Press. 545.

15. James J, Sutton TB (1982) Environmental factors influencing pseudothecial development and ascospore maturaion of *Venturia inaequalis*. Phytopathology 72: 1073–1080.

16. Boric B (1985) Influence of temperature on germability of spores of *Venturia inaequalis* (Cooke) Winter, and their viability as affected by age. Zast Bilja 36: 295–302.

17. Li B, Zhao H, Xu X-M (2003) Effects of temperature, relative humidity and duration of wetness period on germination and infection by conidia of the pear scab pathogen (*Venturia nashicola*). Plant Pathol 52: 546–552.

18. Li BH, Xu XM, Li JT, Li BD (2005) Effects of temperature and continuous and interrupted wetness on the infection of pear leaves by conidia of *Venturia nashicola*. Plant Pathol 54: 357–363.

19. Umemoto S (1990) Dispersion of ascospores and conidia of causal fungus of japanese pear scab, *Venturia nashicola*. Ann Phytopathol Japan Soc 56: 468–473.

20. Rossi V, Salinari F, Pattori E, Giosuè S, Bugiani R (2009) Predicting the dynamics of ascospore maturation of *Venturia pirina* based on environmental factors. Phytopathology 99: 453–461.

21. Spotts RA, Cervantes A (1991) Effect of temperature and wetness on infection of pear by *Venturia pirina* and the relationship between preharvest inoculation and storage scab. Plant Dis 75: 1204–1207.

22. Spotts RA, Cervantes A, Cervantes LA (1994) Factors affecting maturation and release of ascospores of *Venturia pirina* in oregon. Phytopathology 84: 260–264.

23. Villalta O, Washington WS, Rimmington GM, Taylor PA (2000) Influence of spore dose and interrupted wet periods on the development of pear scab caused by *Venturia pirina* on pear (*Pyrus communis*) seedlings. Australas Plant Pathol 29: 255–262.

24. Villalta ON, Washington WS, Rimmington GM, Taylor PA (2000) Effects of temperature and leaf wetness duration on infection of pear leaves by *Venturia pirina*. Aust J Agric Res 51: 97–106.

25. Lan Z, Scherm H (2003) Moisture sources in relation to conidial dissemination and infection by *Cladosporium carpophilum* within peach canopies. Phytopathology 93: 1581–1586.

26. Lawrence E, Zehr E (1982) Enviromental effects on the development and dissemination of *Cladosporium carpophilum* on peach. Phytopathology 72: 773–776.

27. Gottwald TR, Bertrand PF (1982) Patterns of diurnal and seasonal airborne spore concentration of *Fusicladium effusum* and its impact on a pecan scab epidemic. Phytopathology 72: 330–335.

28. Gottwald TR (1985) Influence of temperature, leaf wetness period, leaf age, and spore concentration on infection of pecan leaves by conidia of *Cladosporium caryigenum*. Phytopathology 75: 190–194.

29. Latham AJ (1982) Effect of some weather factors and *Fusicladium effusum* conidium dispersal on pecan scab occurrence. Phytopathology 72: 1339–1345.

30. De Marzo L, Frisullo S, Lops F, Rossi V (1993) Possible dissemination of *Spilocaea oleagina* conidia by insects (*Ectopsocus briggsi*). EPPO Bull 23: 389–391.

31. Lops F, Frisullo S, Rossi V (1993) Studies on the spread of the olive scab pathogen, *Spilocaea oleagina*. EPPO Bull 23: 385–387.

32. Obanor FO, Walter M, Jones EE, Jaspers MV (2008) Effect of temperature, relative humidity, leaf wetness and leaf age on *Spilocaea oleagina* conidium germination on olive leaves. Eur J Plant Pathol 120: 211–222.

33. Obanor FO, Walter M, Jones EE, Jaspers MV (2010) Effects of temperature, inoculum concentration, leaf age, and continuous and interrupted wetness on infection of olive plants by *Spilocaea oleagina*. Plant Pathol 60: 190–199.

34. Viruega JR, Moral J, Roca LF, Navarro N, Trapero A (2013) *Spilocaea oleagina* in olive groves of southern Spain: survival, inoculum production, and dispersal. Plant Dis 97: 1549–1556.

35. Viruega JR, Roca LF, Moral J, Trapero A (2011) Factors affecting infection and disease development on olive leaves inoculated with *Fusicladium oleagineum*. Plant Dis 95: 1139–1146.

36. Eikemo H, Gadoury DM, Spotts RA, Villalta O, Creemers P, et al. (2011) Evaluation of six models to estimate ascospore maturation in *Venturia pyrina*. Plant Dis 95: 279–284.

37. Li B, Yang J, Dong X, Li B, Xu X (2007) A dynamic model forecasting infection of pear leaves by conidia of *Venturia nashicola* and its evaluation in unsprayed orchards. Eur J Plant Pathol 118: 227–238.

38. Rossi V, Bugiani R (2007) A-scab (Apple-scab), a simulation model for estimating risk of *Venturia inaequalis* primary infections. EPPO Bull 37: 300–308.

39. Xu X, Butt DJ, Santen VAN (1995) A dynamic model simulating infection of apple leaves by *Venturia inaequalis*. Plant Pathol 44: 865–876.

40. Roubal C, Regis S, Nicot PC (2013) Field models for the prediction of leaf infection and latent period of *Fusicladium oleagineum* on olive based on rain, temperature and relative humidity. Plant Pathol 62: 657–666.

41. Payne AF, Smith DL (2012) Development and evaluation of two pecan scab prediction models. Plant Dis 96: 1358–1364.

42. Trapman M, Jansonius PJ (2008) Disease management in organic apple orchards is more than applying the right product at the correct time. Ecofruit-13th International Conference on Cultivation Technique and Phytopathological Problems in Organic Fruit-Growing: Proceedings to the Conference from 18th February to 20th February 2008 at Weinsberg/Germany. 16–22.

43. Holb IJ, Jong PF, Heijne B (2003) Efficacy and phytotoxicity of lime sulphur in organic apple production. Ann Appl Biol 142: 225–233.

44. Jamar L, Cavelier M, Lateur M (2010) Primary scab control using a "during-infection" spray timing and the effect on fruit quality and yield in organic apple production. 14: 423–439.

45. Giosuè S, Bugiani R, Caffi T, Pradolesi GF, Melandri M, et al. (2010) Used of the A-scab model for rational control of apple scab. IOBC WPRS Bull 54: 345–349.

46. Rossi V, Caffi T, Salinari F (2012) Helping farmers face the increasing complexity of decision-making for crop protection. Phytopathol Mediterr 51: 457–479.

47. Gent DH, Mahaffee WF, McRoberts N, Pfender WF (2013) The use and role of predictive systems in disease management. Annu Rev Phytopathol 51: 267–289.

48. Boller EEF, Avilla J, Gendrier JP, Jörg E, Malavolta C (1998) Integrated Production in Europe: 20 years after the declaration of Ovronnaz. IOBC Bull 21: 1–34.

49. Alavanja MCR, Hoppin JA, Kamel F (2004) Health effects of chronic pesticide exposure: cancer and neurotoxicity. Annu Rev Public Health 25: 155–197.

50. Brent KJ, Hollomon DW (2007) Fungicide resistance in crop pathogens: How can it be managed? FRAC Monog 2. Fungicide Resistance Action Committee.

51. Shtienberg D (2013) Will decision-support systems be widely used for the management of plant diseases? Annu Rev Phytopathol 51: 1–16.

52. Leffelaar P (1993) On Systems Analysis and Simulation of Ecological Processes. Kluwer. London.

53. Rossi V, Giosuè S, Caffi T (2010) Modelling plant diseases for decision making in crop protection. In: Oerke E-C, Gerhards R, Menz G, Sikora RA, editors. Precision Crop Protection-the Challenge and Use of Heterogeneity.

54. Hui C (2006) Carrying capacity, population equilibrium, and environment's maximal load. Ecol Modell 192: 317–320.

55. Townsend C, Begon M, Harper J (2008) Essentials of ecology. John Wiley and Sons. New York. 510.

56. Zadoks J, Schein R (1979) Epidemiology and plant disease management. Oxford University Press, New York. 427.

57. Bennett JC, Diggle A, Evans F, Renton M (2012) Assessing eradication strategies for rain-splashed and wind-dispersed crop diseases. Pest Manag Sci 69: 955–963.

58. Caffi T, Rossi V, Bugiani R, Spanna F, Flamini L, et al. (2009) A model predicting primary infections of *Plasmopara viticola* in different grapevine-growing areas of Italy. J Plant Pathol 91: 535–548.

59. Ghanbarnia K, Dilantha Fernando WG, Crow G (2009) Developing rainfall- and temperature-based models to describe infection of canola under field conditions caused by pycnidiospores of *Leptosphaeria maculans*. Phytopathology 99: 879–886.

60. Gilligan CA, van den Bosch F (2008) Epidemiological models for invasion and persistence of pathogens. Annu Rev Phytopathol 46: 385–418.

61. Buck AL (1981) New equations for computing vapor pressure and enhancement factor. J Appl Meteorol 20: 1527–1532.

62. Teng P (1981) Validation of computer models of plant disease epidemics: a review of philosophy and methodology. J Plant Dis Prot 88: 49–63.

63. Madden L V, Hughes G, van den Bosch F (2007) The study of plant disease epidemics. APS press. St. Paul. 421.

64. Nash J, Sutcliffe J (1970) River flow forecasting through conceptual models part I. J Hidrol 10: 282–290.

65. González-Domínguez E, Rodríguez-Reina J, García-Jiménez J, Armengol J (2014) Evaluation of fungicides to control loquat scab caused by *Fusicladium eriobotryae*. Plant Heal Prog Accepted.

66. De Wolf ED, Isard SA (2007) Disease cycle approach to plant disease prediction. Annu Rev Phytopathol 45: 203–220.

67. Krause RA, Massie LB (1975) Predictive systems: modern approaches to disease control. Annu Rev Phytopathol 13: 31–47.

68. Fourie P, Schutte T, Serfontein S, Swart F (2013) Modeling the effect of temperature and wetness on *Guignardia pseudothecium* maturation and ascospore release in citrus orchards. Phytopathology 103: 281–292.

69. Gadoury DM, Machardy WE (1982) A model to estimate the maturity of ascospores of *Venturia inaequalis*. Phytopathology 72: 901–904.

70. Holtslag QA, Remphrey WR, Fernando WGD, Ash GHB (2004) The development of a dynamic disease- forecasting model to control *Entomosporium mespili* on Amelanchier alnifolia. Can J Plant Pathol 313: 304–313.

71. Legler SEE, Caffi T, Rossi V (2013) A Model for the development of Erysiphe necator chasmothecia in vineyards. Plant Pathol. DOI:10.1111/ppa.12145.

72. Luo Y, Michailides TJ (2001) Risk analysis for latent infection of prune by *Monilinia fructicola* in California. Phytopathology 91: 1197–1208.

73. Rossi V, Caffi T, Giosuè S, Girometta B, Bugiani R, et al. (2005) Elaboration and validation of a dynamic model for primary infections of *Plasmopara viticola* in North Italy. Riv Ital di Agrometeorol 13: 7–13.

74. Gadoury D, Machardy WE (1986) Forecasting ascospore dose of *Venturia inaequalis* in commercial apple orchards. Phytopathology 76: 112–118.

75. Gent DH, De Wolf E, Pethybridge SJ (2011) Perceptions of risk, risk aversion, and barriers to adoption of decision support systems and integrated pest management: an introduction. Phytopathology 101: 640–643.

76. Schut M, Rodenburg J, Klerkx L, van Ast A, Bastiaans L (2014) Systems approaches to innovation in crop protection. A systematic literature review. Crop Prot 56: 98–108.

77. Mills W, Laplante A (1954) Diseases and insect in the orchard. Cornell Ext Bull 711.

78. GVA (2013) Octubre-Noviembre 2013. Butlletí d'avisos 13.

79. Machardy WE, Gadoury DM (1989) A revisions of Mills's criteria for predicting apple scab infection periods. Phytopathology 79: 304–310.

80. González-Domínguez E, Armengol J, García-Jiménez J, Soler E (2013) El moteado del níspero en la Marina Baixa. Phytoma 247: 50–52.

81. Teng P (1981) Validation of computer models of plant disease epidemics: a review of philosophy and methodology. J Plant Dis Prot 88: 49–63.

Determination of 17 Organophosphate Pesticide Residues in Mango by Modified QuEChERS Extraction Method Using GC-NPD/GC-MS and Hazard Index Estimation in Lucknow, India

Ashutosh K. Srivastava[1], Satyajeet Rai[1], M. K. Srivastava[1], M. Lohani[2], M. K. R. Mudiam[1], L. P. Srivastava[1]*

1 Pesticide Toxicology Laboratory, CSIR-Indian Institute of Toxicology Research, (Council of Scientific and Industrial Research, Govt. of India), Mahatma Gandhi Marg, Lucknow, India, 2 Department of Biotechnology, Integral University, Lucknow, India

Abstract

A total of 162 samples of different varieties of mango: Deshehari, Langra, Safeda in three growing stages (Pre-mature, Unripe and Ripe) were collected from Lucknow, India, and analyzed for the presence of seventeen organophosphate pesticide residues. The QuEChERS (Quick, Easy, Cheap, Effective, Rugged and Safe) method of extraction coupled with gas chromatography was validated for pesticides and qualitatively confirmed by gas chromatography- mass spectrometry. The method was validated with different concentrations of mixture of seventeen organophosphate pesticides (0.05, 0.10, 0.50 mg kg^{-1}) in mango. The average recovery varied from 70.20% to 95.25% with less than 10% relative standard deviation. The limit of quantification of different pesticides ranged from 0.007 to 0.033 mg kg^{-1}. Out of seventeen organophosphate pesticides only malathion and chlorpyriphos were detected. Approximately 20% of the mango samples have shown the presence of these two pesticides. The malathion residues ranged from ND-1.407 mg kg^{-1} and chlorpyriphos ND-0.313 mg kg^{-1} which is well below the maximum residues limit (PFA-1954). In three varieties of mango at different stages from unpeeled to peeled sample reduction of malathion and chlorpyriphos ranged from 35.48%-100% and 46.66%-100% respectively. The estimated daily intake of malathion ranged from 0.032 to 0.121 µg kg^{-1} and chlorpyriphos ranged from zero to 0.022 µg kg^{-1} body weight from three different stages of mango. The hazard indices ranged from 0.0015 to 0.0060 for malathion and zero to 0.0022 for chlorpyriphos. It is therefore indicated that seasonal consumption of these three varieties of mango may not pose any health hazards for the population of Lucknow, city, India because the hazard indices for malathion and chlorpyriphos residues were below to one.

Editor: Aditya Bhushan Pant, Indian Institute of Toxicology Research, India

Funding: The financial assistance for the publication of manuscript is provided by GAP-168 MPRNL (Monitoring of Pesticide Residues at National Level) Government of India, as well as the CSIR-SRF Contingency. The funders had no role in study design, data collection and analysis, decision to publish, or preparation of the manuscript.

Competing Interests: The authors have declared that no competing interests exist.

* E-mail: laxmanprasad13@gmail.com

Introduction

Mango (*Mangifera indica*) is one of the most common and highly consumable tropical fruits of India. It is rich in carotenoid, minerals, carbohydrates and vitamins. India ranked first in mango production in the world during 2010-11 [1]. Lucknow, the capital of Uttar Pradesh, India, is the largest producer of mangoes producing around 3469.5 metric tons, the productivity being about 12.8 tones/hectare [2]. There are many insect pest pressures for mangos grown in this region of India requiring the use of pesticides to increase the productivity [3]. Therefore, for obtaining good quality and high productivity of mango fruits, the commercial cultivation of mango receive frequent application of various contact and systemic pesticides throughout the cropping season [4]. Most pesticide residues find their way into the human body through fruits, vegetables, cereals, water and other food commodities. Thus, analysis of pesticide residues in food commodities and other environmental samples have become an essential requirement for consumers, producers and food quality control authorities [5]. Due to increased use of pesticide in the orchard, pesticide residues may remain in the raw fruits and their products such as juices, nectar, jellies and ice cream 'pose to be poisonous hazards to human health owing to their toxicity' [6–7]. To increase foreign trade under WTO regime, it is imperative to produce pesticide free mango [8]. Among the various pesticides, organophosphates (OPs), are the most extensively used insecticides in many crops including mango. Due to low persistence and high bio-efficiency of organophosphates, many farmers regularly use this group of pesticides for various vegetables and fruit crops. The continuous use of pesticides has caused deleterious effects the ecosystem [9]. Because of wide spread use of pesticides, the presence of their toxic residues have been reported in various environmental component/commodities [5,10–11]. Public awareness of health hazards posed by pesticide residues in fruits and vegetables has led to the development of many analytical methods. [5,7,8,12–16]. Method validation is an important requirement in chemical analysis. The analyst must generate information to show that a method intended for this purpose is capable of providing adequate specificity, accuracy and precision at relevant analyte concentrations in appropriate matrices. In the present study, an attempt has been made to validate modified QuEChERS method using ethyl acetate (EtOAc) for the extraction. In the method QuEChERS is reported that the acetonitrile is not compatible with system of gas chromatography due to high volume spray, which significantly increases the internal pressure of the chromatographic system. So we adapt the QuEChERS method employing the ethyl acetate solvent.

Seventeen organophosphate pesticides like dichlorvos, phorate, phorate-sulfone, phorate-sulfoxid, dimethoate, diazinon, methyl-parathion, chlorpyrifos-methyl, fenitrothion, malathion, chlorpyrifos, chlorfenvinfos, profenofos, ethion, edifenophos, anilophos and phosalone in mango fruits were analysed by Gas chromatography using Nitrogen Phosphorus Detector (GC-NPD). Majority of these pesticides are being used in mango orchards during spraying [17–18]. The validated method has been applied to determined 17 OPs residues in three delicious varieties of mango like Deshehari, Langra and Safeda of Malihabad, Lucknow, Uttar Pradesh, India as these varieties are prone to insect pests and consumed largely (Fig. 1), determining estimated daily intake (EDI) and hazard index of detected OPs residues for the consumption of mangoes by local population of Lucknow, India.

Materials and Methods

Ethics Statement

No specific permissions were required for these locations/activities. There is no requirement for ethical permission for this study.

Chemicals and Pesticide Standards

All solvents like n-hexane, acetone and ethyl acetate (HPLC grade) were purchased from Sigma-Aldrich. Co. USA, Spectrochem Pvt. Ltd. India and were glass distilled before use. Acetone was refluxed over potassium permanganate for 4 hr and then distilled. Sodium chloride (NaCl), anhydrous sodium sulphate (Na_2SO_4) and magnesium sulphate ($MgSO_4$) were procured from Himedia Pvt. Ltd. India. Before use sodium sulphate (Na_2SO_4) and magnesium sulphate ($MgSO_4$) were purified with acetone and baked for 4 hr @ 600°C in muffle Furness to remove possible phthalate impurities. Primary secondary amine (PSA) bondasil 40 μm part 12213024 of Varian was used.

Pesticide standards (Dichlorvos 98.9%, phorate 96.0%, phorate-sulfone 96.8%, phorate-sulfoxid 94.8%, Dimethoate 99.6%, diazinon, methyl-parathion 99.7%, chlorpyrifos-methyl 99.9%, fenitrothion 95.2%, Malathion 97.2%, chlorpyriphos 99.9%, chlorofenvinfos 98.7%, profenofos 95.0%, Ethion 97.8%, Edifenphos 99.9%, anilophos 97.5% and phosalone 95.2%) were purchased from Supelco Sigma Aldrich USA, Fluka Sigma-Aldrich Schweis and Rankem Pvt. Ltd. New Delhi, India.

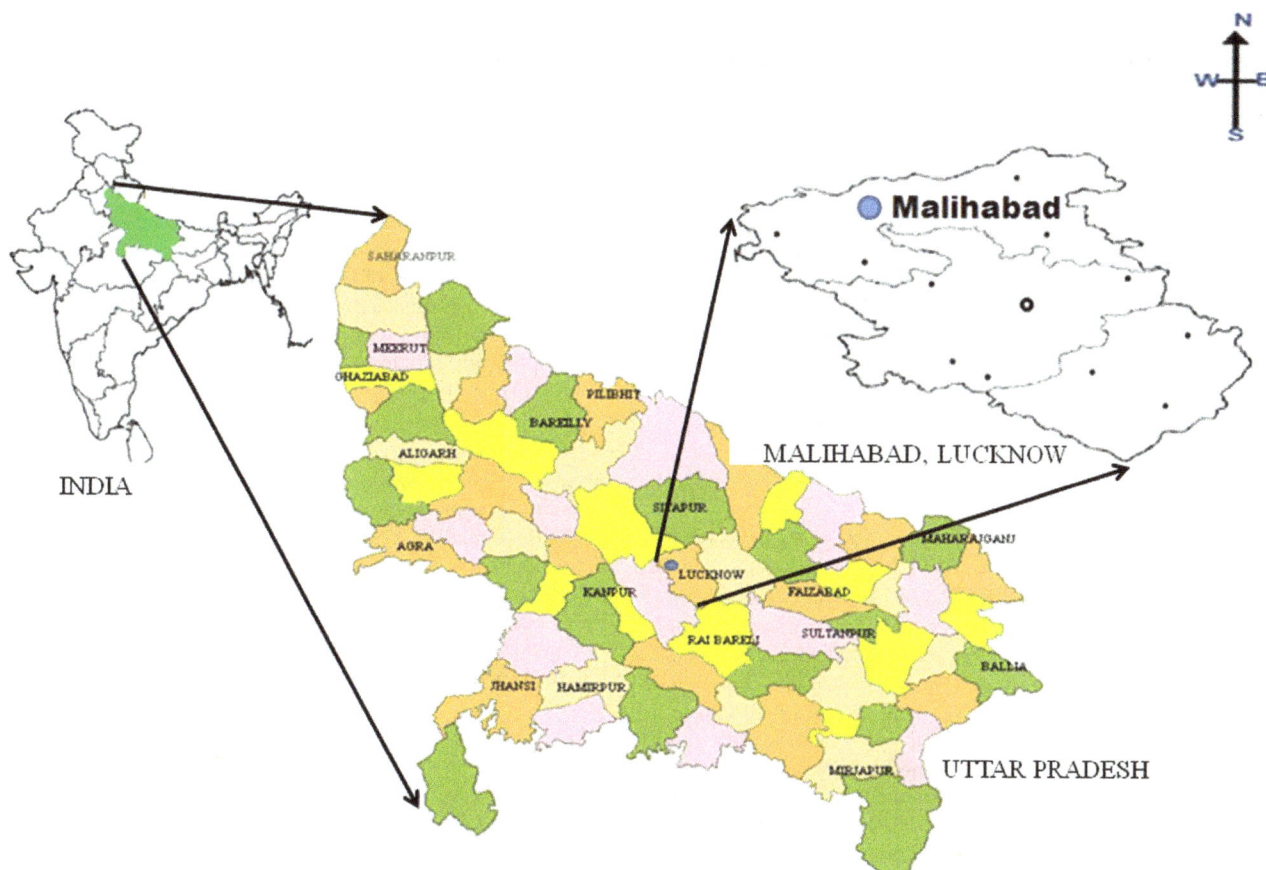

Figure 1. Map showing sampling site Malihabad, Lucknow, Uttar Pradesh, India.

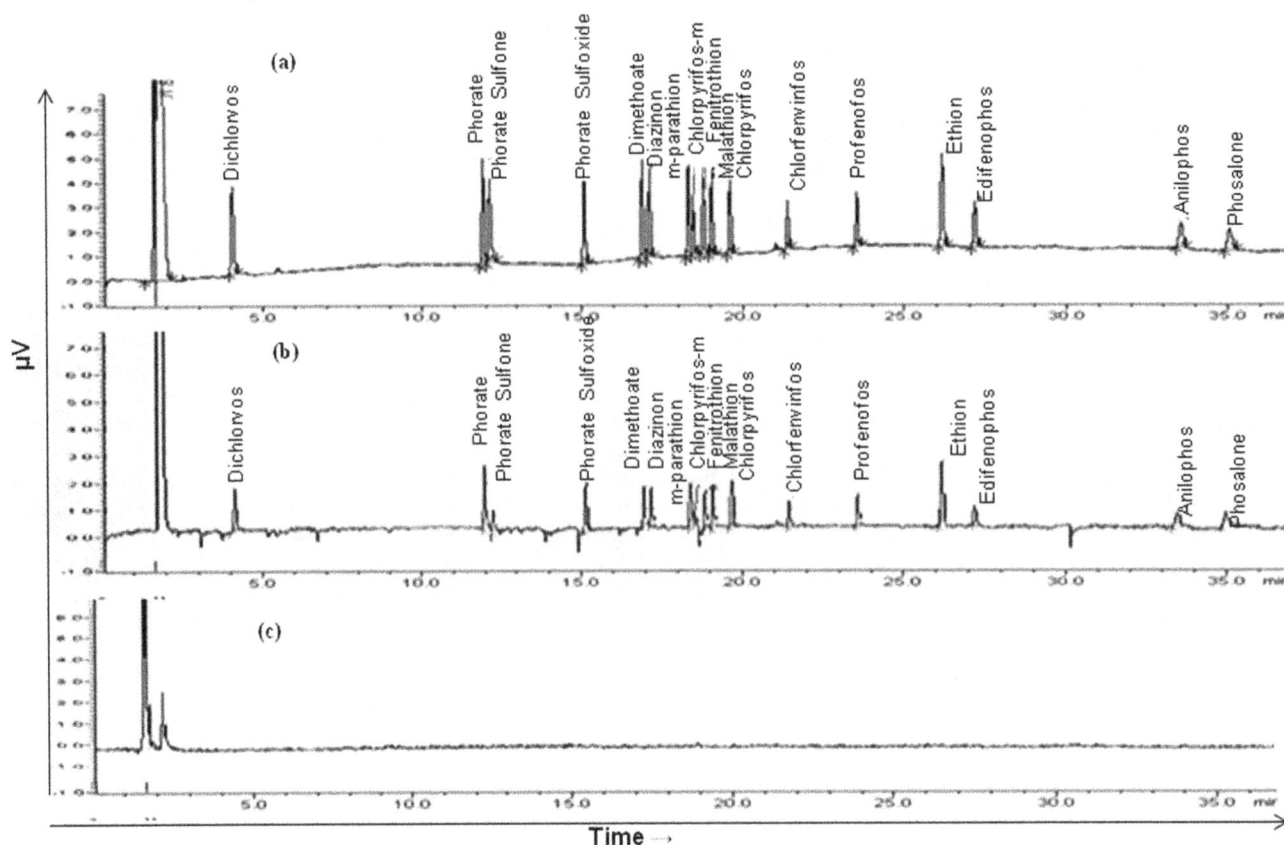

Figure 2. GC-NPD Chromatogram of (a) standard solution of 17 pesticides (0.5 mg l^{-1}). (b) spiked mango sample (0.10 mg kg^{-1}) and (c) blank mango sample.

Sample collection

A total of 162 mango samples like Deshehari, Langra and Safeda of pre mature, unripe and ripe, from three different orchards of Malihabad, Lucknow, Uttar Pradesh, India were collected. (Fig. 1). Three batches of pre-mature, unripe and ripe mangoes (6 samples per batch) of individual varieties were taken for analysis. Samples of Mangoes (6 samples x 3 conditions/ batches x 3 varieties x 3 orchards = 162 samples) were brought to the laboratory and analyzed as soon as possible or stored in refrigerator at 4 ± 2°C until analysis.

QuEChERS Sample Preparations

The unpeeled and peeled mango samples (50 g) of each variety were chopped and grinded in Warring blander. Macerated sample of each mango in triplicates were extracted and cleaned by QuEChERS method as follows: 10 ± 0.1 g macerated sample was mixed with 10 ml ethyl acetate, 4 g activated anhydrous $MgSO_4$, 1.0 g activated NaCl in centrifuge tube and shaken for 10 min. at 50 rpm on rotospin. The extract was centrifuged for 10 min at 8,000 rpm. 1 ml supernatant of extract was cleaned with the mixture of 50 mg PSA, 150 mg anhydrous $MgSO_4$ and 10 mg activated charcoal (application of activated charcoal in the stage of clean up to remove pigments from matrix). The extract was again shaken for 10 min at 50 rpm on rotospin and centrifuged for 10 min at 8,000 rpm. The supernatant was collected in GC vial for the analysis. Steps of extraction and clean up using a longer agitation related to the use of ethyl acetate solvent, which is less

efficient than the acetonitrile with regard to power extraction of pesticides in the partitioning process is documented [12]. The use of larger time in the agitation may have been an attempt to maximize the extraction of pesticides, but promoted a significant increase in analysis time compared with the original method. 1 μl clean extract was injected in gas chromatography equipped with– nitrogen phosphorus detector (NPD) for the analysis OP pesticide residues. (Fig. 2)

GC-NPD Analysis

Residues were analyzed on Shimadzu GC-2010 equipped with fused silica capillary column, DB-1 (30 mt.×0.25 mm. id) coated with 100% dimethylpolysiloxane (0.25 μm film thickness) using NPD. General operating conditions were as follows; Injector port temperature: 250°C; detector temperature 280°C; carrier gas Nitrogen (N_2); flow 1.46 ml min^{-1}; Hydrogen (H_2) makeup is 30 ml min^{-1} and zero air 60 ml min^{-1}, (zero air has less than 0.1 ppm hydrocarbons, to decreases the background noise level and gives the baseline much better stability, considerably increasing detector sensitivity and ensuring precise analytical results) column temperature program: initially 130°C hold 2 min, increase at 5°C/min to 170°C hold 3 min, then increase 220°C min^{-1} at 5°C min^{-1}, hold for 14 min; injection volume: 1 μl split ratio 1:5. The total run time was 37 min and Shimadzu, GC Solution software was used for instrument control and data analysis. Quantification of the pesticides was done by peak area using the standard method.

Table 1. Fortification experiment of organophosphate pesticides residues from spiked mango at different levels (recovery and repeatability), limit of detection (LOD), limit of quantification (LOQ), maximum residues limits (MRL) and acceptable daily intake (ADI).

P Pesticides	Fortification level (mg kg⁻¹)	Recovered	(%) Recovery (n = 3)	LOD (mg kg⁻¹)	LOQ (mg kg⁻¹)	%RSD[a]	MRL[b] (mg kg⁻¹)	ADI[c] (mg kg⁻¹ day⁻¹)
Dichlorvos	0.05	0.039±3.22	78.90					
	0.10	0.080±3.22	80.23	0.003	0.010	4.42	0.1	0.004
	0.50	0.423±3.22	84.57					
Phorate	0.05	0.036±3.22	72.33					
	0.10	0.075±3.22	75.45	0.002	0.007	3.17	0.05	0.0005
	0.50	0.391±3.22	78.20					
Phorate Sulfone	0.05	0.035±3.22	70.22					
	0.10	0.071±3.22	71.45	0.010	0.033	3.84	0.05	0.0005
	0.50	0.373±3.22	74.60					
Phorate Sulfoxide	0.05	0.040±3.22	80.35					
	0.10	0.080±3.22	80.00	0.003	0.010	5.35	0.05	0.0005
	0.50	0.426±3.22	85.25					
Dimethoate	0.05	0.039±3.22	78.23					
	0.10	0.080±3.22	80.45	0.003	0.010	6.95	2.0	0.002
	0.50	0.425±3.22	85.10					
Diazinon	0.05	0.043±3.22	85.28					
	0.10	0.086±3.22	86.30	0.001	0.004	9.50	NA	0.02
	0.50	0.450±3.22	90.00					
m-Parathion	0.05	0.045±3.22	91.11					
	0.10	0.088±3.22	88.35	0.001	0.004	3.23	0.2	0.003
	0.50	0.467±3.22	93.45					
Chlorpyrifos-m	0.05	0.043±3.22	85.25					
	0.10	0.084±3.22	83.70	0.003	0.010	4.91	NA	0.01
	0.50	0.444±3.22	88.90					
Fenitrothion	0.05	0.036±3.22	72.90					
	0.10	0.070±3.22	70.20	0.010	0.033	5.37	0.5	0.005
	0.50	0.372±3.22	74.50					
Malathion	0.05	0.041±3.22	81.25					
	0.10	0.082±3.22	82.50	0.002	0.008	8.49	4.0	0.02
	0.50	0.428±3.22	85.55					
Chlorpyrifos	0.05	0.045±3.22	90.57					
	0.10	0.092±3.22	92.40	0.001	0.004	6.70	0.5	0.01
	0.50	0.476±3.22	95.25					
Chlorfenvinfos	0.05	0.044±3.22	88.10					

Table 1. Cont.

P Pesticides	Fortification level (mg kg^{-1})	Recovered	(%) Recovery (n = 3)	LOD (mg kg^{-1})	LOQ (mg kg^{-1})	%RSD[a]	MRL[b] (mg kg^{-1})	ADI[c] (mg kg^{-1} day^{-1})
	0.10	0.091±3.22	90.75	0.003	0.010	6.48	NA	NA
	0.50	0.466±3.22	93.25					
Profenofos	0.05	0.042±3.22	83.25					
	0.10	0.080±3.22	80.25	0.004	0.013	8.68	NA	NA
	0.50	0.417±3.22	83.45					
Ethion	0.05	0.040±3.22	80.25					
	0.10	0.080±3.22	80.40	0.005	0.017	8.23	2.0	NA
	0.50	0.428±3.22	85.70					
Edifenophos	0.05	0.046±3.22	92.34					
	0.10	0.090±3.22	90.40	0.001	0.005	7.23	NA	NA
	0.50	0.467±3.22	93.50					
Anilophos	0.05	0.039±3.22	77.72					
	0.10	0.079±3.22	79.45	0.002	0.007	8.18	NA	NA
	0.50	0.416±3.22	83.20					
Phosalone	0.05	0.042±3.22	84.22					
	0.10	0.080±3.22	80.50	0.003	0.010	9.61	5.0	0.02
	0.50	0.431±3.22	86.25					

[a]RSD-Relative Standard Deviation.
[b]PFA - Prevention of Food Adulteration Act Govt. of India, 1954 [31].
[c][32].
NA- Not Available.

Table 2. Level of malathion and chlorpyrifos residues (mg kg^{-1}) in different species of mango fruits and their percent reduction after peeling.

Mango	Pre mature n = 6		Unripe n = 6		Ripe n = 6		Samples [A]
	Malathion Mean ± SD	Chlorpyrifos Mean ± SD	Malathion Mean ± SD	Chlorpyrifos Mean ± SD	Malathion Mean ± SD	Chlorpyrifos Mean ± SD	Detected/Analysed
Deshehari							14/54
Unpeeled	0.969±0.050{3} (ND-1.185)	0.090±0.035{4} (ND-0.223)	0.062±0.020{3}*** (ND-0.197)	0.028±0.005{2}** (ND-0.113)	0.021±0.010{2}*** (ND-0.304)	ND	
Peeled	0.011±0.005{3}### (ND-0.046)	0.048±0.050{4} (ND-0.142)	0.040±0.025{3}* (ND-0.128)	0.014±0.025{2} (ND-0.057)	ND	ND	
% reduction	98.86	46.66	35.48	50.00	100		
Langra							11/54
Unpeeled	0.049±0.030{2} (ND-0.294)	0.160±0.070{2} (ND-0.313)	0.679±0.125{5}*** (ND-1.407)	ND	0.213±0.050{2}*** (ND-0.853)	ND	
Peeled	ND	ND	0.272±0.015{5}### (ND-0.932)	ND	ND	ND	
% reduction	100	100	59.94		100		
Safeda							7/54
Unpeeled	ND	ND	0.480±0.015{5} (ND-0.908)	ND	0.147±0.080{2} (ND-0.591)	ND	
Peeled	ND	ND	0.139±0.025{5}### (ND-0.239)	ND	0.035±0.650{2} (0.105)	ND	
% reduction			71.04		76.19		
Total Samples Detected/ Analysed							32/162

[A] Total samples of each variety of mango.

Values within small parentheses ( ) indicates range of residues and middle {} indicates number of samples detected with pesticides.

* p<0.05, ** p<0.01, *** p<0.001, ### p<0.001.

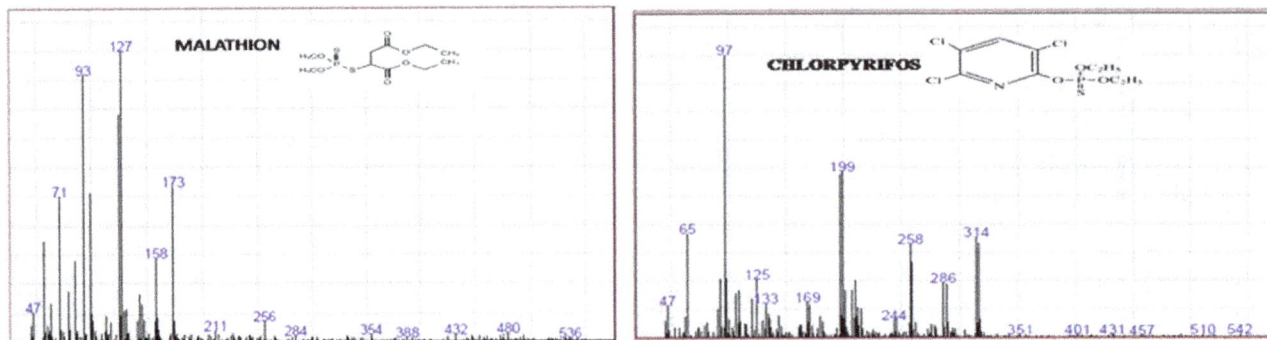

Figure 3. GC-MS chromatogram showing confirmation of malathion and chlorpyrifos.

GC-MS Confirmation

A Perkin Elmer GC-MS consisting of Auto system XL Gas Chromatograph with a Turbo Mass Spectrometer was used for analysis. The column used in this study is Elite-5MS fused-silica capillary column (30 m×0.32 mm I.D., 0.25 mm film thickness). Carrier gas used was helium (purity 99.999%) with a flow rate of 1.6 ml min^{-1}. A 1 µl aliquot of the extract was injected using the splitless mode. The oven temperature program is 100°C for 1 min and then @ 20°C min^{-1} to 210°C and hold for 1 min; then @45°C min^{-1} to 300°C and hold for 1 min. The total runtime of the GC is 9.5 min. Base peaks 173, 158, 127 m/z for malathion as target ion as well as 314, 286, 258 m/z for chlorpyriphos were noticed as qualifiers in selective ion mode (SIM) for analysis. The injector temperature was set at 300°C. The transfer line and source temperature was set at 280°C and 230°C respectively. Solvent delay for MS is 5 min.

Method validation

The validation of the analytical method was performed by the accuracy, precision, linearity and limit of detection (LOD), quantification (LOQ). All the analysis was carried out using the same blank samples of mango fruit.

Accuracy and precision data were obtained with recovery studies by spiking samples with pesticide standards at levels of 0.05, 0.1, and 0.5 mg kg^{-1}. The spiked and control samples were analyzed in five replicates. Precision of the method was evaluated through the relative standard deviations (%RSD) associated with pesticides measurements during recovery.

Linearity was determined by plotting calibration curve with standard solutions in n-hexane containing five different concentrations (0.025, 0.05, 0.1, 0.5 and 1.0 mg L^{-1}). Five injections were made at each of the five concentration levels.

The limits of detection (LOD) and Quantification (LOQ) were calculated according standard guidelines [19–20]. Five independent analysis of mango samples spiked with mixture of 17 OPs at the level of 0.05, 0.1, and 0.5 mg kg^{-1} were performed for the percentage recovery of each pesticides. The LOD and LOQ were calculated at the spiking level of 0.1 mg kg^{-1} from the standard deviation of this determination. Table 1

In order to maintain analytical quality control for each sample batch a spiked sample (similarly used in the recovery study) was analyzed simultaneously. Batch results were considered unsatisfactory when the sample used as quality control had low recovery.

Hazard index estimation

Health risk estimations were performed on the basis of pesticide analysis data obtained in present study and annual fruit intake per person. The estimated daily intake (EDI) of pesticides is the toxicological criteria for their exposure. It is calculated as per international guidelines [21–23] using the equation: EDI = C x F/ D x W where C is the mean of individual pesticide concentration (mg kg^{-1}); F is mean annual intake of fruit per person (kg); D is days in a year (365 days) and W is the mean human body weight (60 kg). The annual intake as the total fruits per person is 9.5 kg per year according to Indian survey perform in years 2005–2006 [24–26]. The EDI (mg kg^{-1} day^{-1}) as obtained were used to estimate the hazards index by dividing them to their corresponding value of known acceptable daily intake (Hazard index = EDI/ ADI).

Statistical Analysis

The data were statically analyzed by using one way ANOVA. Criterion of significance was taken as P<0.05, P<0.01 and P< 0.001. All statistical calculations have been done using IBM SPSS statistics version 20.

Results and Discussion

Validation of QuEChERS method

Fig. 2 (a,b,c) showing the representative chromatogram for standard of OP mixture, spiked and blank mango fruit samples. Adequate separation of the 17 OPs was achieved. No interference peaks were obtained in the blank sample chromatogram at the same retention time of the target compounds.

All OPs showed linearity ranged (0.025 to 1.0 mg kg^{-1}) with the coefficient co-relation more than 0.997 (r). The relative standard deviation (RSD) of the three replicate injections ranged from 3.17 to 9.50% showing good repeatability.

Table 1, shows that recovery data and repeatability of 17 OPs analyzed at three different spiking levels. The recovery ranged from 70.20 to 93.50%. The overall recovery was more than 70% (except phorate sulfone and fenitrothion) at lower spiking level 0.05 mg kg^{-1} with RSD below 10% which represent satisfactory repeatability of the method for all pesticides (Barakat et al. 2007). It is observed that LOD and LOQ for 17 pesticides ranged from 0.001 to 0.010 mg kg^{-1} and 0.007 to 0.033 mg kg^{-1} respectively. It is interesting to note that, LOQ analyzed of all 17 OPs was lower than their respective MRL's established by PFA 1954, Govt. of India (Table 1). Few studies have been reported for the presence of organophosphate residues in mango using various extraction technique used in gas chromatographic analysis [13,20,27–28].

In the present study, the QuEChERS sample preparation and GC analysis using ethyl acetate solvent has been validated. The results indicate that the QuEChERS sample preparation method

Table 3. Estimated daily intake and hazard index of malathion and chlorpyriphos residues in mango.

Mango	Average Pesticide concentration in three varieties of Mangos (ΣC) (µg kg⁻¹)		Mean annual intake of fruit [a,b] per-person/day (kg) (F)	No. of Days in year (D)	Average weight of person (W)	ADI (µg kg⁻¹ bw daily)		EDI (µg kg⁻¹ bw daily)		Hazard index		Hazard Risk	
	Mala	CPF				Mala	CPF	Mala	CPF	Mala	CPF	Mala	CPF
Pre-Mature	172	50	9.5	365	60	20	10	0.075	0.022	0.0037	0.0022	No	No
Unripe	279	7	9.5	365	60	20	10	0.121	0.003	0.0060	0.0003	No	No
Ripe	69	0	9.5	365	60	20	10	0.030	0.00	0.0015	0.00	No	No

Mala = Malathion; CPF = Chlorpyriphos.
[a]From National Nutrition Monitoring Board (2008).
[b]National Sample Survey Organization (2000).

coupled with GC-NPD analysis is suitable for the determination of 17 OPs viz; dichlorvos, phorate, phorate-sulfone, phorate-sulfoxide, dimethoate, diazinon, methyl-parathion, chlorpyrifos-methyl, fenitrothion, malathion, chlorpyrifos, chlorofenvinfos, profenofos, ethion, edifenophos, anilophos and phosalone. Successful use of QuEChERS method has been reported for the analysis of pesticide residues in high sugar content matrices like honey and sugarcane juice [27,29]. Similarly the present study revealed good result by applying QuEChERS method for the analysis of various organophosphate in ripe mango fruit which are enriched in high sugar content.

Pesticide residue in various stages of the mango

Pesticides are commonly sprayed to mango trees at pre-mature, unripe and one month before the harvest. Therefore, mangoes of all three stages were studied. Analysis was done in samples of three different varieties of mangos (deshehari, langra and safeda) in triplicates for the presence of pesticides residues and values are given in Table 2. Out of seventeen analyzed pesticide only two pesticides malathion and chlorpyrifos were detected in three varieties of mangoes.

Deshehari

Comparison between peeled and unpeeled samples. The results of Table 2 showed the presence of malathion and chlorpyrifos residues in different stages of mango. The mean concentration of malathion and chlorpyrifos was 0.969 (ND-1.185) and 0.090 (ND-0.223) mg kg⁻¹ respectively in pre-mature unpeeled samples. However, malathion and chlorpyrifos residues were 0.011(ND-0.046) and 0.048 (ND-0.142) mg kg⁻¹ in the pre-mature peeled mango. It is observed that there was a reduction of 98.86% malathion and 46.66% chlorpyrifos after the peeling of samples. The level of malathion residue in unpeeled to peeled mango samples was significantly (P<0.001) reduced and no significant variation was observed for chlorpyrifos. However in unripe unpeeled mangoes the mean concentration of malathion was 0.062 (ND-0.197) mg kg⁻¹ and chlorpyrifos was 0.028 (ND-0.113) mg kg⁻¹. The residues of malathion and chlorpyrifos were 0.040(ND-0.128) mg kg⁻¹ and 0.014 (ND-0.057) mg kg⁻¹ in unripe peeled mangoes. The percent reduction of malathion was 35.48 and chlorpyrifos was 50.00 through peeling in terms of mean values. The variation in level of malathion and chlorpyrifos was not significant. In ripe unpeeled mangoes, the residue of malathion 0.021 mg kg⁻¹ (ND-0.304) was observed while chlorpyrifos was below to detectable level. None of these two pesticides were detected after the peeling of ripe mangoes. It is interesting to note that in ripe mangoes, the OP residues were least as compared to unripe or pre-mature mangoes. It may be because of judicious use of pesticides in the latent harvesting period of mangoes.

Comparison between pre-mature, unripe and ripe mango samples. The variation of malathion residues in unpeeled unripe and ripe samples was significant (P<0.001) in comparison to pre-maturate samples. Similarly, residues of chlorpyrifos was also significant (P<0.05) in unripe Deshehari mango. Whereas only malathion was significant (P<0.01) in unripe peeled mango samples.

Langra

Comparison between peeled and unpeeled samples. The mean concentration of malathion and chlorpyrifos was 0.049 (ND-0.294) mg kg⁻¹ and 0.160 (ND-0.313) mg kg⁻¹, respectively in pre-mature unpeeled langra mango samples. The cent percent malathion and chlorpyrifos residues were reduced from unpeeled to peeled samples. However in unripe unpeeled mango the mean concentration of malathion was 0.679 (ND-1.407) mg kg^{-1} and 0.272 (ND-0.932) mg kg^{-1} in peeled mango samples. The percent reduction of malathion was 59.94 through peeling with respect to their mean values. The reduction in the level of malathion was significant (P<0.001) in unpeeled to peeled mango samples. Chlorpyrifos was not detected in unripe, unpeeled and peeled mango samples. In ripe mango malathion was detected 0.213 (ND-0.853) mg kg^{-1} in unpeeled sample with 100% reduction from unpeeled to peeled samples. However, chlorpyrifos was again not detected in ripe unpeeled and peeled samples.

Comparison between pre-mature, unripe and ripe mango samples. The variation of malathion residues in unpeeled samples of pre-maturate, unripe and ripe mangoes was significant (P<0.001), if compared with premature Langra mango samples.

Safeda

Comparison between peeled and unpeeled samples. Malathion and chlorpyrifos residues were detected in unripe and ripe stages of mangoes. In unripe unpeeled mango malathion residue was 0.480(ND-0.908 mg kg^{-1}) and in peeled samples the residue was 0.139(ND-0.239 mg kg^{-1}) with 71.04% reduction of their mean values after peeling of the samples. The reduction of malathion was significant (P<0.001) in unpeeled to peeled mango samples. In ripe mangoes, malathion was 0.147(ND-0.591 mg kg^{-1}) in unpeeled samples and 0.035(ND-0.105 mg kg^{-1}) in peeled samples with 76.19% reduction is seen in their mean values after peeling. Malathion and chlorpyrifos residues were not detected in unpeeled and peeled pre-mature mangoes. However, chlorpyrifos was below detection limit in all three stages of safeda mango.

Comparison between pre-mature, unripe and ripe mango samples. The malathion and chlorpyrifos residues in unpeeled and peeled pre-maturate, unripe and ripe mangoes was not significant in safeda mango samples.

The texture, sugar and water content of mango may vary on the maturity, varieties and cultivars of the fruits. The texture and water content play an important role in trapping of pesticides and recovery efficiency.

The results of the method validation indicated that the QuEChERS sample preparation coupled with the GC-NPD analysis is suitable for the determination of 17 OPs in mango samples. In the present study none of pesticide residues accept malathion and chlorpyriphos were detected in 162 mango samples. Therefore, malathion and chlorpyrifos were further confirmed on GC-MS (Fig. 3).

Table 3 shows average estimated daily intake and hazard index for malathion and chlorpyrifos in pre-mature, unripe and ripe mango fruits. Hazard indices were calculated 0.0037, 0.0060 and 0.0015 for malathion and 0.0022, 0.0003 and zero for chlorpyrifos respectively in pre-mature, unripe and ripe mangoes. It is therefore indicated that the consumption of mango may not pose health hazards for the population of Lucknow city, India as hazard indices for malathion and chlorpyrifos in all three stages of mangoes were below one [30]. The total concentration of malathion and chlorpyrifos residues were 0.520 and 0.057 mg kg^{-1} which are below to maximum residues limits, 4.0 mg kg^{-1} and 0.5 mg kg^{-1} [31]. It is further stated that there is no health risk associated with malathion and chlorpyrifos residues after the consumption of mango fruits.

Conclusion

The validated QuEChERS method applied in the present study fulfils the established criteria for sensitivity and confident identification of organophosphorus pesticides at low level in matrix with high sugar content like mango. The results revealed that none of OPs pesticides except traces of malathion and chlorpyrifos residues were present in mango fruits. It is also observe that peeling has great influence in reduction of pesticide residues in the pulp. To avoid adverse effects on public health it is necessary to set up control measures so as to make sure that each pesticide residues should be below MRL in the fruits to be marketed. Therefore the study has explored the significant information regarding the analysis of 17 OPs residues in different varieties of mango fruits of Malihabad, Lucknow region of India. It is further observed that mango fruits appear to be safe from OP pesticides residues as revealed by EDI and hazard index. In this manner one can assume that there is no apparent risk to the health of consumers of mango, as two detected pesticides have hazard index below to one. It is therefore, suggested that regular evaluation of pesticide residues should be carried out on mango fruits at national level for the planning of future policies about the use of pesticides in mango orchards and enables pesticide free fruits.

Acknowledgments

The authors are grateful to the Dr. K.C. Gupta Director, CSIR-Indian Institute of Toxicology Research (IITR), Lucknow for his keen interest and providing research facilities. The financial assistance to Ashutosh Kumar Srivastava funded by Council of Scientific and Industrial Research, New Delhi as SRF is also acknowledged. IITR communication no. is 3050.

Author Contributions

Conceived and designed the experiments: AKS SR LPS. Performed the experiments: AKS SR MKS. Analyzed the data: AKS SR MKS ML LPS. Contributed reagents/materials/analysis tools: LPS MKRM. Wrote the paper: AKS SR MKS LPS. Performed the GC-MS confirmation of pesticides: MKRM.

References

1. Food and Agricultural Organization of United Nations: Economic and Social Department: The Statistical Division, 2010.http://en.wikipedia.org/wiki/Mango.
2. Biswas BC, Kumar L (2011) Revolution in mango production: Success Stories of some Farmers. New Delhi, *Fertilizer Marketing News* 1–3.
3. Mohapatra S, Ahuja AK, Deepa M, Sharma D (2010) Residues of acephate and its metabolite methamidophos in/on mango fruit (Mangiferae indica L.). Bull Environ Contam Toxicol 86: 101–104.
4. Banerjee K, Oulkar DP, Patil SB, Jadhav MR, Dasgupta S, et al. (2009) Multiresidue determination and uncertainty analysis of 87 pesticides in mango by liquid chromatography-tandem mass spectrometry. J Agri Food Chem 57: 4068–4078.
5. Srivastava AK, Trivedi P, Srivastava MK, Lohani M, Srivastava LP (2011) Monitoring of pesticide residues in market basket samples of vegetable from Lucknow City, India: QuEChERS method. Environ Monit Assess 176: 465–472.
6. Filho AM, Santos FND, Pereira PADP (2011) Multi-residue analysis of pesticide residues in mangoes using solid-phase microextraction coupled to liquid chromatography and UV–Vis detection. J Sepr Sci DOI 10.1002/jssc.201100341.
7. Perret D, Gentili A, Sergi SM, Dascenzo G (2002) Validation of a method for the determination of multiclass pesticide residues in fruit juices by liquid chromatography/tandem mass spectrometry after extraction by matrix solid-phase dispersion. J AOAC Int 85: 724–730.
8. Mukherjee I, Singh S, Sharma PK, Jaya M, Gopal M, et al. (2007) Extraction of multi-class pesticide residues in mango fruits (mangiferae indica L.): application of pesticide residues in monitoring of mangoes. Bull Environ Contam Toxicol 78:380–383.

9. Sharma D, Nagpal A, Pakade YB, Katnoria JK (2010) Analytical methods for estimation of organophosphorus pesticide residues in fruits and vegetables. A review Talanta 82: 1077–1089.

10. Bhanti M, Taneja A (2007) Contamination of vegetables of different seasons with organophosphorous pesticides and related health risk assessment in northern India. Chemosphere 69: 63–68.

11. Wang L, Liang Y, Jiang X (2008) Analysis of eight organophosphorus pesticide residues in fresh vegetables retailed in agricultural product market of Nanjing, China. Bull of Environ Contam Toxicol 81: 377–382.

12. Anastassiades M, Lehotay SJ, Stajnbaher D, Schenck FJ (2003) Fast and easy multiresidue method employing acetonitrile extraction/partitioning and "dispersive solid phase extraction" for the determination of pesticides residues in produce. J AOAC Int 86: 412–431.

13. Aysal P, Ambrus AR, Lehotay SJ, Andrew C (2007) Validation of an efficient method for the determination of pesticide residues in fruits and vegetables using ethyl acetate for extraction. J Environ Sci and Health Part B42: 481–490.

14. Lehotay SJ (2007) Determination of pesticide residues in foods by acetonitrile extraction and partitioning with magnesium sulphate: Collaborative study. J AOAC Int 90: 485–520.

15. Payá P, Anastassiades M, Mack D, Sigalova I, Tasdelen B, et al. (2007) Analysis of pesticide residues using the Quick Easy Cheap Effective Rugged and Safe (QuEChERS) pesticide multiresidue method in combination with gas and liquid chromatography and tandem massspectrometric detection. Anal and Bio anal Chem 389: 1697–1714.

16. Furlani RPZ, Marcilio KM, Leme FM, Tfouni SAV (2010) Analysis of pesticide residues in sugarcane juice using QuEChERS sample preparation and gas chromatography with electron capture detection. Food Chem 126: 1283–1287.

17. Pathak MK, Fareed M, Bihari V, Reddy MMK, Patel DK, et al. (2011) Nerve conduction studies in sprayers occupationally exposed to mixture of pesticides in a mango plantation at Lucknow, North India. Toxicological & Environmental Chemistry 93: 1, 188–196

18. Pathak MK, Fareed M, Srivastava AK, Pangtey BS, Bihari V, et al. (2013) Seasonal variations in cholinesterase activity, nerve conduction velocity and lung function among sprayers exposed to mixture of pesticides. Environmental Science and Pollution Research DOI 10.1007/s11356-013-1743-5.

19. Taylor JK (1987). In Quality assurance of chemical measurements. Chelsea, USA: Lewis Publishers, Inc. 31. PFA (1954) Prevention of Food Adulteration Act 1954. Act No. 37 with Prevention of Food Adulteration Rules 1955 and Notification and Commodity Index 16th edition. Eastern Book Company Publishing Pvt. Ltd. Lucknow, India.

20. Saadati N, Pauzi Abdullah MD, Zakaria Z, Belin S, Sany T, et al. (2013) Limit of detection and limit of quantification development procedures for organochlorine pesticides analysis in water and sediment matrices. Chemistry Central Journal 7: 63–73.

21. World Health Organization: WHO (1997) Guideline for predicting dietary intake of pesticide residues (revised). Global Environment Monitoring System-Food Contamination and Assessment Programme (GEMS/Food) in collaboration with codex committee on pesticide residues.

22. Food and Agriculture Organization: FAO (2002) Submission and evaluation of pesticide residues data for the estimation of maximum residues levels in food and feed (Ist edition). Rome: Food and Agriculture Organization. <http://www.fao.org/ag/AGP/AGPP/Pesticide/p.htm>.

23. Osman KA, Al-Humaid AI, Al-Rehiayani A, Al-Redhaiman KN (2011) Estimated daily intake of pesticide residues exposure by vegetables grown in greenhouse in Al-Qassim region Saudi Arabia. Food Control 22: 947–953.

24. National Sample Survey Organization (NSSO). 1975–2000; http://mospi.nic.in/mospi_nsso_rept_pubn.htm; last accessed on 24/0907, Ministry of Statistics and Programme Implementation, Govt. of India.

25. European Commission (2007) Method validation and quality control procedures for pesticide residues analysis in food and feed, Document No. SANCO/2007/3131. Availablefrom:<www.ec.europa.eu/plant/prtection/resources/qualcontrol_en.pdf>

26. National Nutrition Monitoring Bureau (NNMB). 1979–2008. *NNMB Report (7.7)*: National Institute of Nutrition Hyderabad India.

27. Barakat AA, Badaway HMA, Salama E, Attallah E, Maatook G (2007) Simple and rapid method of analysis for determination of pesticide residues in honey using dispersive solid phase extraction and GC determination. J Food Agri & Environ 5: 97–100.

28. Raju MB, Rao CN, Kumar GVR, Rao BT, Krishna PM, et al. (2011) Determination of multiclass pesticide residues in mangoes by Liquid chromatography–tandem mass spectrometry. J App Bio and Pharm Tech2: 279–289.

29. Prati P, Camargo GA (2008) Characteristics of sugarcane juice and your influence in the beverage stability. Bio Eng2: 37–44.

30. Darko G, Akoto O (2008) Dietary intake of organophosphorous pesticide through vegetables from Kumasi, Ghana. Food and Chem Toxicol 46: 3703–3706.

31. PFA (1954) Prevention of Food Adulteration Act 1954. Act No. 37 with Prevention of Food Adulteration Rules 1955 and Notification and Commodity Index (16th ed.). Lucknow: Eastern Book.

32. Sharma KK (2007) Pesticide residues Analysis Manual. Directorate of Information and Publications of Agriculture, New Delhi, 2007.

Spatial and Temporal Patterns of Carbon Storage in Forest Ecosystems on Hainan Island, Southern China

Hai Ren[1][*][◊], Linjun Li[1][◊], Qiang Liu[2][◊], Xu Wang[3][◊], Yide Li[4][◊], Dafeng Hui[5][◊], Shuguang Jian[1][◊], Jun Wang[1][◊], Huai Yang[4][◊], Hongfang Lu[1][◊], Guoyi Zhou[1][◊], Xuli Tang[1], Qianmei Zhang[1], Dong Wang[6], Lianlian Yuan[1], Xubing Chen[1]

1 Key Laboratory of Vegetation Restoration and Management of Degraded Ecosystems, South China Botanical Garden, Chinese Academy of Sciences, Guangzhou, China, 2 College of Life Science, Hainan Normal University, Haikou, China, 3 College of Environment and Plant Protection, Hainan University, Haikou, China, 4 Research Institute of Tropical Forestry, Chinese Academy of Forestry, Guangzhou, China, 5 Department of Biological Sciences, Tennessee State University, Nashville, TN, United States of America, 6 School of Life Sciences, Central China Normal University, Wuhan, China

Abstract

Spatial and temporal patterns of carbon (C) storage in forest ecosystems significantly affect the terrestrial C budget, but such patterns are unclear in the forests in Hainan Province, the largest tropical island in China. Here, we estimated the spatial and temporal patterns of C storage from 1993–2008 in Hainan's forest ecosystems by combining our measured data with four consecutive national forest inventories data. Forest coverage increased from 20.7% in the 1950s to 56.4% in the 2010s. The average C density of 163.7 Mg C/ha in Hainan's forest ecosystems in this study was slightly higher than that of China's mainland forests, but was remarkably lower than that in the tropical forests worldwide. Total forest ecosystem C storage in Hainan increased from 109.51 Tg in 1993 to 279.17 Tg in 2008. Soil C accounted for more than 70% of total forest ecosystem C. The spatial distribution of forest C storage in Hainan was uneven, reflecting differences in land use change and forest management. The potential carbon sequestration of forest ecosystems was 77.3 Tg C if all forested lands were restored to natural tropical forests. To increase the C sequestration potential on Hainan Island, future forest management should focus on the conservation of natural forests, selection of tree species, planting of understory species, and implementation of sustainable practices.

Editor: Ting Wang, Wuhan Botanical Garden, Chinese Academy of Sciences, Wuhan, China, China

Funding: This research was supported by the "Strategic Priority Research Program" of the Chinese Academy of Sciences, grant No. XDA05050206. The funders had no role in study design, data collection and analysis, decision to publish, or preparation of the manuscript.

Competing Interests: The authors have declared that no competing interests exist.

* Email: renhai@scbg.ac.cn

◊ These authors contributed equally to this work.

Introduction

Carbon (C) storage in forest ecosystems is one of the largest and most active components of C cycling in terrestrial ecosystems and plays an important role in global C cycling and climate change [1,2]. Information on the spatial distribution of C sources and sinks and their temporal changes is critical for understanding C cycle mechanisms and is essential for formulating climate change policies [3]. As a result, estimation of C budgets at large spatial scales has received increasing attention in recent years [4].

While occupying only 6% of land area, tropical forests contain about 40% of the stored C in the terrestrial biosphere, with vegetation accounting for 58% and soil accounting for 41% [5]. However, there is substantial uncertainty about the estimates of C storage. Conflicting results on tropical forest C storage have been reported. Houghton et al. (1992), for example, indicated that tropical forests are a C source (from 1.2 to 2.2 Pg C/yr) because of deforestation and forest degradation [6]. Malhi and Grace (2000), in contrast, reported that tropical forests are C sinks (1–3 Pg C/yr) while northern forests are C sources [7]. Further studies on C storage in tropical forests at large scales are still needed.

Hainan, the largest tropical island and the second largest island province in China, is part of the Indo-Burma biodiversity hotspot and harbors large areas of tropical forests. Several studies have been conducted on forest resources and C storage on Hainan Island, but produced remarkably varying results. For example, Fang et al. (1996) reported that the total biomass of forests on Hainan Island was 59.79 Tg during 1984–1988 [8]. Zhao and Zhou (2004) found that the forest C storage on the island was 30.92 Tg during 1989–1993 [9]. After considering forest age and vegetation types, Wang (2001) reported that the forest C storage was only 23.21 Tg [10]. Cao et al. (2002) reported that forest C stored in vegetation increased from 30.45 Tg in 1979 to 37.74 Tg in 1993 [11]. Li and Lei (2010) estimated that the total C storage was as high as 50.83 Tg in 2004–2008, while Guo et al. (2013) recently reported the total forest C storage was 37.3 Tg [12,13].

The large discrepancies among those studies are probably due to differences in the methods used to calculate C storage. While all studies used the data from national forestry inventories (seven inventories have been conducted since 1973) conducted by the Sate Forest Agency on Hainan Island, the studies used different inventory datasets, different components of C storage, C concentration coefficients (i.e. the proportion of carbon contained in dry mass of plant organs), or age structures. For example, Cao et al. (2002), used a C concentration coefficient of 0.50 while Wang et al. (2001) used a coefficient of 0.45 [11,14]. Although C storage in ecosystems includes both biomass C and soil C, all of the previous studies considered only the C stored in tree vegetation and failed to consider that stored in the understory or soil. In addition, the spatial distribution of C storage on Hainan Island has not been reported. Thus, it remains unclear how the spatial and temporal patterns of C storage have changed in forest ecosystems during 1993–2008 on Hainan Island, Southern China.

The goal of this study was to examine the spatial and temporal patterns of C storage in forest ecosystems on Hainan Island, China. The specific objectives were to determine: 1) changes in C density of forest vegetation on Hainan Island from 1993–2008; 2) the temporal and spatial patterns of C storage in forest ecosystems on Hainan Island during this period; and 3) how the potential for C storage can be increased.

Materials and Methods

Ethics Statement

This study was based on forest inventory data and our field measurements. For the field study, all necessary permits were obtained from Hainan Bureau of Forestry. The field study did not involve endangered or protected species.

Description of Hainan Island

Hainan Island has a land area of 33,920 km^2 and is located at the northern edge of the tropics (latitude 18°10′–20°10′N, longitude 108°37′–111°03′E). Its tropical monsoon climate includes distinct dry and wet seasons and typhoons. Average annual rainfall is 1500–2500 mm, and average annual temperature is 22–26°C. The soil type is mainly laterite. The main zonal vegetation types include tropical rain forest and tropical mountain rain forest. The island contains more than 4200 plant species (259 families, 100 genera) including about 2000 tropical species [15].

Vegetation classification based on remote sensing and image processing

We collected the Landsat TM satellite images (November 2008), 1:250,000 Digital Elevation Model (DEM), Hainan forest maps (1:500,000), and administrative maps. The images were processed using ERDAS IMAGINE 8.31 [4]. This included geometric correction processing, unsupervised classification method, vegetation information extraction, image classification, and determination of area statistics [16]. The image contained a total of 17 spectral clusters of land cover of which nine were vegetation. These nine spectral clusters were merged into six vegetation types based on the Chinese vegetation taxonomy system [17]: tropical natural rain forest, *Eucalyptus* plantation, rubber plantation, *Casuarina* plantation, coniferous plantation, and orchard. The spatial location of the six vegetation types was overlaid with the Hainan forest maps to show the actual geographical distribution of the studied vegetation types and created the distribution map of forests on Hainan Island in 2008. Finally, we selected the field control points to verify and correct the distribution map, and overlaid the digital map of the administrative boundary onto the processed TM image to estimate the area of each forest type [4,18,19].

Forest inventory data

Forest area and timber volume for each age class and forest type have been inventoried in China once every five years since 1973 [20]. The systematic inventorying of forests on Hainan Island began in 1989 after the island became a province split from Guangdong Province. The forest inventory database used in this study included four inventories, each of which covered a 5-year period: 1989–1993, 1994–1998, 1999–2003, and 2004–2008. The inventory data included statistical report data, a plot database, and a sample trees database. The plot database contained more than 60 factors including plot number, name of dominant species, average tree diameter at breast height (DBH), average tree height, stand volume, number of standing trees (or bamboo), and litter thickness. The sample trees database contained 11 factors including the number of sampled trees, stand type, plot number, DBH, and volume. For the plot database, plots were established using a systematic sampling method. The southwest crossing point of each grid was used as a reference point to establish a 25.82-m×25.82-m plot within a 4-km×6-km grid in 1989 (1421 plots in total). Grid size was changed to 4-km×3-km in 1994 (2829 plots in total) [21].

Field survey plots in 2012 (field sampling data)

To verify the accuracy of the forest inventory data and to estimate C storage in the understory, litter, and soil layers, we established 100 field survey plots in 2012. The plots were selected based on forest type, spatial distribution, forest area, stand volume, and age class on the island. The number of plots for each forest type was as follows: 50 for natural forest (tropical rain forest), 24 for rubber plantation, 8 for eucalyptus plantation, 3 for *Acacia* plantation, 3 for *Pinus* plantation, 2 for *Casuarina* plantation, 1 for mixed coniferous and broad-leaved species forest, 3 for mango orchard, 3 for betel nut orchard, 2 for lychee orchard, 1 for longan orchard, and 1 for other hardwood forest. There were three replicated quadrats in each plot. The area per quadrat was 3600 m^2 for natural forest, 800 m^2 for plantation, and 400 m^2 for orchard. The measured variables were the same as in forest inventory. In addition, for each quadrat, we sampled plant tissue in the tree and understory layer, litter, and soil for laboratory analysis.

Estimation of C storage in forest ecosystems

The C in forest ecosystems includes C stored in the tree layer (tree C, including tree root C), shrub layer, herb layer, litter layer, and soil layer. C storage in the tree layer was estimated by forest inventory data and validated by our field sampling data in 2012. C storage in the shrub layer, herb layer, litter layer, and soil layer in 2012 was calculated using our field sampling data. The methods for estimating C storage in these layers were described below. Since the data of C storage in the shrub layer, herb layer, litter layer, and soil layer were not included in the forest inventories, we estimated these data using the relationships between measurements of shrub, herb litter, soil layer C and tree layer C biomass measurements developed using the measurements in 2012. While C storage in shrub, herb, litter and soil layer, and tree layer C biomass varied among years, we assumed that the relationships did not change.

Estimation of C storage in the tree layer based on forest inventory

The biomass of trees was calculated using the Biomass Expansion Factor (f_{BEF}) method [4].

$$f_{\text{BEF}} = a + b/V \qquad (1)$$

where V is forest stand volume (V, m^3 ha^{-1}, for the measurement method see reference [21]), and a and b are parameters of the conversion factor of a specific tree species from volume to biomass. The conversion factor values for different dominant tree species were obtained from previous studies on Hainan Island (Table 1).

The tree biomass at the forest stand scale (B, Mg ha^{-1}) was calculated using the following formula:

$$B = \sum_{i=1}^{k} A_i \times f_{BEF_i} \times V_i \qquad (2)$$

where i is the dominant species of the forest type, A_i is the forest stand area, V_i is the average storage volume, and $f_{\text{BEF}i}$ is the corresponding conversion factor of the i dominant species in the forest type.

The data of C storage in trees were also calculated at the city scale (Hainan Island has 18 cities. Each city represents an administration area). The biomass of the j-th plot in the i-th city (Bij) can be calculated using the following formula:

$$B_{ij} = aV_{ij} + b \qquad (3)$$

where the units of B_{ij} and V_{ij} are Mg ha^{-1} and m^3 ha^{-1}, respectively, and a and b are conversion factors of the dominant species (Table 1).

The formula for determining the average biomass of trees in the i-th city (B_i, Mg ha^{-1}) was:

$$B_i = \frac{1}{n} \sum_{j=1}^{n} B_{ij} \qquad (4)$$

where n is the total number of plots in the i-th city. The formula for determining the total biomass of trees in the i-th city (Ti) was:

$$T_i = 100 * A_i * C_i * B_i \qquad (5)$$

where A_i is the land area (unit: km^2) in the i-th city, C_i is the percentage of forest coverage in the i-th city, B_i is the average biomass of tree in the i-th city (Mg ha^{-1}), and 100 is the unit conversion factor.

The total tree biomass in Hainan Province (T) was summed for all 18 county/city-level cities as:

$$T = \sum_{i=1}^{18} T_i \qquad (6)$$

Tree C storage on Hainan Island was calculated by multiplying forest biomass (T) by the C concentration. The C concentration was measured in 2012.

Estimation of C storage in the understory layer based on field sampling and laboratory analysis

The understory layer included a shrub layer (0.5 to 1.5 m tall) and a herb layer (<0.5 m tall). To estimate C storage in the understory, we collected all plant individuals including seedlings from three 5-m×5-m subquadrats in each quadrat. The collected material was dried and weighed, and 30% of the dried material per subquadrat was used for determination of C concentration by the potassium dichromate oxidation method [28]. C storage in understory layers was estimated by multiplying the dry mass of the ground layer collected from each plot and the corresponding ground layer C concentration [29].

Estimation of C storage in the litter layer based on field sampling and laboratory analysis

To determine litter layer C, we collected all litter from three 1-m×1-m subquadrats is each quadrat. The methods used for

Table 1. The conversion formulas used in previous studies for estimating the biomass of dominant tree species on Hainan Island.

Dominant species	Biomass expansion factor (f_{BEF}) formula	n*	R²**	Reference
Eucalyptus	$f_{\text{BEF}} = 0.8873 + 4.5539/V$	20	0.80	Han et al., 2010 [22]
Rubber	$f_{\text{BEF}} = 0.7975V + 0.4204$	18	0.87	Cao et al., 2009 [23]
Pinus	$f_{\text{BEF}} = 0.5101V + 1.0451$	12	0.92	Fang et al., 2001 [24]
Cunninghamia lanceolata	$f_{\text{BEF}} = 0.3999V + 22.541$	56	0.95	Fang et al., 1996, 2001 [8,24]
Native broad-leaved species plantation (soft wood#)	$f_{\text{BEF}} = 0.7564V + 8.3103$	12	0.91	Cao et al., 2009 [23]
Native broad-leaved species plantation (hard wood#)	$f_{\text{BEF}} = 0.6255V + 91.0013$	19	0.86	Li, 1993 [25]
Tropical rain forest species	$f_{\text{BEF}} = 1.0357V + 8.0591$	17	0.89	Li et al., 1995 [26]
Acacia	$f_{\text{BEF}} = 0.6255V + 91.0013$	19	0.86	Zhou et al., 2008 [27]
Fruit species	$f_{\text{BEF}} = 0.3154V + 3.4171$	6	0.76	Cao et al., 2009 [23]
Casuarina equisetifolia	$f_{\text{BEF}} = 0.7441V + 3.2377$	10	0.95	Fang et al., 2001; Zhou et al., 2008 [24,27]
Mixed coniferous and broad-leaved tree species	$f_{\text{BEF}} = 0.8136V + 18.4660$	10	0.99	Fang et al., 2001; Zhou et al., 2008 [24,27]

*: n is the number of trees used in developing the regression model.
**: R² is the coefficient of determination. All the regression models are significant (P<0.05).
#: hard wood (wood density >0.7); soft wood (wood density <0.7).

collection, analysis, and calculation were the same as those used for the understory.

Estimation of C storage in the soil layer based on field sampling and laboratory analysis

For determination of C in the forest soil, we collected three soil cores (4 cm diameter and 100 cm deep) per subquadrat with a soil auger. We separated each 10-cm layer for the top 70 cm (seven layers), while the 70–100 cm depth was sampled as one layer because of its relatively constant C concentration. Soil depth varied among subquadrats, and we collected cores to the maximum depth in each case. The soil bulk density was measured in accordance with the soil layers of every 1 meter soil profile [30]. The samples were processed by the potassium dichromate oxidation method for determination of the organic matter [28].

C storage in the soil of the j-th plot of the i-th city (SOC$_{ij}$) was calculated as:

$$SOC_{ij} = 0.58 * 100 * W_{ij} * D_{ij} * R_{ij} \qquad (7)$$

where the units for SOC$_{ij}$ are Mg ha^{-1}; W_{ij} is soil bulk density (g cm^{-3}); D_{ij} is soil depth (cm, soil depth ranged from 60 to 100 cm for different soil types, which depended on the soil layer depth in the field); R_{ij} is the average soil organic matter content (%) of the j-th plot in the i-th city; 0.58 is the conversion coefficient from organic matter to organic C [4]; and 100 is the unit conversion factor. The mean SOCi of the i-th city was calculated as:

$$SOC_i = \frac{1}{18} \sum_{j=1}^{18} SOC_{ij} \qquad (8)$$

where 18 represent that there are 18 cities in Hainan.

The total ecosystem C storage of i-th city (Total C$_i$, Mg ha^{-1}) was summed by vegetation C and SOC. Therefore, we used the same calculation method as above to obtain C storage data for different cities on Hainan Island.

Mapping methods

Based on the estimation of total C storage (Vegetation C and SOC) in each city, we produced the spatial distribution map of C storage on the administration map in Hainan. The spatial distribution maps of forest ecosystem C storage on Hainan Island in 2008 were created by overlaying the spatial distribution map of tree biomass C storage in 2008 and the spatial distribution map of C storage in the shrub, herb, litter, and soil layers in 2012.

Uncertainty analysis

There were three major sources of uncertainty in C storage estimation in forest ecosystems on Hainan Island: the uncertainty in estimation of C storage in tree layer, uncertainty in relationships used to estimate C storage in the shrub, herb, litter, and soil layers from C storage in tree layer and uncertainty in forest area estimation in our research. The uncertainty of estimations was conducted by analysis of the different error sources. The error of estimation on C storage in tree layer mainly came from the input data such as inventory of forest area and volume and model parameters associated with regression coefficients used for estimation of dominant tree biomass. The Monte-Carlo method [4] was used to calculate the uncertainty in estimation of C storage in tree layer and uncertainty in forest area estimation. The uncertainty in relationships used to calculate C storage in the

shrub, herb, litter, and soil layers from C storage in tree layer in 2012 could come from two sources. One was the modeling fitting of C storage in tree layer with C storage in other layer. Another source was the application of these relationships developed in 2012 to other years. We estimated these uncertainty using error propagation method following the Guide to the Expression of Uncertainty in Measurement [31,32]. The law of the propagation of uncertainty or the Taylor method [32] was also used in this analysis.

Results

Change in forest coverage from the 1940s to the 2010s and the spatial distribution of forests on Hainan in 2008

Forest coverage (defined as the percentage of total land area in a region that is covered by any kind of natural or artificial forest) on Hainan Island increased from 20.7% in the 1950s to 56.4% in the 2010s. However, the natural forest coverage (defined as the percentage of total land area in a region that is covered by natural forests) decreased from 49.9% in the 1940s to 6.9% in the 2010s (Fig. 1). According to the forest inventory reports in 2008, Hainan Island had six types of forest ecosystems in 1993 and 11 types in 2008 (Table 2. Five new forest types were counted). Among them, natural tropical rain forests occupied the largest area, and followed with *Eucalyptus* (Table 2).

The area occupied by plantations was much larger than that occupied by natural forests in 2008 (Fig. 2). The natural tropical rain forests mainly grew in the mountainous areas of the central south of Hainan, while the plantations were distributed in the northern hilly land and the surrounding coastal plateau.

Change in C density of the forest ecosystems from 1993 to 2008

The average C density across all forest types in Hainan in 2008 was 163.7 Mg C/ha. Among the layers of tree, shrub, herb, litter, and soil, C density in the soil layer was the largest and accounted for most of the C in each forest ecosystem (Table 2). The C density in the soil layer was 121.4 Mg C/ha, which accounted for about 74% of the total C density. The vegetation C density was about

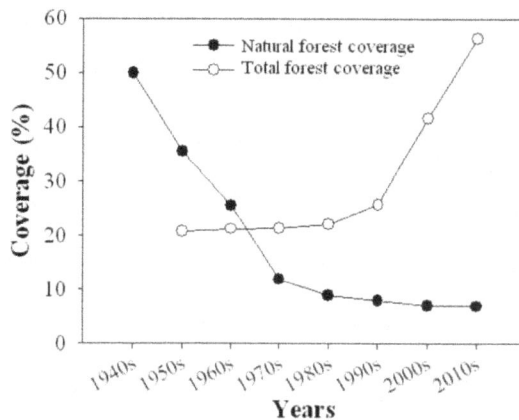

Figure 1. Total forest coverage and natural forest coverage on Hainan Island from 1940s to 2010s. Data are from the National Forest Resources Inventory.

Table 2. Carbon (C) density, forest area, and C storage of forest ecosystems on Hainan Island.

Year and forest type	Area (100 ha)	C density (t/ha)						Total forest ecosystem C storage (Tg)
		Tree layer	Soil layer	Shrub layer	Herb layer	Litter layer	Forest ecosystem	
1993								
Casuarina plantation	551	12.89	116.44	2.12	0.76	1.03	133.25	7.34
Native broad-leaved species plantation (soft wood**)	179	49.59	124.06	3.30	0.52	1.13	178.60	3.20
Eucalyptus plantation	1596	9.51	114.79	2.04	0.82	1.01	128.16	20.45
Tropical rain forest (natural+secondary)	3539	81.58	126.99	4.84	0.43	1.17	215.01	76.09
Pinus plantation	144	7.56	113.56	1.99	0.86	0.99	124.96	1.80
Cunninghamia lanceolata plantation	48	10.83	115.49	2.07	0.80	1.02	130.21	0.62
Total	6057							109.51
1998								
Casuarina plantation	588	12.73	116.38	2.12	0.77	1.03	133.02	7.82
Native broad-leaved species plantation (soft wood**)	204	67.31	125.85	4.08	0.47	1.15	198.86	4.06
Eucalyptus plantation	1703	13.88	116.85	2.15	0.75	1.04	134.66	22.93
Tropical rain forest (natural+secondary)	5169	71.19	126.18	4.27	0.46	1.16	203.26	105.06
Pinus plantation	216	8.00	113.86	2.00	0.85	1.00	125.72	2.72
Cunninghamia lanceolata plantation	48	19.60	118.76	2.30	0.69	1.06	142.41	0.68
Acacia plantation	240	49.20	124.01	3.28	0.52	1.13	178.14	4.28
Total	8168							147.55
2003								
Casuarina plantation	516	15.06	117.30	2.18	0.74	1.04	136.31	7.03
Native broad-leaved species plantation (soft wood**)	239	57.73	124.95	3.64	0.49	1.14	187.95	4.49
Eucalyptus plantation	1667	14.96	117.26	2.18	0.74	1.04	136.17	22.70
Tropical rain forest (natural+secondary)	5756	69.87	126.07	4.21	0.46	1.16	201.76	116.13
Pinus plantation	264	15.03	117.29	2.18	0.74	1.04	136.28	3.60
Cunninghamia lanceolata plantation	71	20.94	119.13	2.34	0.68	1.06	144.15	1.02
Native broad-leaved species plantation (hard wood**)	23	43.97	123.36	3.08	0.54	1.12	172.07	0.40
Acacia plantation	384	52.77	124.42	3.43	0.51	1.13	182.26	7.00
Total	8920							162.37
2008								
Casuarina plantation	312	16.30	117.73	2.21	0.72	1.05	138.01	4.31
Native broad-leaved species plantation (soft wood**)	48	46.06	123.63	3.16	0.53	1.12	174.51	0.84

Table 2. Cont.

Year and forest type	Area (100 ha)	C density (t/ha)						Total forest ecosystem C storage (Tg)
		Tree layer	Soil layer	Shrub layer	Herb layer	Litter layer	Forest ecosystem	
Eucalyptus plantation	1930	12.55	116.30	2.11	0.77	1.03	132.75	25.62
Tropical rain forest (natural+secondary)	5133	73.82	126.40	4.41	0.45	1.16	206.23	105.86
Pinus plantation	324	15.08	117.30	2.18	0.74	1.04	136.34	4.42
Cunninghamia lanceolata plantation	11	33.73	121.83	2.73	0.59	1.10	159.98	0.18
Native broad-leaved species plantation (hard wood**)	107	79.21	126.82	4.71	0.44	1.17	212.33	2.27
Acacia plantation	480	51.28	124.25	3.37	0.52	1.13	180.54	8.67
Mixed coniferous and broad-leaved tree species plantation.	72	50.71	124.19	3.34	0.52	1.13	179.89	1.30
Rubber plantation	5754	18.17	118.34	2.26	0.70	1.05	140.52	80.86
Orchard	3203	17.82	118.23	2.25	0.71	1.05	140.05	44.86
Total	17374*							279.17

*In 2008, rubber plantations, orchards, and mixed coniferous and broad-leaved tree species plantations were included in the statistics. The rubber plantation and orchard were first accounted, and some *Pinus* plantation developed into mixed coniferous and broad-leaved tree species plantation.
**hard wood (wood density >0.7); soft wood (wood density <0.7).

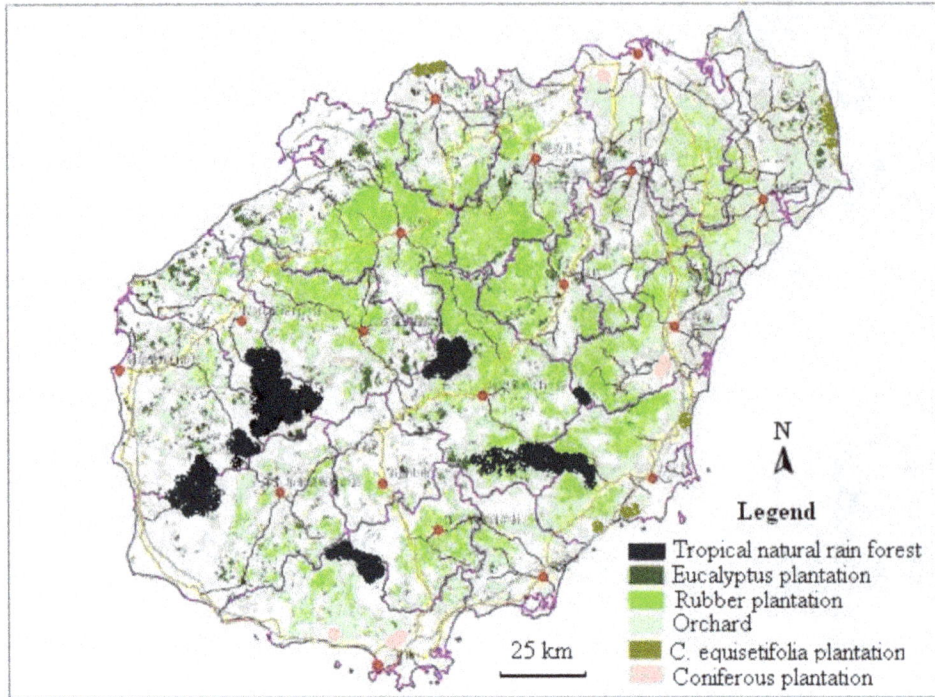

Figure 2. The distribution of forests on Hainan Island in 2008.

41.2 Mg C/ha, and the C density in litter layer was only 1.1 Mg C/ha (Table 2).

The C density of most forest ecosystems on Hainan Island gradually increased from 1993 to 2008 (Table 2). The C density of forest ecosystems in 1993 varied with forest type and ranged from 125.0 Mg C/ha in *Pinus* plantations to 215.0 Mg C/ha in natural tropical rain forests. The C density of native broad-leaved species plantations (hard wood) was the highest and followed by natural tropical rain forests. In 2008, the lowest C density was in

Eucalyptus plantations. Overall, C density was higher in natural tropical rain forests and native broad-leaved species plantations (hard wood) than in more artificial systems such as rubber and *Eucalyptus* plantations. The C storage was higher in forest types with natural regeneration (e.g., mixed coniferous and broad-leaved species plantation) than in plantations.

The average carbon density of forest ecosystems on Hainan Island increased about 2.11 Mg C/ha from 1993 to 2008 (Excluding the statistics on rubber plantations and orchards in

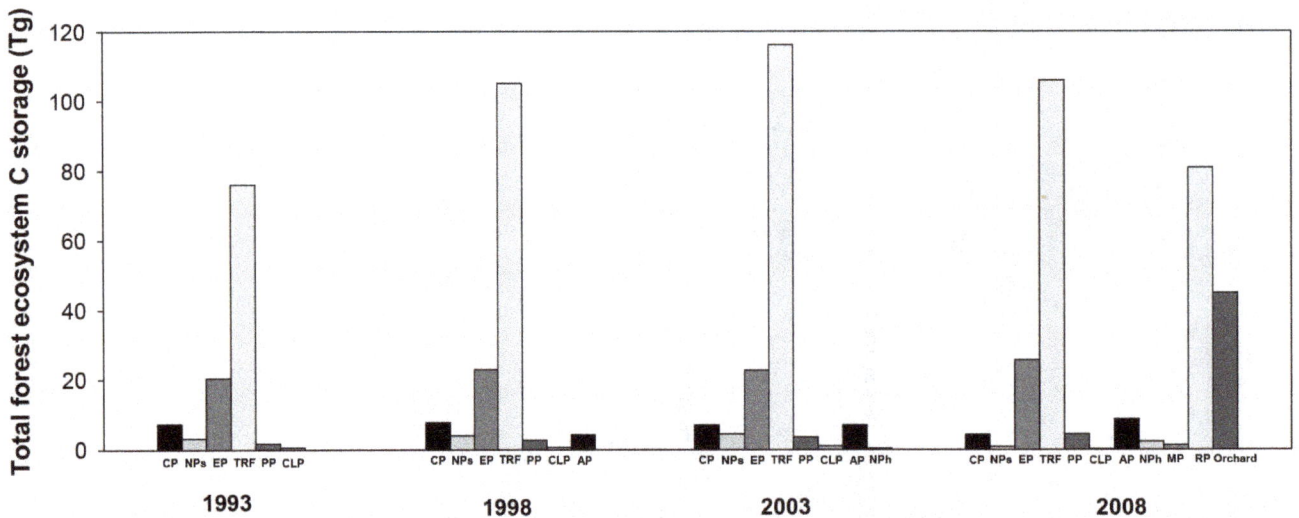

Figure 3. Total C storage of different forest ecosystems on Hainan Island during 1993–2008. CP: *Casuarina* plantation; NP (soft wood): Native broad-leaved species plantation (soft wood); EP: *Eucalyptus* plantation; TRF: Tropical rain forest (natural+secondary); PP: *Pinus* plantation; CLP: *Cunninghamia lanceolata* plantation; AP: *Acacia* plantation; NP (hard wood): Native broad-leaved species plantation (hard wood); MP: Mixed coniferous and broad-leaved tree species plantation; RP: Rubber plantation.

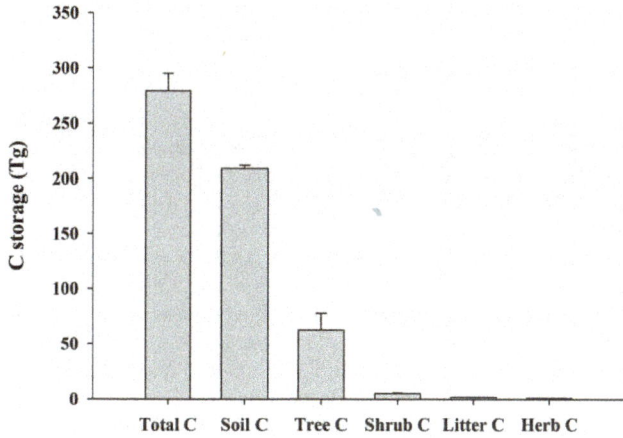

Figure 4. The C storage in different layers of forest ecosystems on Hainan Island in 2008. Tree C includes C in above- and below-ground biomass. Values are means ± SE.

2008). Although the coverage of natural forest with higher carbon density decreased, the average carbon density across all forest types increased. Since other types of forests accounted for a large area and also continued to accumulating C. The results meant that the average carbon density was strongly dependent on the spatial extent of the region and types of land uses included. The average carbon density had changed along with the shifts in forest type and forested area (Table 2; Fig. 1).

Change in C storage in forest ecosystems from 1993 to 2008

Over the past 15 years, the total forest C storage on Hainan Island gradually increased 1.55 times from 109.51 in 1993 to 279.17 in 2008. The C storage of most forest ecosystems increased from 1993 to 2009. Among them, the C storage in *Pinus* and *Eucalyptus* plantations increased 35% from 1993 to 2008. In *Casuarina* plantations and natural tropical rain forests, however, C storage increased from 1993 to 2003 but decreased from 2003 to 2008 (Fig. 3).

The forest ecosystems on Hainan island in 2008 stored about 279.17 Tg C, with 209.07 Tg in the soil layer, 62.19 Tg in the tree

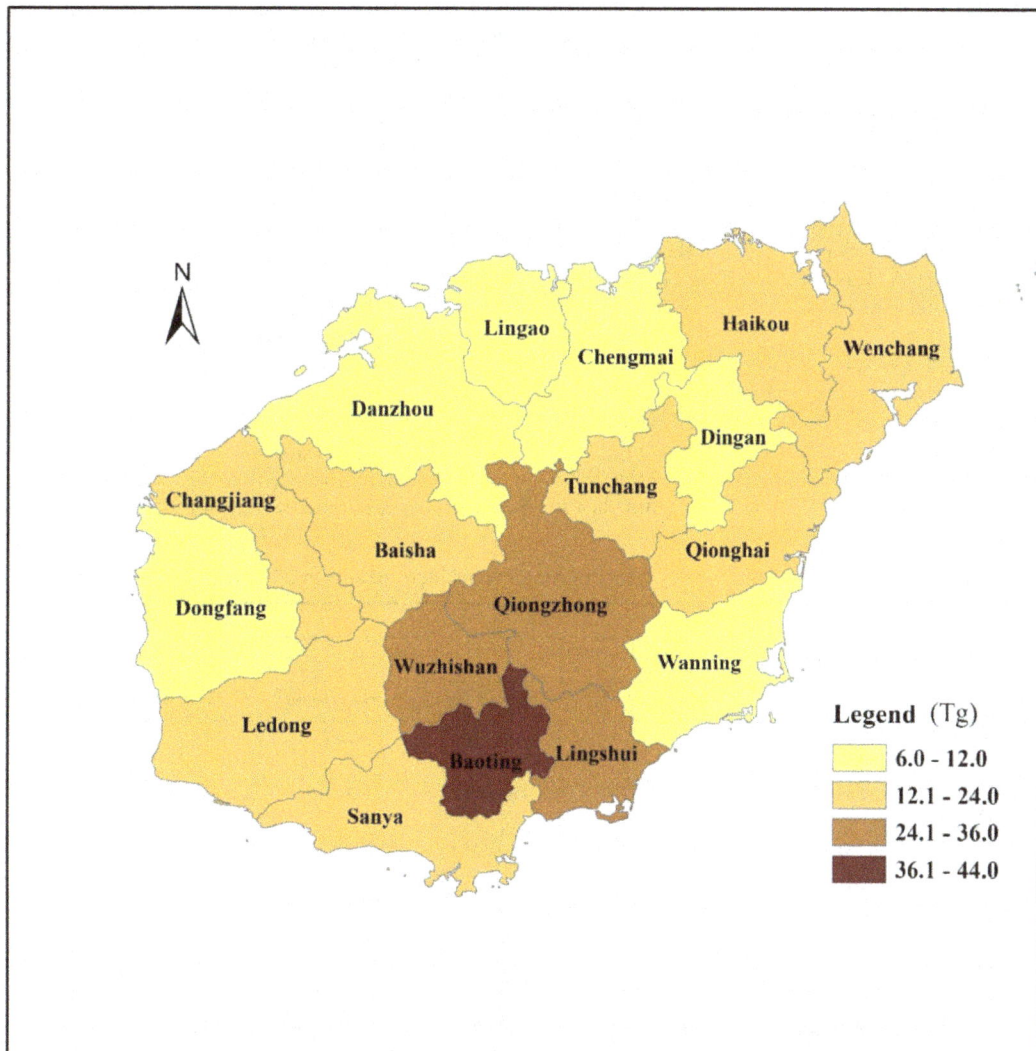

Figure 5. The spatial distribution of forest ecosystem C storage on Hainan Island in 2008.

layer, 5.06 Tg in the shrub layer, 1.06 Tg in the herb layer, and 1.79 Tg in the litter layer. Soil C accounted for 74.9% of the total C storage (Fig. 4).

The spatial distribution of forest ecosystem C storage on Hainan Island in 2008

The spatial distribution of forest ecosystem C storage (combination of tree biomass C storage and C storage in the understory, litter, and soil layers) in 2008 was not homogenous across the province (Fig. 5). The forest ecosystem C storage was highest in the south central region (24.1–44.0 Tg C), lowest in the north (6.0–12.0 Tg C), and intermediate in other regions (12.1–24.0 Tg C).

Uncertainty analysis of total forest ecosystems C density

The results of uncertainty analysis of forest ecosystem C density indicated that forest ecosystem C density errors were partly came from the uncertainty in relationships used to calculate from C storage in tree layer to that in shrub, herb, litter and soil layers among three sources, it accounted for an average 3.2% ($\pm$5.05 Mg C/ha). The uncertainty on the estimation of C storage in tree layer and uncertainty in forest area estimation accounted for an average 6.1% ($\pm$10.13 Mg C/ha) of the total error (Table 3). The uncertainty of different forest types varied remarkably (Table 3).

Discussion

The C storage in forest ecosystems is closely related to ecosystem area and forest health. Both forest ecosystem area and health have declined on Hainan Island since 1900s [15]. The tropical forest area has decreased at a rate of 2.02% per year since 1950 [15]. The main causes of the tropical forest loss were excessive lumbering, planting of rubber trees, slash-and-burn cultivation, and unrestricted deforestation for fuels and other usages. Change from natural forests to artificial plantations has caused an obvious decrease in forest quality. Fortunately, the government realized the importance of protecting natural forests in the 1990s and prohibited further deforestation on Hainan Island. Rubber plantations, *Eucalyptus* plantations, and orchards, however, remain abundant. Because of rapid population growth and economic development, natural tropical forests on the plains and hilly land and along the coast have mostly been destroyed, and only remain in the mountain areas. The current status reflects a long history of human disturbance and of persistent conflict between development and conservation.

The average C density of 163.7 Mg C/ha in Hainan's forest ecosystems as estimated in this study was slightly higher than the average in China's mainland forests. For example, Wang et al. (2001) and Ren et al. (2013) reported that China's mainland forests contain a total of 141.3–147.5 Mg C/ha, with an average of 36–42 Mg C/ha in the vegetation and 105.3 Mg C/ha in the soil [4,14]. Among those forest types in Hainan in 2008, the average C density of natural forest was the highest and was 206.23 Mg/ha in total, 78.68 Mg/ha in vegetation layer, 1.16 Mg/ha in litter layer and 126.40 Mg/ha in soil layer. The average C density of other plantation forests varied from 132.75 to 180.54 Mg C/ha. Hainan could not provide more land for planting trees [21], therefore, the potential carbon sequestration scenario was that all forested lands were restored to natural tropical forests. The potential carbon sequestration would be 77.3 Tg C. In addition, our estimation was lower than the C density in tropical forests worldwide (279 Mg/ha in total, 157 Mg/ha in vegetation and 122 Mg/ha in soil) without considering the influence of climate, fertility and other limiting factors to the forest growth [2]. The C density in forest soils on Hainan Island was close to the average C density in soil of tropical forests worldwide, but the C density in the vegetation layer in forests on the island was far less than that in tropical forests worldwide. The C storage on Hainan Island could be increased by selecting tree species with high C densities and by improving community structure. The C density in the forest ecosystems on Hainan Island was high in the soil and vegetation layers and was low in the litter layer. The low C density in the litter was reasonable because litter decomposition and nutrient cycling should occur at rapid rates under the high temperate and moisture conditions on Hainan Island.

The total forest ecosystem C storage on Hainan Island increased from 109.51 Tg in 1993 to 279.17 Tg C in 2008, with a total increase of 169.66 Tg. The increase was partially due to a 30% net increase in forest coverage during this period (Fig. 1) and partially due to the shifts in forest types. This increase was similar to the average increase in forest ecosystems in China [24]. It is worth noting that, if the C stored in rubber plantations and orchards was removed from the calculation, the total forest C storage on Hainan Island would increase by only 43.94 Tg. Another finding was that

Table 3. Estimations of carbon density in different layers of major forest ecosystems in 2008 with uncertainty analysis (Mean$\pm$SE).

Forest type	C density (t/ha)					
	Tree layer	Soil layer	Shrub layer	Herb layer	Litter layer	Forest ecosystem
Tropical rain forest (natural+ secondary)	67.67$\pm$20.35	125.88$\pm$1.78	4.10$\pm$1.00	0.47$\pm$0.05	1.15$\pm$0.02	199.27$\pm$20.45
Native broad-leaved species plantation (hard wood**)	66.34$\pm$20.19	125.76$\pm$1.80	4.03$\pm$0.98	0.47$\pm$0.05	1.15$\pm$0.02	197.75$\pm$20.30
Acacia plantation	54.56$\pm$8.13	124.61$\pm$0.87	3.50$\pm$0.34	0.50$\pm$0.03	1.14$\pm$0.01	184.32$\pm$8.19
Native broad-leaved species plantation (soft wood**)	47.32$\pm$4.80	123.78$\pm$0.59	3.21$\pm$0.18	0.53$\pm$0.02	1.13$\pm$0.01	175.96$\pm$4.84
Cunninghamia lanceolata plantation	24.89$\pm$0.00	120.10$\pm$0.00	2.45$\pm$0.00	0.65$\pm$0.00	1.08$\pm$0.00	149.17$\pm$0.00
Pinus plantation	24.20$\pm$8.65	119.94$\pm$2.02	2.43$\pm$0.25	0.65$\pm$0.06	1.08$\pm$0.03	148.30$\pm$8.89
Eucalyptus plantation	18.92$\pm$9.48	118.56$\pm$2.79	2.28$\pm$0.26	0.69$\pm$0.09	1.06$\pm$0.04	141.51$\pm$9.89
Casuarina plantation	13.07$\pm$7.80	116.52$\pm$3.27	2.13$\pm$0.20	0.76$\pm$0.11	1.03$\pm$0.04	133.51$\pm$8.46

**hard wood (wood density >0.7); soft wood (wood density <0.7).

C storage from 2003 to 2008 did not increase much or even decreased slightly. This occurred because local farmers planted large areas with rubber plantations, *Eucalyptus* plantations, and orchards for economic reasons. The local government encouraged farmers to convert the existing commercial forest stand such as rubber plantations, *Eucalyptus* plantations, and orchards into ecological forest (i.e. the forests or plantations to provide ecosystem services and social services in important eco-regions or fragile regions). However, the annual compensation fee of ecological forests was only about 25% of the commodity value of plantation such as timber, rubber, and fruits [34,35].

Rubber plantations, pulp plantations (*Eucalyptus* and *Acacia*), and orchards represent a serious threat to Hainan's natural tropical forests and C storage. The regrowth of tropical secondary forests and plantations cannot offset the C that is released as a consequence of forest deforestation, resulting in an overall net C loss on tropical lands. The C density varied among different forest types, and the C density of natural tropical rain forest was higher than that of other forest types on Hainan Island. Although Song et al. (2014) hypothesized that rubber plantations in tropical China may act as a large C sink, they were not a C sink when the deforestation of pre-existing tropical forests was considered during the establishment of rubber plantations [33]. Our previous study of C storage in *Eucalyptus* plantations and orchard on mainland China showed similar results as the rubber plantations [4]. Those studies indicated that the conversion from natural forest to plantation would result in decreasing C storage. However, farmers preferred to cut natural forests, grow the fast growing commercial trees, and sell timber to obtain the immediate economic benefits. They seldom considered the tradeoffs between conservation and agriculture [15]. Conversion of remaining natural forests to plantations would result in a loss of 105 Tg C, thus preservation of remaining natural tropical forests could make an important contribution to carbon sequestration and other ecosystem services on the island. Therefore, we provide the following recommendations to increase C sequestration in forest ecosystems: protection of all natural forests, afforestation in barren lands or waste lands, planting hard wood native species with high C fixation abilities, and restoration of forests from croplands in low productivity areas.

The estimation of forest C storage on Hainan Island varied among studies due to that various methods were used by investigators and the forest ecosystems are complex in nature. To guide climate change studies, Intergovernmental Panel on Climate Change (IPCC) published a methodological and technological guideline [36]. We applied a method that similar to the method recommended by the IPCC. Compared to other methods

in previous studies [8–12], our estimation accounted for additional C storage in the understory layer, litter layer and soil layer and directly measured C concentration coefficient of plant organs. In addition, the uncertainty analysis of ecosystem C density, and the temporal and spatial heterogeneity of C storage were first studied, which provided more useful information for forest management in Hainan.

Conclusions

By combining field measurements with data from forest inventories, we quantified the C storage in tropical forest ecosystems on Hainan Island between 1993 and 2008. The average C density in Hainan's forests in 2008 was 163.7 Mg C/ha, with 121.4 Mg C/ha in the soil, 1.1 Mg C/ha in the litter, and 41.2 Mg C/ha in the vegetation (trees, shrubs, herbs, including their roots). The C density of Hainan's forests was higher than the average C density of terrestrial forest ecosystems in China but lower than the worldwide average for such ecosystems. Hainan's tropical forest ecosystems stored 109.51, 147.55, 162.37, and 279.17 Tg C in total in 1993, 1998, 2003, and 2008, respectively. The total C storage in the above- and below-ground portions of forest ecosystems increased over time because of the increase in forest area and the forest type change. The spatial distribution of forest C storage on Hainan Island has been and remains uneven, and the spatial heterogeneity is related to land use, forest type, soil type and climate factors. With the increase in forest area and forest development on Hainan Island, C storage is expected to continuously increase. The potential carbon sequestration was 77.3 Tg C if all the forest stands were still natural tropical forests in 2008. From C sequestration point of view, future forest management should focus on the selection of tree species, the rational planting of understory vegetation at plantations, and implementation of sustainable practices.

Acknowledgments

The authors are indebted to Prof. Bruce Jaffee for English editing, and the anonymous reviewers for their constructive comments.

Author Contributions

Conceived and designed the experiments: HR QL XW YL GZ. Performed the experiments: HR QL XW SJ JW HY HL XT QZ LY XC. Analyzed the data: HR LL DH DW. Contributed to the writing of the manuscript: HR LL DH JW HL.

References

1. Ciais P (1995) A large northern hemisphere terrestrial CO_2 sink indicated by the $^{13}C/^{12}C$ ratio of atmospheric CO_2. Science 269: 1098–1102.
2. Lal R (2005) Forest soils and carbon sequestration. For Ecol Manag 220: 242–258.
3. Houghton RA (2005) Aboveground forest biomass and the global C balance. Glob Change Biol 11:945–958.
4. Ren H, Chen H, Li L, Li P, Hou C, et al. (2013) Spatial and temporal patterns of carbon storage from 1992 to 2002 in forest ecosystems in Guangdong, Southern China. Plant Soil 63: 123–138.
5. Ashton MS, Tyrrell ML, Spalding D, Gentry B (2012) Managing forest carbon in a changing climate. London: Springer. 397p.
6. Houghton RA (1992) Tropical forests and climate. Paper presented at the International workshop on ecology, conservation and management of Southeast Asian rainforests, October 12–14, Kuching, Sarawak.
7. Malhi Y, Grace J (2000) Tropical forests and atmospheric carbon dioxide. Trends Ecol Evol 15: 332–337.
8. Fang JY, Liu GH, Xu SL (1996) Biomass and net production of forest vegetation in China. Acta Ecol Sin 16: 497–508.
9. Zhao M, Zhou GS (2004) Carbon storage of forest vegetation and its relationship with climatic factors. Sci Geogr Sin 4: 50–54.
10. Wang XK, Feng ZW, Ouyang ZY (2001) Vegetation carbon storage and density of forest ecosystems in China. Chinese J Appl Ecol 12: 13–16.
11. Cao J, Zhang YL, Liu YH (2002) Changes in forest biomass carbon storage in Hainan Island over the past 20 years. Geogr Res 21: 551–560.
12. Li HQ, Lei YC (2010) Estimation and evaluation of forest biomass carbon storage in China. Beijing: China Forestry Publishing House. 60p.
13. Guo ZD, Hu HF, Li P, Li NY, Fang JY (2013) Spatial-temporal changes in biomass carbon sinks in China's forests during 1977–2008. Sci China Life Sci 43: 421–431.
14. Wang XK, Feng ZW, Ouyang ZY (2001) Vegetation carbon storage and density of forest ecosystems in China. Chinese J Appl Ecol 12: 13–16.
15. Zhou G (1995) Influences of tropical forest changes on environmental quality in Hainan province, P.R. of China. Ecol Eng 4: 223–229.
16. Jobin B, Beaulieu J, Grenier M, Bélanger L, Maisonneuve C, et al. (2003) Landscape changes and ecological studies in agricultural regions, Québec, Canada. Landscape Ecol 18: 575–590.
17. Hou XY (2001) The vegetation atlas of China. Beijing: Science Press. 280p.
18. Achard F, Eva H, Mayaux P (2001) Tropical forest mapping from coarse spatial resolution satellite data: production and accuracy assessment issues. Int J Remote Sens 22: 2741–2762.

19. Lee TM, Yeh HC (2009) Applying remote sensing techniques to monitor shifting estuary mangrove communities, Taiwan. Ecol Eng 35: 487–496.

20. Guo Z, Fang J, Pan Y, Birdsey R (2010) Inventory-based estimates of forest biomass carbon stocks in China: A comparison of three methods. For Ecol Manag 259: 1225–1231.

21. Hainan Bureau of Forestry (1999) Forest resource statistics of China (1994–1998). Department of Forest Resource and Management, Hainan Bureau of Forestry, Haikou, China.

22. Han FY, Zhou QY, Chen SX, Chen WP, Li TH, et al. (2010) Study on biomass and energy of two different-aged Eucalyptus stands. Forest Res 23: 690–696.

23. Cao JH, Jiang JS, Lin WF, Xie GS, Tao ZL (2009) Biomass of *Hevea* clone PR107. Chinese J Trop Agr 29: 1–8.

24. Fang J, Chen A, Peng C, Zhao S, Ci L (2001) Changes in forest biomass carbon storage in China between 1949 and 1998. Science 292: 2320–2322.

25. Li YD (1993) Comparative analysis for biomass measurement of tropical mountain rain forest in Hainan Island, China. Acta Ecol Sin 13: 314–320.

26. Li YD, Wu ZM, Zeng QB, Zhou GY, Chen BF, et al. (1998) Carbon pool and carbon dioxide dynamics of tropical mountain rain forest ecosystem at Jianfengling, Hainan Island. Acta Ecol Sin 18: 371–378.

27. Zhou C, Wei X, Zhou G, Yan J, Wang X, et al. (2008) Impacts of a large-scale reforestation program on C storage dynamics in Guangdong, China. For Ecol Manag 255:847–854.

28. Liu GS, Jiang NH, Zhang LD, Liu ZL (1996) Soil physical and chemical analysis and description of soil profiles. Beijing: Standards Press of China. 266p.

29. Liu H, Ren H, Hui D, Wang W, Liao B, et al. (2014) Carbon stocks and potential carbon storage in the mangrove forests of China. J Environ Manag 133: 86–93.

30. Zhang JP, Shen CD, Ren H, Wang J, Han WD (2011) Estimating change in sedimentary organic carbon content during mangrove restoration in Southern China using carbon isotopic measurements. Pedosphere 22: 58–66.

31. Cox M, Harris P, Siebert BPL (2003) Evaluation of measurement uncertainty based on the propagation of distributions using Monte Carlo simulation. Meas Tech 46: 824–833.

32. Krouwer JS (2003) Critique of the guide to the expression of uncertainty in measurement method of estimating and reporting uncertainty in diagnostic assays. Clin Chem 49: 1818–1821.

33. Song QH, Tan ZH, Zhang YP, Sha LQ, Deng XB, et al. (2014) Do the rubber plantations in tropical China act as large carbon sinks? Forest 7: 42–47.

34. Deng F, Chen Q, Chen X (2007) Comparison of ecological service among natural forest, rubber and Eucalyptus plantations. J South China Univer of Trop Agr 13:19–23.

35. Ren H, Shen W, Lu H, Wen X, Jian S (2007) Degraded ecosystems in China: Status, causes, and restoration efforts. Landscap Eco Engine 3:1–13.

36. IPCC (2000) Land use, land-use change, and forestry. Cambridge: Cambridge University Press. 375p.

A Melting Pot of Old World Begomoviruses and Their Satellites Infecting a Collection of *Gossypium* Species in Pakistan

Muhammad Shah Nawaz-ul-Rehman[1¤], **Rob W. Briddon**[2], **Claude M. Fauquet**[1]*

1 Danforth Plant Science Center, St. Louis, Missouri, United States of America, **2** Agricultural Biotechnology Division, National Institute for Biotechnology and Genetic Engineering, Jhang Road, Faisalabad, Pakistan

Abstract

CLCuD in southern Asia is caused by a complex of multiple begomoviruses (whitefly transmitted, single-stranded [ss]DNA viruses) in association with a specific ssDNA satellite; Cotton leaf curl Multan betasatellite (CLCuMuB). A further single ssDNA molecule, for which the collective name alphasatellites has been proposed, is also frequently associated with begomovirus-betasatellite complexes. Multan is in the center of the cotton growing area of Pakistan and has seen some of the worst problems caused by CLCuD. An exhaustive analysis of the diversity of begomoviruses and their satellites occurring in 15 *Gossypium* species (including *G. hirsutum*, the mainstay of Pakistan's cotton production) that are maintained in an orchard in the vicinity of Multan has been conducted using φ29 DNA polymerase-mediated rolling-circle amplification, cloning and sequence analysis. The non-cultivated *Gossypium* species, including non-symptomatic plants, were found to harbor a much greater diversity of begomoviruses and satellites than found in the cultivated *G. hirsutum*. Furthermore an *African cassava mosaic virus* (a virus previously only identified in Africa) DNA-A component and a *Jatropha curcas mosaic virus* (a virus occurring only in southern India) DNA-B component were identified. Consistent with earlier studies of cotton in southern Asia, only a single species of betasatellite, CLCuMuB, was identified. The diversity of alphasatellites was much greater, with many previously unknown species, in the non-cultivated cotton species than in *G. hirsutum*. Inoculation of newly identified components showed them to be competent for symptomatic infection of *Nicotiana benthamiana* plants. The significance of the findings with respect to our understanding of the role of host selection in virus diversity in crops and the geographical spread of viruses by human activity are discussed.

Editor: Baochuan Lin, Naval Research Laboratory, United States of America

Funding: This project was supported by the Pak-US linkage program (Pakistan-US Science and Technology Cooperation Program) under the contract PGA-7251-05-007. Dr. Briddon is supported by the Higher Education Commission (Pakistan) under the Foreign Faculty Hiring Program. The funders had no role in study design, data collection and analysis, decision to publish, or preparation of the manuscript.

Competing Interests: The authors have declared that no competing interests exist.

* E-mail: iltab@danforthcenter.org

¤ Current address: Molecular Virology Laboratory, Center of Agricultural Biochemistry and Biotechnology (CABB), University of Agriculture, Faisalabad, Pakistan

Introduction

Geminiviruses are plant-infecting viruses with circular single-stranded (ss)DNA genomes that are encapsidated in twinned icosahedral (geminate) particles. Based on genome organization, insect vector and host range, the family *Geminiviridae* is classified into four genera: *Begomovirus, Curtovirus, Mastrevirus* and *Topocovirus* [1,2]. The genus *Begomovirus* encompasses the majority of the known, as well as economically the most important, geminiviruses that are transmitted exclusively by the whitefly *Bemisia tabaci* [3,4]. Begomoviruses native to the New World (NW) have genomes consisting of two components, known as DNA-A and DNA-B. Although genetically distinct, bipartite begomoviruses have been identified in the Old World (OW). However, some of the emerging OW begomoviruses are monopartite, consisting of a component homologous to the DNA-A component of the bipartite viruses. Recently it has become clear that the majority of monopartite begomoviruses are associated with betasatellites that are important for infecting some hosts [5].

Both components of bipartite begomoviruses are required for infectivity of plants [6]. The DNA-A component encodes all viral proteins required for replication, control of gene expression and transmission between plants, whereas the DNA-B component encodes two proteins required for intra- and intercellular movement in plants [7]. The integrity of the bipartite genome is maintained by virtue of both components sharing a sequence, known as the common region (CR), which contains the origin of replication (*ori*) [8]. The *ori* consists of the ubiquitous nonanucleotide sequence (TAAT/GATTA/CC) that forms part of a predicted hairpin structure which is nicked by the DNA-A-encoded replication-associated protein (Rep: a rolling circle initiator protein) to initiate replication, and repeated sequence motifs (known as "iterons"), which are sequence specific Rep binding sites and are distinct for each species [9,10].

Betasatellites are approximately half the size (~1350 nt) of their helper begomoviruses, which they require for replication and movement in host plants, as well as transmission between plants [5]. Many begomoviruses that associate with betasatellites are wholly dependent on their satellites to efficiently and symptomatically infect some hosts. However, some begomoviruses have a more relaxed relationship, being able to infect plants and induce symptoms in both the presence and absence of the satellite [11]. The mechanism of *trans*-replication of betasatellites by their helper begomoviruses remains unclear. Betasatellites do not encode the iterons of their helper begomoviruses, although they have a predicted hairpin structure (containing a nonanucleotide motif) with similarity to the *ori* of geminiviruses. In most cases betasatellites are capable of being *trans*-replicated by several different begomoviruses. This indicates that their interaction with begomoviruses is distinct from the interaction of DNA-B components with their cognate DNA-A components [12,13].

Betasatellites encode a single product (known as βC1) that mediates all functions so far ascribed to these molecules. The βC1 protein is a pathogenicity determinant, a suppressor of post-transcriptional gene silencing (PTGS: a host defense mechanism against foreign nucleic acids), possibly binds DNA, may up-regulate viral DNA levels in plants and may provide virus movement functions [14,15,16].

Many begomovirus-betasatellite complexes also associate with a further class of satellite-like molecules that are collectively known as alphasatellites. The alphasatellites are also approximately half the size of a typical begomovirus component (~1380 nt) and encode a single product with similarity to the Rep proteins of another family of ssDNA viruses, the nanoviruses [5,17]. Alphasatellites are capable of autonomous replication in host plants but require the helper begomovirus for movement in plant tissues and transmission between plants. These molecules apparently perform no essential function for infectivity of plants. However, their almost ubiquitous presence in plants infected with begomovirus-betasatellite complexes suggests they perform some useful, if subtle, function which may provide a selective advantage to the helper begomovirus [5].

Cotton is the major source of fiber and has been produced on the sub-continent since prehistoric times [18]. Fiber in Asia was initially produced from a native cotton species, *Gossypium arboreum*, but is now produced from *G. hirsutum*, which was introduced from Mexico in 1818. Cotton leaf curl disease (CLCuD) was a sporadic problem across southern Asia prior to 1986. In 1986, in the vicinity of Multan (Pakistan), the disease became epidemic and rapidly spread to virtually all cotton growing regions of the country, as well as eastwards into India during the 1990s [19]. During the late 1990s losses due to the "Multan strain" of CLCuD were finally overcome by the introduction of resistant cotton varieties [20]. However, in 2001, resistant cotton varieties in the vicinity of Burewala (Pakistan) began to exhibit symptoms of CLCuD [21]. This signaled the beginning of the second CLCuD epidemic, known as the "Burewala strain", which now affects all cotton growing areas of Pakistan and northwestern India.

CLCuD in Pakistan during the 1990s was caused by a begomovirus-betasatellite complex that was associated with representatives of at least 6 begomovirus species - *Cotton leaf curl Multan virus* [CLCuMuV], *Cotton leaf curl Rajasthan virus* [CLCuRaV], *Cotton leaf curl Kokhran virus* [CLCuKoV], *Cotton leaf curl Alabad virus* [CLCuAlV], *Papaya leaf curl virus* [PaLCuV] and *Tomato leaf curl Bangalore virus*, either as single or multiple infections [22,23,24,25]. In contrast only a single betasatellite (Cotton leaf curl Multan betasatellite [CLCuMuB]) was isolated [26]. The Burewala epidemic, at least at the present time, is associated with only a single virus, *Cotton leaf curl Burewala virus* (CLCuBuV), which is a recombinant consisting of sequences derived from two of the earlier viruses (CLCuMuV and CLCuKoV) [27]. The betasatellite

associated with CLCuBuV is also recombinant, with most of the sequence derived from CLCuMuB [27,28].

The genus *Gossypium* L. (*Malvaceae*) comprises approximately 50 species of shrubs and small trees which originate from the tropics and sub-tropics. Of these, four species (*G. arboreum* and *G. herbacium*, having diploid genomes and originating from Africa-Asia, and *G. hirsutum* and *G. barbadense*, having tetraploid genomes and originating from the NW) have been cultivated as fiber and oilseed crops for at least 5000 years [29]. It is interesting to note that the native sub-continent species *G. arboreum* is immune to CLCuD, whereas the exotic introduced species *G. hirsutum* and *G. barbadense* are highly susceptible [30]. It is presumed that the native species have had a long association with the viruses causing CLCuD and have evolved resistance.

We have analysed the diversity of begomoviruses (which we shall collectively refer to as "cotton leaf curl geminiviruses" [CGs]) and begomovirus-associated DNA satellites occurring in a collection of mostly non-cultivated *Gossypium* spp. that have been maintained in an orchard in Multan, Pakistan, for over 40 years. The results show the presence of a surprising diversity of components, including new virus species and components, as well as viruses and components that have previously been identified in cotton and other plant species. The significance of these findings to our understanding of the evolution of begomoviruses and the selection pressures exerted upon them by agricultural crops are discussed.

Results

Rolling-circle amplification, cloning and sequencing

Leaf samples were collected from symptomatic and asymptomatic plants of the cotton species indicated in Table 1. The symptoms exhibited by plants of selected species are shown in Figure 1 and described in Table 1. DNA was extracted from leaf samples and used to amplify circular DNA molecules using rolling-circle amplification (RCA). RCA yielded a high molecular DNA product from 11 of the 15 *Gossypium* leaf samples. Since RCA exponentially amplifies only circular DNA molecules, the absence of a product for *G. arboreum*, *G. herbaceum* and *G. therburi* is a good indication that these species were not infected with begomoviruses. However, the presence of a high molecular weight DNA product is not necessarily indicative of the presence of a circular DNA virus, since non-viral molecules, such as mitochondrial plasmids, can be amplified by RCA [31]. Following restriction digestion, a total of 34 molecules of ~2800 nt, 87~1400 nt and 4 smaller clones were obtained and sequenced in their entirety, in both orientations, with no ambiguities remaining.

Sequence comparisons showed the ~2800 nt products to consist of molecules with similarity to the genomes (or DNA-A components) and DNA-B components of begomoviruses. The ~1400 nt clones were shown to have similarity with beta- and alphasatellites. The presence of each type of molecule in the individual leaf samples of plants each of the *Gossypium* species is indicated in Table 1.

Diversity of begomoviruses in the *Gossypium* species

The complete sequences of 34 potentially full-length molecules, originating from 10 non-cultivated cotton species and one cultivated species (*G. hirsutum*), homologous to the DNA-A components of bipartite begomoviruses were obtained. The features of these sequences and the accession numbers under which they are available in the databases are given in Table S1 and S2.

The begomovirus genome (or DNA-A component) sequences obtained were used in a phylogenetic analysis based upon an

Table 1. List of *Gossypium* species sampled, symptoms and virus/satellite components identified in each.

Cotton Species	Geographic Origin	Ploidy level	Agricultural status@	Genome status	Symptoms^S	Begomovirus	DNA-B	Betasatellite	ACMV DNA-A	GPML-CuV	CLCu-BuV	CLCu-MuV	CLCu-RaV	CLCu-KoV	GDar-SLA	GMus-SLA	GDav-SLA
G. arboreum	Asia	Diploid	C	A	NS	-	-	-	-	-	-	-	-	-	-	-	-
G. herbaceum	Africa	Diploid	C	A	NS	-	-	-	-	-	-	-	-	-	-	-	-
G. therburi	Arizona	Diploid	W	D	NS	-	-	-	-	-	-	-	-	-	-	-	-
G. davidsonii	California	Diploid	W	D	M	Yes	Yes	Yes	-	Yes	Yes	-	-	Yes	Yes	Yes	Yes
G. gossypioides	Mexico	Diploid	W	D	VS	Yes	Yes	-	Yes	Yes	Yes	-	-	-	-	Yes	-
G. hirsutum	Mexico	Tetraploid	C	AD	VS	Yes	-	Yes	-	-	-	Yes	-	Yes	-	-	-
G. hirsutum	Mexico	Octaploid	+	AADD	VS	Yes	-	Yes	-	-	Yes	-	-	-	Yes	-	-
G. barbadense	Bolivia, Peru	Tetraploid	C	AD	VS	Yes	-	Yes	-	-	-	-	-	-	-	-	-
G. darwinii	Galapagos Island	Tetraploid	W	AD	MS	Yes	Yes	Yes	Yes	-	-	-	Yes	-	Yes	Yes	-
*G. punctatum**	Mexico	Tetraploid	W	AD	S	Yes	Yes	Yes	Yes	Yes	Yes	-	-	-	Yes	Yes	-
G. latifolium	Mexico	Tetraploid	W	AD	MS	Yes	-	Yes	-	-	-	-	Yes	-	-	Yes	Yes
G. mustelinum	Brazil	Tetraploid	W	AD	MS	Yes	Yes	Yes#	-	-	-	-	-	Yes	Yes	Yes	Yes
G. stocksii	Pakistan	Diploid	W	E	M	Yes	Yes	-	Yes	-	-	-	-	-	-	-	-
G. somalense	Africa	Diploid	W	E	M	Yes	Yes	-	Yes	-	-	-	Yes	-	-	-	-
G. lobatum	Mexico	Diploid	W	D	M	Yes	Yes	-	Yes	Yes	Yes	-	-	-	Yes	Yes	-

*A sub-species of *G. hirsutum*.
#Only betasatellite deletion mutants lacking the βC1 gene were identified.
@Species are indicated as either cultivated (C) or wild (W).
SSymptoms are indicated as non-symptomatic (NS), mild (M), moderately severe (MS), severe (S) and very severe (VS).
+Octaploid *G. hirsutum* is non-cultivated, and is maintained for research purposes.

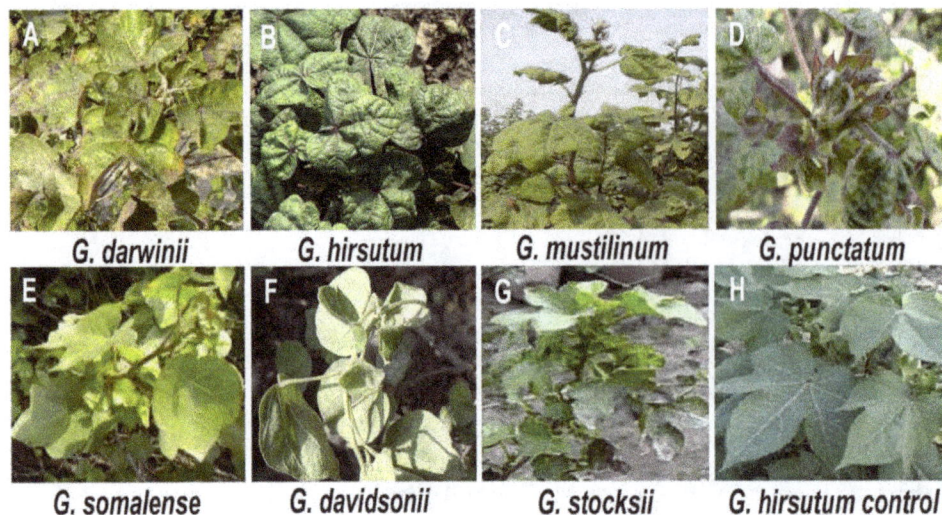

Figure 1. Symptoms displayed by selected *Gossypium* species. Shown are *Gossypium darwinii, G. hirsutum, G. mustilinum* and *G. punctatum*, which showed mild to severe symptoms. In contrast, *G. somalense, G. davidsonii,* and *G. stocksii* showed relatively mild symptoms. A photo of a healthy *G. hirsutum* plant is shown for comparison.

alignment with the full-length genome (or DNA-A) sequences of selected begomovirus isolates available from the databases, representing the majority of begomovirus species so far identified in cotton on the Indian subcontinent (Figure 2) [1]. The tree shows the majority of the sequences obtained here to segregate with previously identified virus species which have been shown to infect cotton in southern Asia, specifically CLCuMuV, CLCuRaV, CLCuBuV and CLCuKoV. The assignment of each sequence to a specific species is indicated in Table S1. Pairwise sequence comparisons confirmed the assignment of each clone to a species based on the presently applicable species demarcation threshold for begomoviruses (89%; Table 2) [32].

In addition to the isolation of begomoviruses species previously identified in cotton, the phylogenetic tree showed an unusual group of isolates (indicated as "GPMLCuV" in Figure 2). These isolates are most closely related to, and segregate with, CLCuAlV. They were isolated from four cotton species; *G. davidsonii, G. gossypioides, G punctatum* and *G. lobatum*. The four clones show between 91 and 100% nucleotide sequence identity (the clones from *G. davidsonii* and *G punctatum* having identical sequences), but less than 87% identity to all other geminivirus sequences available in the databases. The highest levels of sequence identity were to isolates of CLCuAlV (between 81 and 87%) and CLCuMuV (between 77 and 84%; Table 2). To isolates of CLCuRaV available in the databases, the sequences showed only between 75 and 84% identity (Table 2). Based on these results the four clones represent isolates of a new begomovirus species, for which we propose the name *Gossypium punctatum mild leaf curl virus* (GPMLCuV).

Surprisingly the DNA-A component of *African cassava mosaic virus* (ACMV), a bipartite begomovirus not previously identified in Asia, was identified in six of the cotton species (Table 1). However, no evidence for the presence of ACMV DNA-B was found, either by PCR amplification with specific primers or by Southern blot hybridization (results not shown). A total of 9 potentially full-length ACMV DNA-A clones were obtained and sequenced (Table S2). The sequences of these clones show between 76% and 99% identity. To the sequences of ACMV isolates available in the databases they show between 92% and 99% identity, with the highest levels of identity to an isolate originating from Cameroon

(ACMV-[CM:03], AY211884). The predicted amino acid sequences of each of the gene products showed the ACMV isolates from Pakistan to have the highest identity levels to ACMV isolates originating from Cameroon and Ivory Coast (results not shown). Many of the ACMV DNA-A clones obtained are defective, containing frame-shift mutations of the virion-sense genes (as detailed in Figure S1) but with the complementary sense containing only very few single nucleotide exchanges. For many of the clones it is unlikely, even if the cognate DNA-B were present, that they would be able to infect plants autonomously, suggesting that they are maintained by *trans*-complementation.

Recombination between begomoviruses in *Gossypium* species

Recombination is a common feature in the evolutionary history of many geminiviruses [33,34]. To determine whether the begomoviruses identified here show evidence of recombination, RDP3 analysis was conducted based on alignments with full-length sequences of selected begomoviruses available in the databases. The results of this are shown in Figure 3 with the details, including p-values, given in Table S3. The analysis showed CLCuBuV to consist of the virion-sense sequences of CLCuKoV and the complementary-sense sequences of CLCuMuV (Figure 3), as reported previously [27].

Recombination analysis for CLCuMuV clones obtained here shows that each host in which this virus was identified harbors isolates with distinct recombination patterns. Isolates from *G. hirsutum* ([PK:Hir1:08], FJ218486), *G. somalense* ([PK:Mul:Som:08], FJ218487) and *G. darwinii* ([PK:Mul:Dar1:06]) were overall very similar, with sequences just downstream of the hairpin-loop apparently originating from *Bhendi yellow vein mosaic virus* (BYVMV), although the isolate [PK:Hir1:08] also contains a fragment derived from CLCuAlV and resembles an isolate originating from India ([PK:K802a:96]). Interestingly, isolates [PK:Hir1:08] and [PK:Mul:Som:08] contain a recombinant fragment originating from ACMV (Figure 3). However, *G. hirsutum* was not found to harbor ACMV, indicating that either the recombination event yielding CLCuMuV-[PK:Hir1:08] occurred in another plant/species, with the virus subsequently being transmitted to *G.*

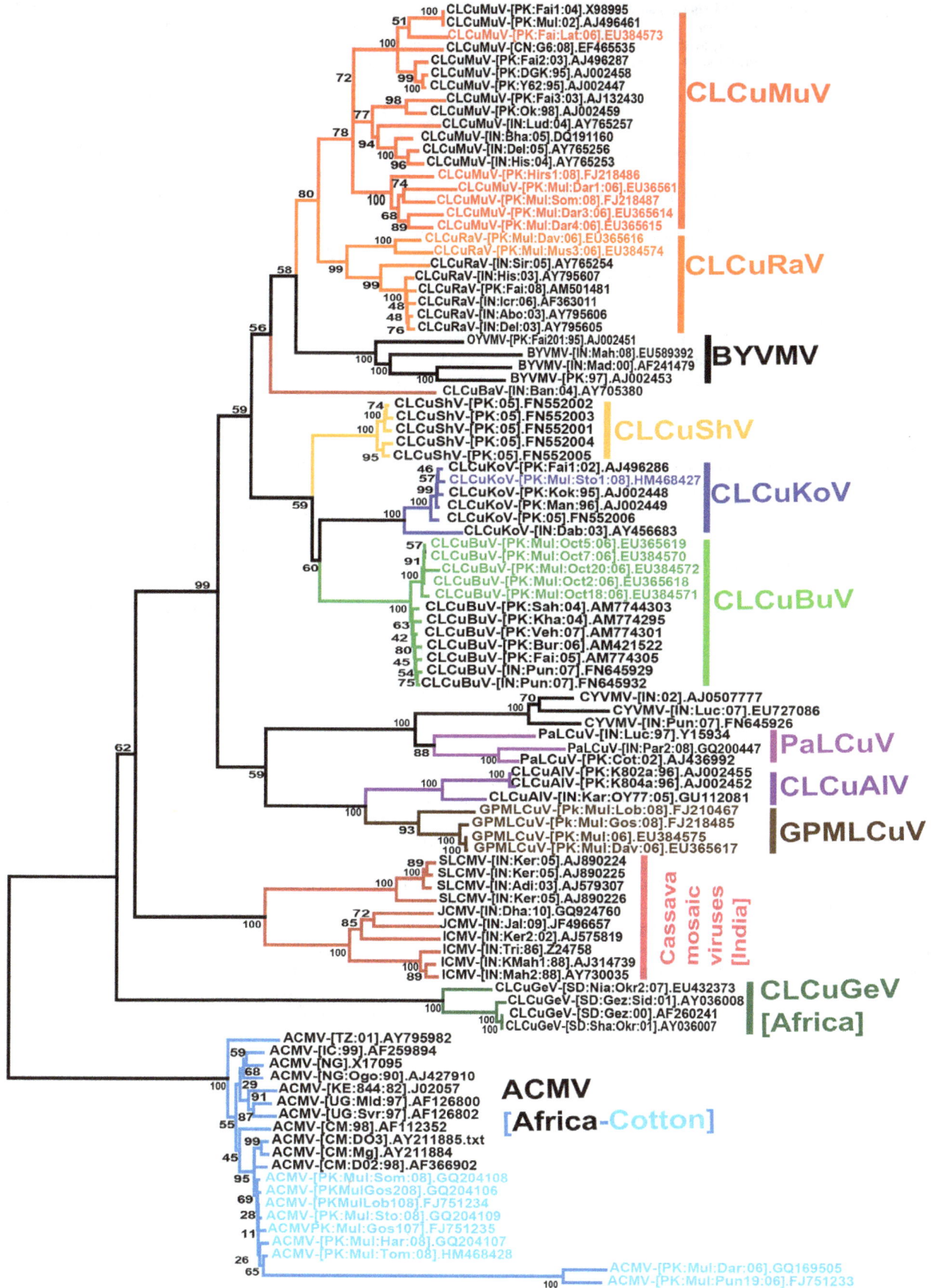

Figure 2. Phylogenetic analysis of virus genome (or DNA-A component) sequences. Neighbor-joining dendrogram based upon a Clustal W alignment all begomovirus genome (or DNA-A component) sequences determined here with sequences of the genomes (or DNA-A components) of selected begomovirus species occurring in the Old World. The database accession number of each sequence is given. The acronyms and isolate descriptors are as described in Fauquet *et al.* [1]. The numbers at nodes represent percentage bootstrap values (1000 replicates).

hirsutum, or that the *G. hirsutum* plant earlier contained ACMV but has since lost it.

CLCuRaV was identified in two cotton species, *G. davidsonii* and *G. mustilinum*, and these two sequences showed differing recombination patterns. In contrast, CLCuKoV, which was only identified in a single cotton species (*G. stocksii*), showed 99% nucleotide sequence identity to previously characterized CLCuKoV isolates, indicative of little, if any, recombination having occurred since these isolates diverged.

GPMLCuV is unique to the orchard in Multan, having not so far been identified elsewhere, and was isolated from 4 cotton species (*G. punctatum*, *G. lobatum*, *G. davidsonii* and *G. gossypioides*). This virus may thus have evolved in this orchard; a hypothesis that is supported by the high levels of sequence conservation (91 to 100% identity) between isolates obtained from distinct cotton species (Table 2). Recombination analysis showed the sequences of GPMLCuV to exhibit incongruous segregation (Figure 3). For three isolates ([PK:Mul:06], [PK:Mul:Dav:06] and PK:Mul:-Gos:06]) the coat protein (CP) and V2 genes showed recombination with BYVMV. However, the V2 gene GPMLCuV isolates [PK:Mul:Gos:08] and [PK:Mul:Lob:08] showed recombination with CLCuAlV. For the Rep gene all isolates exhibited recombination with *Croton yellow vein mosaic virus* (CYVMV). For the TrAP and REn genes all isolates except [PK:Mul:Lob:08] showed recombination with CLCuRaV. Isolate [PK:Mul:Lob:08] instead contained an additional CLCuAlV fragment in the Rep/TrAP gene overlap region.

Overall the recombination analysis shows that, with the possible exception of CLCuKoV, the viruses identified here have a highly recombinant origin which is distinct from, in most cases, the viruses identified earlier. The analysis also shows that viruses with quite distinct histories of recombination can make up a single begomovirus species. This is one of the drawbacks of a taxonomy that includes, as a major criterion, sequence relatedness.

Identification of a begomovirus DNA-B component in *Gossypium* species

No bipartite begomoviruses infecting cotton have previously been reported from the OW. Thus it came as a surprise that molecules with similarity to the DNA-B components of bipartite begomoviruses were isolated from 8 cotton species (Table 1). Despite an extensive search, however, no DNA-B component was identified in *G. hirsutum*, consistent with previous studies [19,23,27].

The eight DNA-B molecules obtained showed between 88 and 98% nucleotide sequence identity, showing them to be closely related, despite each having been isolated from a different cotton species. Comparisons to sequences available in the databases showed only relatively low percentage identity values to the sequences of DNA-B components of other begomoviruses originating from the OW (<50%) with the exception of DNA-B components of *Sri Lanka cassava mosaic virus* (SLCMV; 65–69%), *Indian cassava mosaic virus* (ICMV; 65–71%) and the recently identified *Jatropha curcas mosaic virus* (JCMV; 68–71%) [35](Table 3). This suggests that the DNA-B components identified in cotton originate from one of these three virus species or a species closely related to these viruses.

A phylogenetic tree, based upon an alignment of all available DNA-B sequences with the sequences of DNA-B obtained here (Figure 4A), shows the cotton components to be most closely related to, but distinct from, the DNA-B components of ICMV, JCMV and SLCMV. It has previously been shown that the DNA-B components of ICMV and SLCMV have a common origin, most likely due to component exchange between the two begomovirus species [36]. Although for the most part consisting of the sequence of ICMV DNA-B, the SLCMV DNA-B contains the *ori* of SLCMV, allowing the SLCMV DNA-A-encoded Rep to *trans*-replicate this DNA-B component. To ascertain if a similar exchange of the *ori* has occurred in the cotton DNA-B components, a phylogenetic tree based upon the DNA-B sequences without the CR sequences was produced (Figure 4B). This shows the JCMV DNA-B sequences to segregate with, and be basal to, the cotton DNA-B sequences and both to be distinct from the ICMV and SLCMV DNA-B components. Additionally the DNA-B components of ICMV and SLCMV co-segregate, supporting the conclusion of Saunders *et al.* [36] that they have a common origin.

Alignments of the GPMLCuV sequences with their cognate DNA-B sequences showed them to contain a shared sequence spanning the *ori*. For example, GPMLCuV-[PK:Mul:Dav:06] (EU365617) and its cognate DNA-B (EU384577) share a sequence of ~280 nt (coordinates 2594 to 144 and 2567 to 142, respectively) that has 91.5% identity; overall the sequences show only 46% identity. This is good evidence of *ori* donation. This is supported by the fact that all the DNA-B isolates share the same predicted iteron sequences (GGGGA) that are also found in GPMLCuV (the presumed donor of the *ori* sequences in the DNA-B molecules isolated from cotton), CLCuAlV and CYVMV. In contrast, the predicted iterons of JCMV (the presumed parent of most of the sequences making up the cotton DNA-Bs), ICMV and SLCMV are GGTA, whereas those of CLCuKoV are GGTA/G.

Interestingly, all the cotton DNA-B sequences contain a unique duplication of the right (3′) leg of the nonanucleotide-containing stem-loop structure (Figure S2). The significance of these duplications and whether they might play a part in component replication is unclear.

Betasatellites identified in *Gossypium* species

The presence of betasatellites was shown in 7 of the 14 cotton species examined (Table 1) and the complete sequences of 27 presumed full-length molecules were obtained (Table S4). The molecules showed between 86 and 99% nucleotide sequence identity, indicating that they are all isolates of a single species of betasatellite (based upon the proposed species demarcation threshold of 78% for betasatellites) [37]. With the exception of isolate [PK:Mul:Lat11:06] and 5 clones from *G. davidsonii*, all are of the typical size of betasatellites (~1350 nt) being between 1349 and 1359 nt in length. These presumed full-length betasatellites have the conserved structure shown previously for this class of satellites, consisting of a single gene (βC1), a region of sequence rich in adenine (A-rich) and a sequence of ~100 nt conserved between all betasatellites, known as the satellite conserved region (SCR) [26]. CLCuMuB-[PK:Mul:Lat11:06], (EU384591) is an unusual mutant with a perfect inverted duplication (coordinates 259–744) repeating 243 nt of the βC1 coding sequence which

Table 2. Highest and lowest percentage nucleotide sequence identity values for pairwise comparisons of the begomovirus genomes (DNA-A components) with selected sequences available in the GenBank database.

Sequences obtained from the databases*

Sequences determined here[#]	CLCuMuV (12)	CLCuRaV (6)	CLCuAlV (4)	CLCuKoV (6)	CLCuBaV (1)	CLCuBuV (6)	PaLCuV (4)	GPMLCuV (−)	CLCuShV (5)	ACMV (11)	ICMV (6)
CLCuMuV (6)	97-86	91-83	83-76	84-72	83-77	85-74	71-60	84-77	89-79	65-61	69-64
CLCuRaV (2)		99-95	79-74	89-74	81-80	80-74	71-61	84-75	75-73	64-60	68-65
CLCuAlV (4)			94-89	68-66	75-73	71-66	73-67	87-81	74-71	62-59	64-62
CLCuKoV (1)				99-89	80-77	88-82	80-72	70-65	93-84	66-62	70-68
CLCuBaV (1)					-	79-75	69-63	75-71	82-81	65-63	70-69
CLCuBuV (5)						100-97	76-69	72-68	88-86	66-63	70-67
PaLCuV (4)							100-85	75-67	74-68	64-57	67-60
GPMLCuV (4)								100-91	75-73	60-57	64-60
CLCuShV (5)									99-98	66-64	70-68
ACMV (9)**										99-92	65-62
ICMV (6)											98-89

*The figure in brackets indicates the numbers of sequences used in the analysis.
#The species highlighted (white text on a black background) are the sequences determined here. The remaining species are sequences from the databases included for comparison.

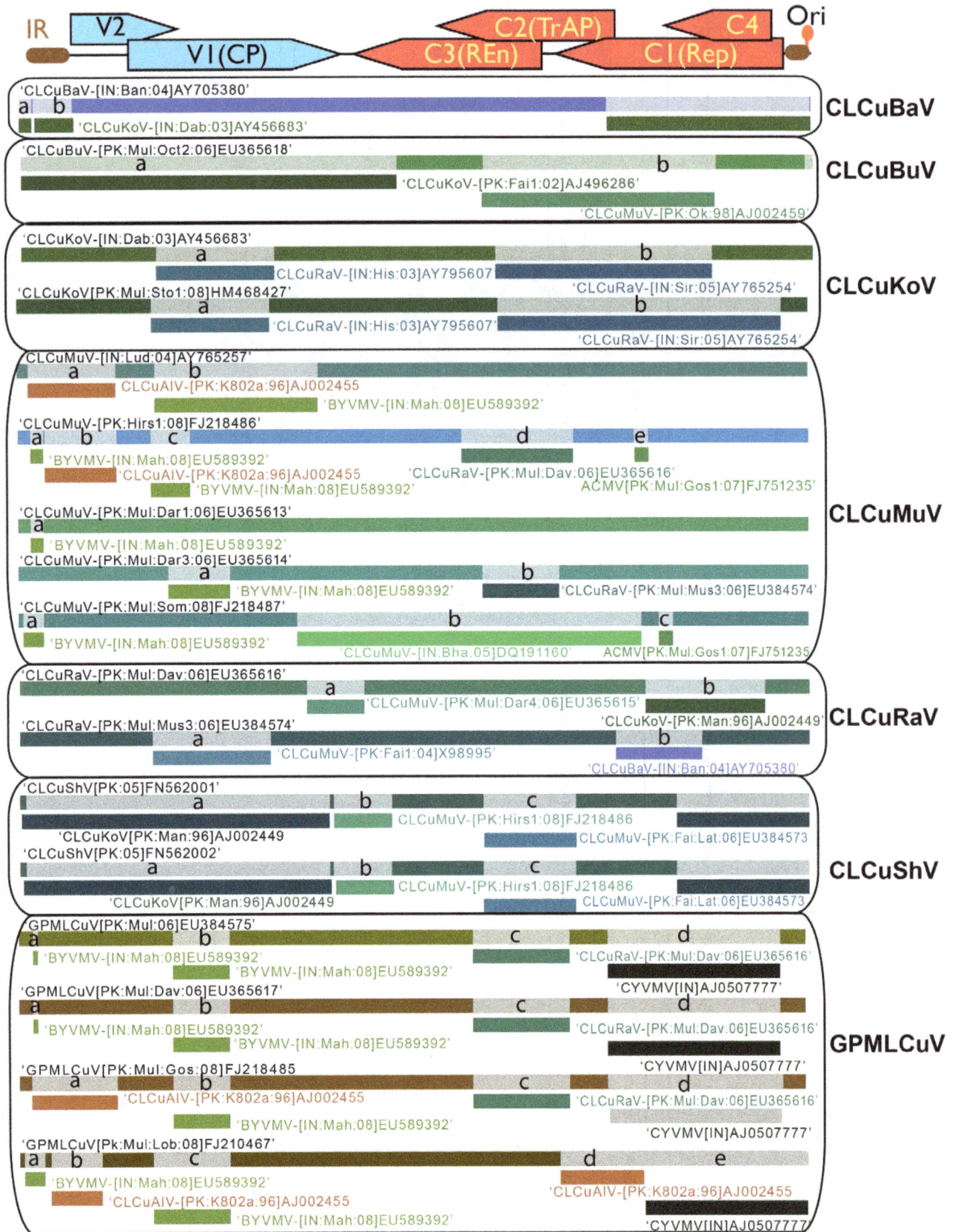

Figure 3. Analysis for recombination among selected begomoviruses associated with CLCuD. Recombinant fragments in the sequences of CLCuD-associated begomoviruses were identified using RDP3 analysis of a Clustal W alignment. The default X-Over settings was used for recombination analysis. Recombinant sequences are identified as colored lines below the sequence (identified above each full-length sequence line)

together with its origin (parent). The sequences analysed are grouped according to species (marked on the right). A liner genome map (with the position of genes and their orientation indicated with arrows; blue for virion-sense and red for complementary-sense) is shown at the top of the figure to indicate the positions of recombinant fragments. The intergenic region (IR; blown bar) and hairpin structure of the *ori* are shown.

replaces the N-terminal end of the *bona fide* βC1 gene, the βC1 promoter and much of the A-rich sequence.

A phylogenetic tree, based upon an alignment of all betasatellite sequences available in the databases with the betasatellite sequences obtained here is shown in Figure S3, whereas a more compact tree with fewer sequences is shown in Figure 5. Both trees show the betasatellites to fall into the two major phylogenetic classes first identified by Briddon *et al.*, (2003); those isolated, for the most part, from hosts in the family *Malvaceae*, and those originating mostly from non-malvaceous hosts. The betasatellites identified here all fall into the malvaceous class and segregate into 4 distinct groups. The first group, consisting of four sequences isolated from *G. davidsonii*, are recombinant, with the SCR replaced by the intergenic sequence derived from CLCuRaV ([PK:Mul:Dav129:06]), or CLCuMuV ([PK:Mul:Dav118:06], [PK:Mul:Dav113:06] and [PK:Mul:Dav85:06]). Such recombinant betasatellites have been identified previously in association with CLCuD affected cotton [19] and other begomovirus-betasatellite complexes [38,39] and are, by convention, not classified as betasatellites [37].

The second group of betasatellite molecules (isolated from *G. punctatum*) segregates with the first CLCuMuB identified, the betasatellite associated with CLCuD in Pakistan during the 1990s [19], which we shall henceforth refer to as CLCuMuB^{Mul}. The third group (isolated from *G. darwinii*) is closely related to CLCuMuB^{Mul} but contains an approximately 95 nt sequence, within the SCR, derived from Tomato leaf curl betasatellite (ToLCuB). This betasatellite, which we shall refer to as CLCuMuB^{Bur}, is associated with CLCuBuV and is the only satellite now prevalent in *G. hirsutum* in most cotton growing areas of Pakistan [27,28]. Interestingly, previously isolated CLCuMuB^{Bur} sequences from Pakistan and India form a cluster that is distinct from CLCuMuB^{Bur} sequences isolated from *G. darwinii*. This may be due to the shorter length of ToLCuB sequence inserted in these CLCuMuBs. Among the betasatellites isolated from *G. darwinii* there are betasatellites ([PK:Mul:Dar17:06], EU384596) which are recombinant with Tomato yellow leaf curl Thailand betasatellite (TYLCTHB) and Ageratum yellow leaf curl betasatellite (AYLCB; [PK:Mul:Dar11:06], EU384599) rather than ToLCB. So far, TYLCTHB has not been identified in Pakistan (Figures 5 and S3).

The fourth group of betasatellites, consisting of two clones isolated from *G. latifolium* (CLCuMuB-[PK:Mul:Lat9:06] and -[PK:Mul:Lat11:06]), do not contain the recombinant SCR of

CLCuMuB^{Bur} and have sequences between the SCR and A-rich region that are distinct from both CLCuMuB^{Bur} and CLCuMuB-^{Mul}, the origin of which remains unclear. Five further CLCuMuB clones, three of which were cloned from *Hibiscus* spp. [40,41,42], segregate with four Kenaf leaf curl betasatellite (KLCuB) isolates which together are basal to all CLCuMuB sequences. A possible explanation is that these CLCuMuB and KLCuB clones were isolated from the far-east Indian state of West Bengal (along the Bangladesh border), and thus are geographically isolated from the areas where CLCuD occurs. Additionally these CLCuMuB clones are also recombinant, containing sequences in the SCR derived from AYLCB and TYLCTHB.

Recently the African CLCuD-associated begomovirus *Cotton leaf curl Gezira virus* (CLCuGeV) has been identified in cotton originating from southern Pakistan in the presence of CLCuMuB [25]. The cognate betasatellite of CLCuGeV, Cotton leaf curl Gezira betasatellite (CLCuGeB), has not so far been identified in Pakistan. The recombination analysis here has shown a small recombinant fragment of CLCuMuB in a CLCuGeB isolate from Sudan ([SD:Gez2:00], AY044143) (Figure 5). This suggests that CLCuMuB may be present in Africa and further highlights the exchange of viruses and associated satellites between Africa and southern Asia.

G. lobatum, *G. gossypoides*, *G. stocksii* and *G. somalense*, despite containing viruses previously classified as betasatellite-requiring monopartite geminiviruses (CLCuBuV or CLCuRaV), did not apparently contain betasatellites. In contrast, only a betasatellite derived from CLCuMuB^{Bur}, with the βC1 gene deleted, was identified in *G. mustilinum* (data not shown). This is the first such molecule derived from CLCuMuB^{Bur}, all previous βC1deletion mutants having been derived from CLCuMuB^{Mul}.

Alphasatellites identified in *Gossypium* species

Alphasatellites were identified in 8 of the 14 *Gossypium* species (Table 1). A total of 60 full-length alphasatellite molecules were cloned and sequenced (Table S4). These range from 1141 to 1373 nt in length, typical of this class of molecule [17]. All the molecules contain a single large gene which encodes a Rep protein homologous to those of previously characterized alphasatellites.

A phylogenetic tree, based upon an alignment of all full-length alphasatellite sequences available in the nucleotide sequence databases with the full-length alphasatellite sequences obtained here is shown in Figure 6. This shows the alphasatellites characterized here to fall into three major groups (indicated as

Table 3. Highest and lowest percentage identity values for pairwise comparisons of the complete nucleotide sequences of the DNA-B components of selected begomoviruses with the DNA-B components identified in cotton species.

	GPMLCuV (8)*	ICMV (6)*	SLCMV (3)*	JCMV (1)*	ToLCNDV (5)*
GPMLCuV (8)*	98-88	71-65	69-65	71-68	35-32
ICMV (6)*	-	98-75	94-91	82-81	34-32
SLCMV (3)*	-	-	98-95	79-78	33-32
JCMV (1)*	-	-	-	100	36-33
ToLCNDV (5)*	-	-	-	-	97-83

*The figures in brackets indicate the numbers of isolates compared.

A- Full-length DNA-B

B- CR deleted DNA-B

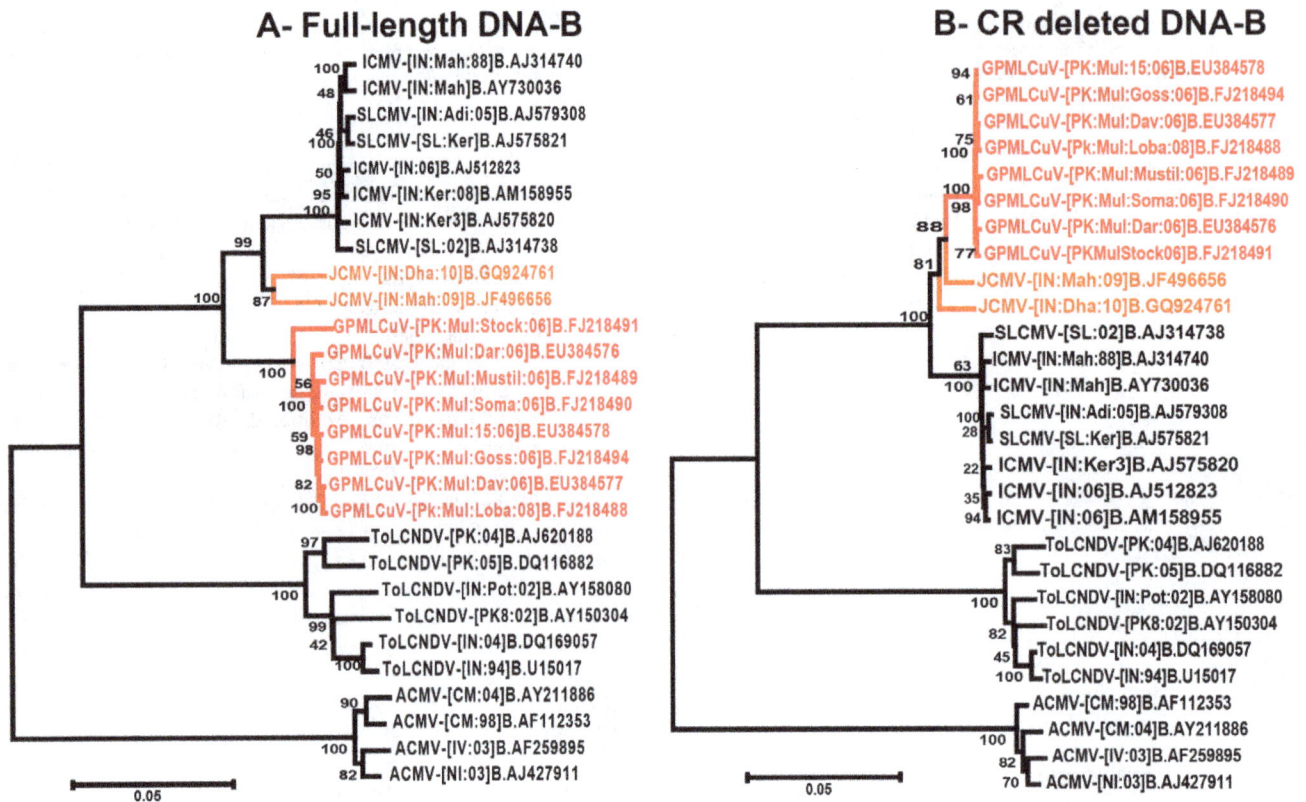

Figure 4. Phylogenetic analysis of DNA-B component sequences. Neighbor-joining dendrogram based upon a Clustal W alignment of all DNA-B components determined here with sequences of selected DNA-B components of begomovirus species occurring in the Old World. The tree in panel A is based upon full-length sequences whereas that in panel B is based upon sequences spanning the BC1 and BV1 region. The database accession number of each sequence is given. The acronyms and isolate descriptors are as described in Fauquet et al. [1]. The numbers at nodes represent percentage bootstrap values (1000 replicates).

"GMusSLA", "GDarSLA" and "GDavSLA"). The alphasatellites identified here are distinct from the two alphasatellite so far identified in cotton in Pakistan; Cotton leaf curl Multan alphasatellite (CLCuMuA) and Cotton leaf curl Shadadpur alphasatellite (CLCuShA) [17,43,44,45].

The 49 alphasatellites labeled as GDarSLA show between 85 and 100% nucleotide identity with each other (Table 4). To all other alphasatellites available in the databases the GDarSLA sequences show less than 67% identity, with the exception of CLCuMuA to which the identity values are between 75 and 86%. Based on the proposed demarcation threshold for distinct aplhasatellites (83%; R.W. Briddon, manuscript in preparation), this indicates that they represent isolates of a single alphasatellite, for which we propose the name Gossypium darwinii symptomless alphasatellite (GDarSLA).

The alphasatellites labelled as GMusSLA, which were isolated from 6 cotton species (Table S5) form a clade with Cotton leaf curl Gezira alphasatellite (CLCuGeA; note that this alphasatellite has so far not been identified in cotton). The GMusSLA sequences share between 88 and 100% nucleotide sequence identity with each other but show less than 57% identity to all other alphasatellites, the highest being to CLCuGeV (51–57%) and Tomato leaf curl alphasatellite (44 to 52%; Table 4). This indicates that they represent a new alphasatellite for which the name *Gossypium mustilinum symptomless alphasatellite* (GMusSLA).

GMusSLA is distinct from all other begomovirus-associated alphasatellites in encoding a Rep consisting of only 295aa with the exception of isolate [PK:Mul:Gos-2:08], that contains a number of

frame shift mutations yielding a gene with the potential to encode a 302aa product and isolates ([PK:Mul:Dav7b:06], (EU384653) and [PK:Mus1:06], (Eu384654) that instead encompass a gene with the capacity to encode a shorter product of 263aa due to a frame shift mutation that C-terminally truncates the coding sequence. Isolates [PK:Mul:Dav7b:06] and [PK:Mus1:06] are significantly smaller (1218 nt) than typical begomovirus-associated alphasatellites (~1380 nt), which may indicate that they are deletion mutants. With the exception of the alphasatellite described by Saunders et al., [46] AYVSGA, (previously named DNA-2), which has recently also been identified in Oman [47], that encodes a 289aa Rep, other full-length begomovirus-associated alphasatellites typically encode a Rep consisting of a predicted 315aa.

Interestingly AYVSGA and the alphasatellites identified in the NW (Cleome leaf crumple alphasatellite [CILCrA] and Euphorbia mosaic alphasatellite [EuMA] [48], are basal to all the other begomovirus-associated alphasatellites and the GMusSLA/CLCu-GeA clade sits between these two groups. This indicates that there are three distinct classes of begomovirus-associated alphasatellites, which are only distantly related to the nanovirus alphasatellites.

Dating estimates for the origins of CGs

To estimate when CGs associated with CLCuD first emerged, time to most recent common ancestor (TMRCA) was estimated from phylogenies of the CP gene using Markov Chain Monte Carlo (MCMC) integrated in BEAST (V1.6). By applying the relaxed uncorrelated relaxed clock model with exponential growth

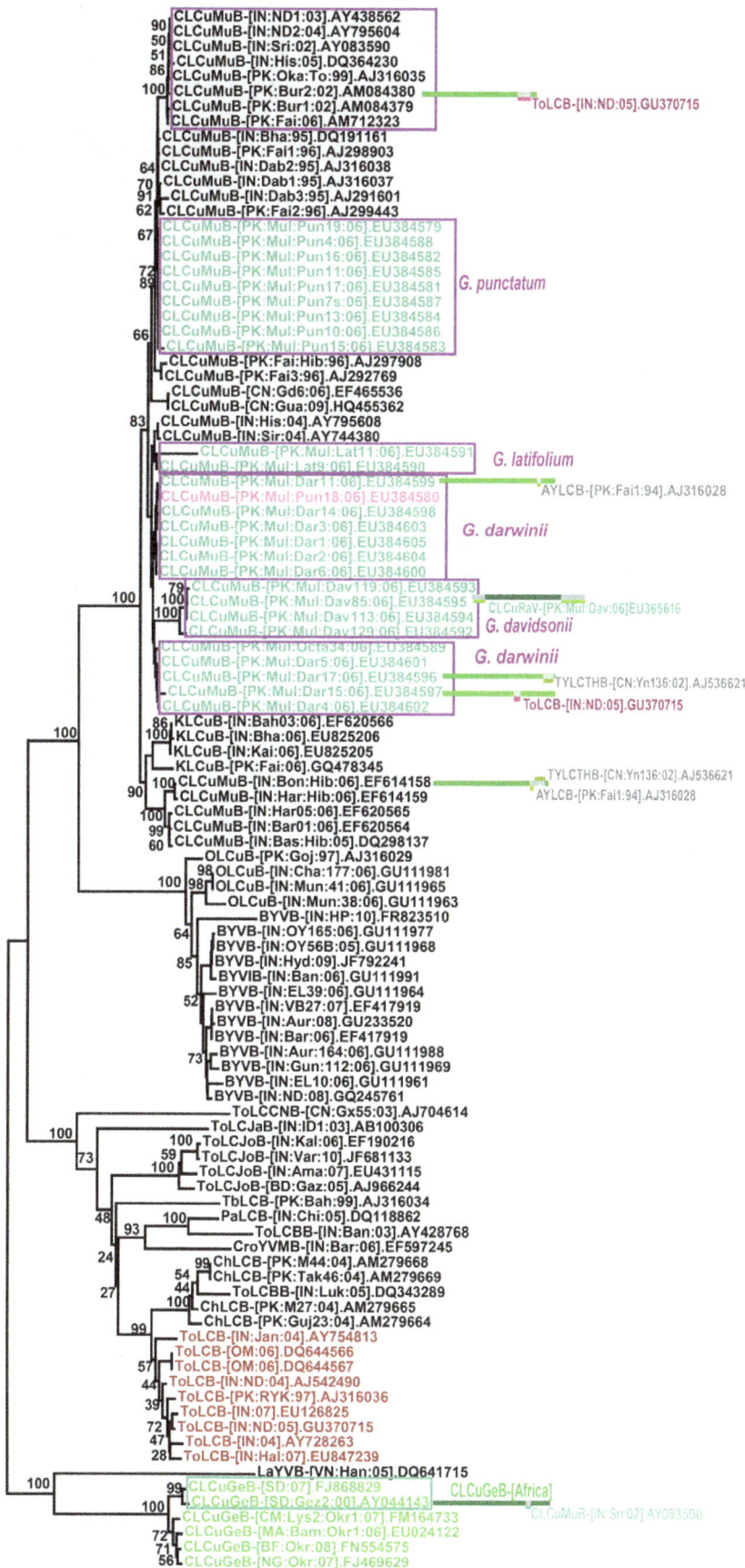

Figure 5. Phylogenetic analysis of betasatellite sequences. Neighbor-joining dendrogram based upon a Clustal W alignment of selected betasatellite sequences available in the databases with the betasatellite sequences determined here (a more complete phylogenetic analysis is shown in Figure S3). The database accession numbers of each sequence are given. The acronyms and isolate descriptors are as described in [37]. The betasatellite isolates obtained from cotton spp. as part of this study are highlighted with colored boxes. The numbers at each node represent the bootstrap values (1000 replicates). The recombinant sequences for different hosts are presented in different rectangles. The representative sequences of recombinant origin are presented with their line diagram generated through RDP3 program. Note that some betasatellite species in this tree are not monophyletic due to the use of the Clustal W algorithm for the alignment (the taxonomy of betasatellites is based upon the Clustal V algorithm).

parameters, the dates for CGs lineages were estimated (Figure 7A). An undated tree, based alignments of the Rep gene sequences, was produced for comparison (Figure 7B).

Interestingly, both trees shown in Figure 7 have two major clades (labeled A and B). In both trees clade A contains CLCuMuV, while the clade B encompasses CLCuKoV. Not surprisingly the phylogenetic tree generated for Rep gene sequences shows CLCuBuV to form a monophyletic group with CLCuMuV, while for the CP tree CLCuBuV segregates with CLCuKoV isolates, confirming the conclusions of Amrao et al. [27] and the recombination analysis here concerning the origins of this species. For the CP tree GPMLCuV segregates with CLCuAlV which form a clade segregates with PaLCuV isolates, indicating the possible emergence of GPMLCuV by recombination between CLCuAlV and PaLCuV. It is possibly significant that CLCuAlV, PaLCuV and GPMLCuV form a distinct group in the Rep tree, suggesting that for the Rep sequences they diverged from the remainder of the cotton viruses some time ago. It is also worth noting that only on one occasion has PaLCuV been identified in cotton [23]. Overall these trees show that, for at least some of the CGs, recombination is a major feature in their origins.

For the CP tree the date estimates have broad confidence intervals (Table S6), indicating that caution needs to be taken in interpreting the divergence time estimates. The CP sequences of nine divergent CGs [Taxa n = 66]) isolated between 1995 and 2010 were used to estimate their times of emergence. The data indicates that CGs could be ~2.5 centuries old and that a major diversification of CGs could have started during the late 19th century (Figure 7A, nodes A and B). CLCuAlV/GPMLCuV and CLCuMuV are estimated to have appeared as long ago as ~1935 and 1928 (nodes F and G), whereas PaLCuV and CLCuKoV are estimated to have diverged in 1945. CLCuD was first noted in Pakistan in 1967 [49]. The TMRCA distribution for DNA-A component shows that 6 out of 9 CGs were potentially co-circulating at that time. However, CLCuBuV and CLCuShV (node E) seem to appear due to recombination after the mid-1980s. In contrast, the nodes for the divergence of GPMLCuV (node F) show that this recombinant is older (mean value 1948) than CLCuBuV. However, since 2001 only CLCuBuV has been found in *G. hirsutum* across most of Pakistan, while GPMLCuV has so far only been identified in the orchard in Multan. This may indicate that GPMLCuV is not well adapted to *G. hirsutum* or at least that CLCuBuV is better adapted.

Dating estimates for the origins of satellites

Due to betasatellites in some cases being recombinant (Figure 5), which might influence the results in the maximum clade credibility (MCC) tree, the βC1 gene (~357 nt) was selected for divergence time estimates. Similarly, due to differences in the sequence lengths of alphasatellites, the Rep gene of alphasatellites (alpha-Rep) gene was used for TMRCA estimates. The TMRCA distributions (Figure 8A and B) suggest that the cotton-associated alphasatellites emerged (mean value 1852) more recently than the cotton-associated betasatellite (mean value 1653). The oldest possible date for the emergence of CLCuMuA is 1968 (clade A;

Figure 8A), when it diverged from its closest relative CLCuShA (AM711116). Interestingly, CLCuMuA isolates from Pakistan and India form separate groups within clade A. This indicates that alphasatellites isolated from these countries are evolutionarily isolated. Importantly, of the alphasatellites in the clade B (Figure 8A), GMusSLA forms a distinct group and has the oldest TMRCA estimates (mean value 1958). It is noteworthy that GDarSLA isolates group according to the host from which they were isolated with the exception of those in group I. Group-I isolates originate from *G. davidsonii* and *G. mustilinum*. This is similar to the pattern observed for the full-length genomes of alphasatellites (Figure 6) and may suggest that the GDarSLA groupings reflect host adapted variants and/or that there is little exchange between the cotton species, with the exception of *G. davidsonii* and *G. mustilinum*.

TMRCA estimates for the βC1 gene of OLCuB, CLCuMuB, and KLCuB (clades A, B and C; Figure 8B) show the same branching pattern as the phylogenetic trees based on full-length sequences (Figure 5). CLCuMuB shares a more recent common ancestor with KLCuB than either does with the betasatellites that infect okra; Bhendi yellow vein betasatellite (BYVB) and Okra leaf curl betasatellite (OLCuB)(node A: Figure 8B). The data suggests that a major diversification of CLCuMuB occurred after 1963, which is just 4 years before CLCuD was first reported in Pakistan. The independent grouping of CLCuMuB isolates from *G. punctatum* and *G. darwinii* suggests that there is little exchange of the betasatellite with these two species.

Estimation of nucleotide substitution rates for CGs and their satellites

The mean nucleotide substitution rates for the CP of CLCuMuV, the βC1 gene of CLCuMuB and the Rep gene of GDarSLA were determined using recombination free datasets with the relaxed clock and Bayesian Skyline Plot (BSP) method (Table 5). For each dataset, the sequences were partitioned into the 3 codons positions. The mean substitute rates for βC1 and the alphasatellite Rep were considerably higher (3.51×10^{-3} and 2.13×10^{-3} substitutions/nucleotide/year, respectively) than those for the CLCuMuV CP (4.24×10^{-4}). This high substitution rate is closer to substitution rate estimated for the *East African cassava mosaic virus* CP (1.37×10^{-3} subst./nt/year), a bipartite cassava infecting begomovirus from Africa [50]. The high nucleotides substitution rate for satellites suggests that they are evolving rapidly. This idea is supported by the fact that satellite clones isolated from a single host segregate within the phylogenetic trees. The substitution rate for the CLCuMuV CP (4.24×10^{-4} subst./nt/year) is similar to that estimated for that estimated for the *Tomato yellow leaf curl virus* CP (4.63×10^{-4} subst./nt/year). This may indicate that these two viruses face the same evolutionary pressure, despite infecting different hosts in different parts of the world. The differences between mutation rates for CP and the satellite genes also indicate that these are likely under different evolutionary (selection) pressures.

To further estimate the selection pressure, we used the 3 position clock model in BSP analysis. Surprisingly, for the

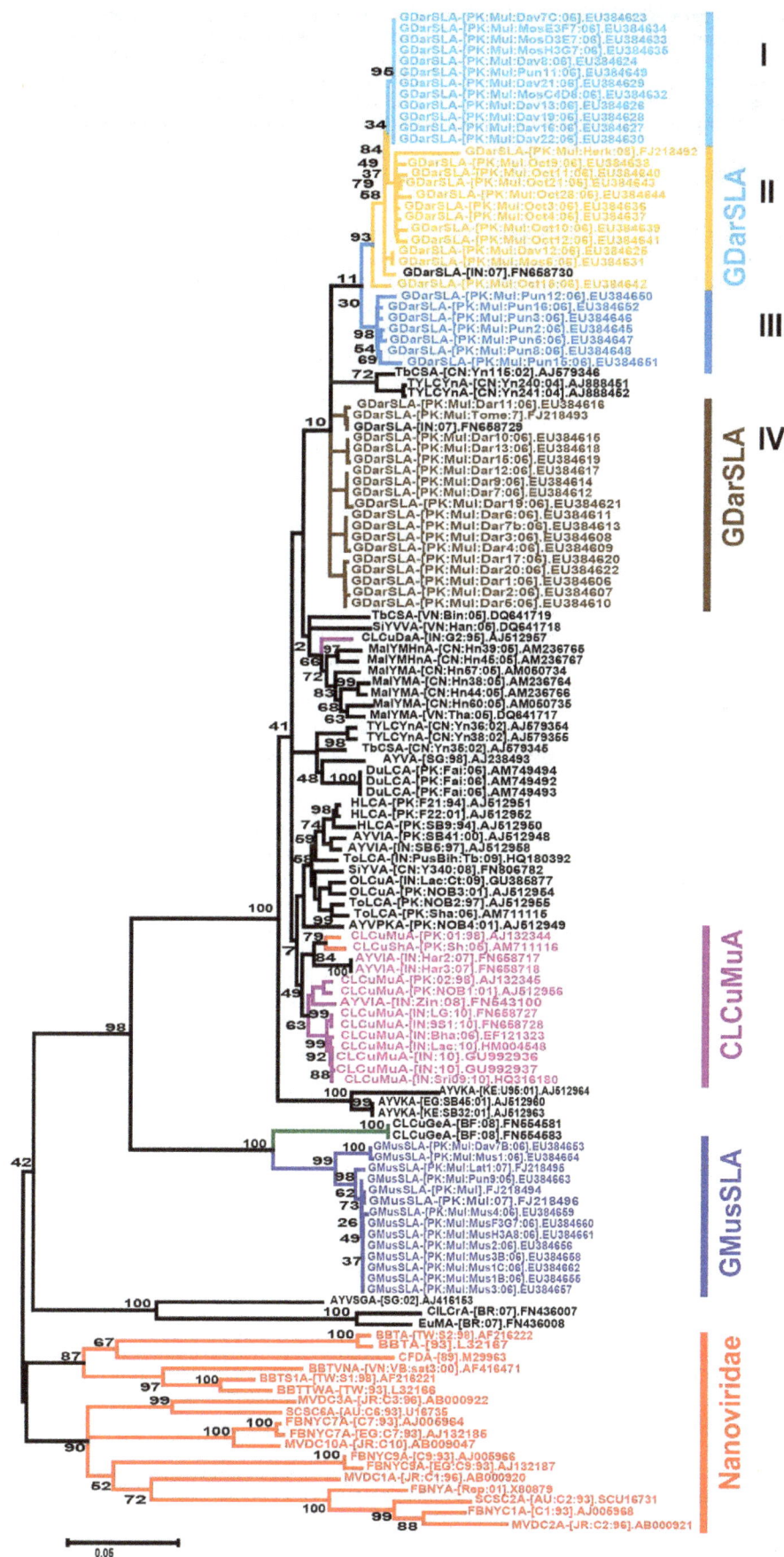

Figure 6. Phylogenetic analysis of alphasatellite sequences. Neighbor-joining dendrogram based upon a Clustal W alignment of the alphasatellite sequences determined here with selected alphasatellite sequences available in the databases. The database accession numbers of each sequence is given. The alphasatellite isolates obtained from cotton (including those not isolated in this study) are highlighted with colored text and indicated on the right. The alphasatellites labeled I to IV are discussed in the text. The alphasatellites associated with nanoviruses are indicated with red text and red branches. The numbers at nodes represent percentage bootstrap values (1000 replicates). Note that some alphasatellite in this tree are not monophyletic due to the use of the Clustal W algorithm for the alignment (the taxonomy of alphasatellites is based upon the Clustal V algorithm).

GDarSLA Rep and the CLCuMuV CP, but not the CLCuMuB βC1, codon position 1 showed a higher substitution rate (1.4 and 1.64, respectively) than codon positions 2 and 3 (0.765, 0.831 and 0.449, 0.909, respectively). This is unexpected since the third codon position (wobble position) normally shows a higher rate of substitution. This likely indicates that the βC1 gene is under a higher selection pressure, preventing sequence change, than the other two genes. Why the CP and alpha-Rep genes might show more rapid sequence change is unclear. Sequence changes (particularly at codon 1) would usually be considered detrimental. For the CP this might interfere with insect transmission, so the data may indicate that GDarSLA is no longer under the stringent selection pressure posed by insect transmission (the viruses possibly no longer requiring insect transmission, since the plants are maintained vegetatively). This cannot, however, be the case for GDarSLA Rep, since mutations here would interfere with replication of the satellite and could lead to extinction. Further studies will be required to investigate this phenomenon.

GPMLCuV can transreplicate both a betasatellite and a DNA-B in *Nicotiana benthamiana*

Biolistic inoculation of GPMLCuV to *Nicotiana benthamiana* resulted in very mild leaf curl symptoms at 14 days post-inoculation (dpi)(Figure 9, panel A). At 21 dpi newly developing leaves showed increasingly milder symptoms and plants recovered from infection (results not shown). Co-inoculation of GPMLCuV ([PK:Mul:Dav06], EU365617) with the cognate DNA-B component ([PK:Mul:Dav06], EU384577) resulted in severe downward leaf curling and vein thickening symptoms at 14 dpi (panel B). However, no recovery was observed for these infections. Interestingly, inoculation of GPMLCuV with either CLCuMuB^{Mul} or CLCuMuB^{Bur} resulted in very severe symptoms (panels C and D, respectively), including leaf enations and infertility of flowers. Symptoms in the presence of either betasatellite were more severe than in the presence of the DNA-B. For each of the inoculations, 10 plants were inoculated and all showed symptoms of infection.

Southern blot analysis showed that in *N. benthamiana* plants GPMLCuV is capable of maintaining both CLCuMuB and the DNA-B component (Figure 9, panels E–G). Infections of GPMLCuV in the presence of the DNA-B raised viral DNA levels above those seen in plants infected with only the virus. However, there was no significant difference in viral DNA levels between infections with either CLCuMuB^{Mul} or CLCuMuB^{Bur}, in the presence of the DNA-B. Both betasatellite variants were efficiently maintained by the virus in *N. benthamiana*.

Discussion

Annual crops such as cotton are re-infected with geminiviruses every growing season from sources that must include other crops and weeds, as well as volunteer (ratoon) cotton. These plants act as reservoirs of both the viruses and insect vectors in the off season. This annual cycle between crop and reservoir hosts is a stringent bottleneck that potentially reduces the genetic diversity of the viruses in the crop. Each year only the best adapted viruses survive

the bottleneck and spread within the crop. It is only recently that researchers have come to realize that weeds may harbor a far greater diversity of viruses and their satellites, than actually appears in the crop, and that it is possible that genetic changes (including for example component exchange and recombination) within weeds plays a major role in virus diversification.

The orchard of *Gossypium* spp. maintained in Multan is a unique resource. Plants here are allowed to grow undisturbed and are re-grown from seeds only when they die. In many cases they represent genotypes that are present nowhere else in the country and the selection pressures they exerted on the viruses they harbor are thus likely to be entirely different to those in the widely grown *G. hirsutum* or in the annual weeds which may harbor the cotton viruses in the off season. Since the orchard has been maintained for some 40 years, it is likely that distinct virus populations have been maintained by insect transmission between plants within the orchard. There will of course have been introductions of viruses from the surrounding ecosystem but the numbers of bottlenecks, due to insect transmission and the death of plants, is likely to have been significantly less. The diversity of begomoviruses and associated components we have identified is wholly consistent with these assumptions. The analysis presented here thus provides an indication of the begomoviruses and their associated satellites that are, or have been, present in the environment, as well as an indication of their potential for evolution by component exchange and recombination. The study also highlights the effect of plant host background and, possibly, agricultural practices can have on plant virus populations. The plants in the orchard are under the same environmental pressures as the cotton plants in the adjacent farmer's fields, with the possible exception that no control measure for viruses or insects vectors (insecticides) are implemented in the orchard, and are inoculated with the same viruses carried by *B. tabaci* as the cotton plants in the farmer's fields, yet they contain an entirely different set of begomoviruses and associated satellites. At the time of sampling, only a single virus species, *Cotton leaf curl Burewala virus* [27], and its betasatellite (CLCuMuB^{Bur} [27,28]) was present in the cultivated cotton surrounding the orchard. Cotton is by far the predominant crop in this region during the summer months when the insect vector, *B. tabaci*, is active.

The isolates of CLCuBuV characterized here differ from those identified earlier identified in having an intact TrAP gene. Amrao *et al.* (2010) showed that CLCuBuV, cloned for the most part from *G. hirsutum* plants carrying the CP-15/2-LRA-5166 derived resistance [20] to viruses of the Multan strain of CLCuD and showing severe CLCuD symptoms, lacked an intact TrAP [27]; a finding since confirmed for CLCuBuV originating from India [51,52]. A hypothesis was put forward suggesting that the TrAP protein may be the avirulence determinant recognized by the resistance gene(s) of CP-15/2-LRA-5166 derived varieties, which could have selected for a virus lacking this gene. The close proximity of the orchard to the area where the resistance breaking virus, CLCuBuV, was first reported may suggest that CLCuBuV originated in the orchard. Certainly there is the distinct possibility that this, and other viruses, could originate from plants, such as the *Gossypium* spp. in the orchard, which harbor a great diversity of

Table 4. Highest and lowest percentage identity values for comparisons of the complete nucleotide sequences of the alphasatellites identified here with selected alphasatellites associated with begomoviruses and nanoviruses obtained from the databases.

Sequences determined in this study	Sequences obtained from the databases						
	GMusSLA (14)*	GDarSLA (51)* [95]#	CLCuMuA (9)* [99-86]@	CLCuGeA (2)* [100]@	ToLCA (3)* [94-80]@	FBNYVC9A (2)* [98]@	BBTA◇ (2)* [92.3]@
GMusSLA (14)*[100-88]▲	-	57-52	56-54	57-51	52-44	41-38	46-45
GDarSLA (49)*[100-85]▲	-	-97-89	78-75	37-33	67-56	40-32	32-24

*The figures in round brackets indicate the numbers of isolates compared.
@The figures in square brackets are the percentage identity values for comparisons between the sequences obtained from the databases.
#There are two GDarSLA sequences available in the databases, originating from India. The figure in brackets is the percentage identity between these two sequences.
▲The figures in square brackets are the percentage identity values for comparisons between the sequences obtained from Gossypium spp.
◇Two nanovirus-associated alphasatellites are included for comparison; Faba bean necrotic yellow vein C9 alphasatellite (FBNYVC9A) and Banana bunchy top alphasatellite (BBTA).

begomoviruses and associated components over a prolonged period – a situation that promotes recombination and components exchange.

Despite the preponderance of CLCuBuV in the environment surrounding the orchard at the time of sampling, the cotton species contained begomoviruses in addition to CLCuBuV. CLCuRaV was identified in two cotton species but has only recently been identified in Pakistan for the first time, infecting tomato in Faisalabad [53]. It has not been identified in cotton in Pakistan but has previously been identified in *G. hirsutum* affected by CLCuD in India [22].

The presence of a begomovirus DNA-B component in cotton is surprising. No DNA-B components have previously been identified in *G. hirsutum* in Pakistan where only betasatellite-requiring begomoviruses have been identified. It is also surprising to find that the DNA-B is closely related to the DNA-B components of ICMV, JCMV and SLCMV. Although several other bipartite begomoviruses occur in Pakistan, such as *Tomato leaf curl New Delhi virus* [54], *Mungbean yellow mosaic India virus* [55] and *Squash leaf curl China virus* [56], ICMV, JCMV and SLCMV have so far not been identified in Pakistan. The origin of the DNA-B component thus remains a mystery. SLCMV is believed to be a monopartite virus which has "captured" a DNA-B component from ICMV [36]. Thus SLCMV DNA-B has most of the sequence of the ICMV DNA-B, but contains the ori of SLCMV DNA-A; presumably obtained by recombination with SLCMV DNA-A – a process referred to as "origin donation" [36]. Closer analysis of the DNA-B components shows them to carry the ori of GPMLCuV. This indicates that origin donation has occurred which is supported by all the DNA-B molecules having the same predicted iteron sequences as GPMLCuV that differ from those of the ACMV DNA-A components isolated here. This is consistent with GPMLCuV being the cognate "DNA-A" of these DNA-B components. However, in three cotton species (*G. darwinii*, *G. stocksii* and *G. somalense*) no begomovirus (or DNA-A component) was identified that has iteron sequences compatible with the DNA-B components. The precise mechanism for maintenance of the DNA-B components in these cotton species thus remains unclear and will require further investigation.

ICMV, JCMV and SLCMV are not known to infect *Gossypium* spp. The DNA-B components of bipartite begomoviruses encode two products (NSP and MP) that mediate intra- and intercellular movement of viruses in hosts, requiring interaction with both the virus (specifically viral DNA) and host factors [57,58,59,60,61,62]. The question thus arises whether the DNA-B component identified here is functional in cotton or is maintained as a non-functional "satellite-like" molecule which makes no contribution to the infectivity of the virus(es) with which it is associated. This will require experimental verification, specifically infectivity studies with clones in cotton, which at this time is not possible. However, maintenance of the NSP and MP coding sequences suggests that there is a purifying selection for their maintenance, which in turn is good evidence that the genes are functional and thus act as virus movement proteins in *Gossypium* spp. If this turns out to be the case, this would be the first bipartite begomovirus identified infecting cotton in the OW. To investigate whether GPMLCuV can trans-replicate DNA-B, clones were inoculated to *N. benthamiana*. The DNA-B was maintained in *N. bethamiana* and together the two components induced symptoms that were distinct from plants infected with only GPMLCuV. Additionally the viral DNA levels were elevated, indicating that GPMLCuV can trans-replicate the DNA-B and the DNA-B contributes to the virus infection (this likely indicating that the genes encoded by the DNA-B are functional). Similarly GPMLCuV was inoculated with the DNA-B and CLCuMuB to *N. benthamiana* plants and the satellite was maintained,

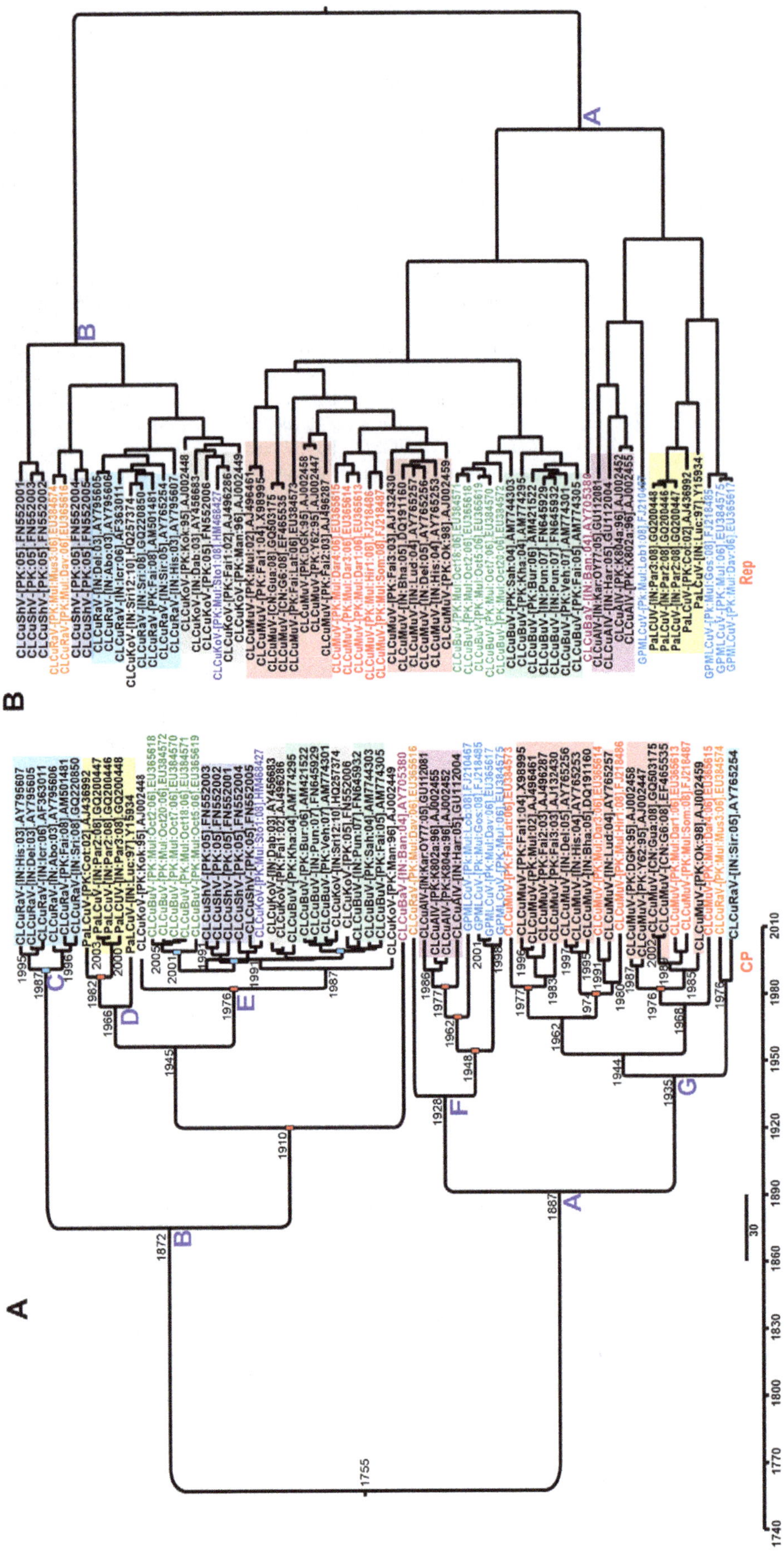
A
B
CP
Rep
1740
1770
1800
1830
1860
1890
1920
1950
1980
2010
30

Figure 7. Dating estimates for the origins of CGs. Dated phylogeny for the coat protein (CP) genes of CLCuD-associated begomoviruses originating from Asia (A). The scaled to time trees were generated by using uncorrelated relaxed LogNormal clock model in the BEAST program (v1.6). The base of each clade is labeled with the mean time to most recent common ancestor (TMRCA) values. Red boxes on nodes indicate viruses which were present before the first appearance of CLCuD in 1967, while blue boxes represent viruses which appeared after the first incidence of the disease. A non-dated phylogeny for the replication-associated protein of CLCuD-associated begomoviruses originating from Asia is shown for comparison (B). Due to recombination the lineages in panel B do not the mirror those of panel A and recombination also leads to some virus species not being monophyletic.

inducing severe symptoms. This indicates that GPMLCuV can potentially act as either a monopartite satellite-associated virus or as a bipartite virus. However, whether this is also the case in *G. hirsutum* remains to be determined and is an important question.

The identification of a JCMV-derived DNA-B component in these cotton species may indicate that ICMV, JCMV, SLCMV or a closely related species, sharing a DNA-B component with them, may be present in Pakistan. Possibly one or more of these viruses remain to be identified in weed hosts. Weeds of the family *Euphorbiaceae* and *Fabaceae* are known to be one of the few alternative hosts/reservoirs of the African cassava-infecting begomoviruses [63,64], thus possibly JCMV has yet to be identified in such weeds.

G. arboreum and *G. herbaceum* are native to Asia and Africa and were found not to contain any begomoviruses and their associated components, confirming earlier reports [19,23]. This has been taken to indicate that these cotton species have had a long association with the viruses present in these regions and have evolved efficient resistance mechanisms against them. This is clearly not the case for the NW cotton species, which are highly susceptible and appear to be selecting for unique viruses and their associated components in this orchard, in comparison to the *G. hirsutum* crop. Nevertheless, there is some inherent capacity for resistance to the OW begomoviruses in the genetic make-up of NW cotton, as seen in the resistance to the CLCuMuV selected from *G. hirsutum* by conventional breeding during the epidemic of the 1990s [20].

In addition to the monopartite begomovirus species which have previously been shown to be associated with CLCuD many of the cotton species examined here also contained a begomovirus which has not been identified previously. In common with the majority of viruses this has a recombinant origin with most of the sequence derived from CLCuMuV. The name *Gossypium punctatum mild leaf curl virus* (GPMLCuV) is proposed for this new species. It seems likely that this begomovirus species evolved in the orchard. The lack of apparent spread of GPMLCuV out of the orchard may suggest that *G. hisutum* is resistant to it, or that the virus is not well adapted to *G. hirsutum*.

Six of the cotton species were shown to contain ACMV DNA-A. In addition, the sequence of DNA-A isolate [PK:Mul:Som:08], (FJ218487), isolated from *G. somalense*, contains a small (approximately 68 nt) fragment derived from ACMV. ACMV occurs across sub-Saharan Africa [65] but has not been reported from elsewhere, including Pakistan and India. The presence of ACMV is surprising since cassava is not grown in Pakistan. There are long-standing trade routes between North Africa and Pakistan and some of the peoples of coastal Pakistan have their ethnic roots in Sudan. Also there are large expatriate south Asian communities in many East African countries. These are all possible routes for transfer of viruses between regions. It has recently been suggested that *Bean yellow dwarf virus* (BeYDV), a dicot-infecting mastrevirus, was introduced into southern Africa by this means: BeYDV occurs in both southern Africa and in Pakistan [66].

ACMV is a bipartite virus and requires the presence of a DNA-B, which encodes movement functions, to infect plants [6]. However, in none of the plants was there any evidence for the presence of ACMV DNA-B. It is noticeable that all plants shown to contain

ACMV DNA-A also contained the JCMV-like DNA-B. However, the predicted iterons of ACMV and those of the DNA-B components detected are distinct, suggesting that the ACMV Rep would be unable to trans-replicate these DNA-B components. However, this will need to be confirmed experimentally, as exceptions are possible. Many of the ACMV DNA-A clones isolated from the cotton species are defective. This may suggest that the ACMV DNA-A component is being maintained as a satellite-like molecule, with movement functions (to complement the missing DNA-B) and insect transmission (for those ACMV components lacking the CP) being provided *in trans* from another begomovirus.

The high levels of sequence identity between the ACMV DNA-A components identified in cotton and ACMV isolates originating from western Africa (particularly Cameroon and Ivory Coast) suggest that this is the geographical origin of the ACMV that has been introduced into Pakistan. This finding also suggests that the introduction has been quite recent; the virus having diverged little from the isolates present in Africa. In order to rule out any possibility of contamination during the cloning process, the orchard in Multan was re-sampled in 2008 and the presence of ACMV DNA-A was confirmed. The finding that some isolates of CLCuMuV and CLCuGeB contain small recombinant fragments of ACMV and CLCuMuB respectively, further strengthens the suggestion that there has been exchange of viruses between Africa and southern Asia.

Only a single species of betasatellite, the CLCuD-associated betasatellite CLCuMuB, was shown to be present in the diverse cotton species. CLCuMuB has been shown to be essential for inducing disease in *G. hirsutum* [19]. In the absence of this betasatellite CLCuMuV was shown to be very poorly infectious to *G. hirsutum* and to induce atypical symptoms [67]. Betasatellites encode a single gene product (βC1) that is a pathogenicity determinant which suppresses PTGS [15,16]. Expression of the CLCuMuB βC1 from a PVX vector has been shown to induce all the symptoms typical of CLCuD [68]. It is thus no surprise to find CLCuMuB in these exotic cotton species, indicating that, for many of them, the begomoviruses they harbor likely also require the additional functions provided by this betasatellite to effectively infect them. However, three cotton species (*G. gossypioides*, *G. somalense* and *G. lobatum*) were identified which contain viruses that have previously been shown to be betasatellite-requiring but in which no betasatellites were detected. For example, in *G. somalense* the presence of CLCuRaV was shown but without a betasatellite. Additionally, for *G. mustelinum*, although the presence of the betasatellite-requiring CLCuKoV was shown, only CLCuMuB mutants lacking the βC1 gene were identified. These results suggest that, in contrast to *G. hirsutum*, in some instances infection of these cotton species does not require the presence of the functions provided by the CLCuMuB. Nevertheless, in most cases, the severe symptom phenotype correlated with the presence of CLCuMuB.

Although for *G. davidsonii* the presence of the betasatellite was shown, this species exhibited only mild symptoms, whereas *G. gossypioides* did not contain the betasatellite, but showed very severe symptoms. These results suggest that disease in some cotton species can be induced by begomoviruses without CLCuMuB.

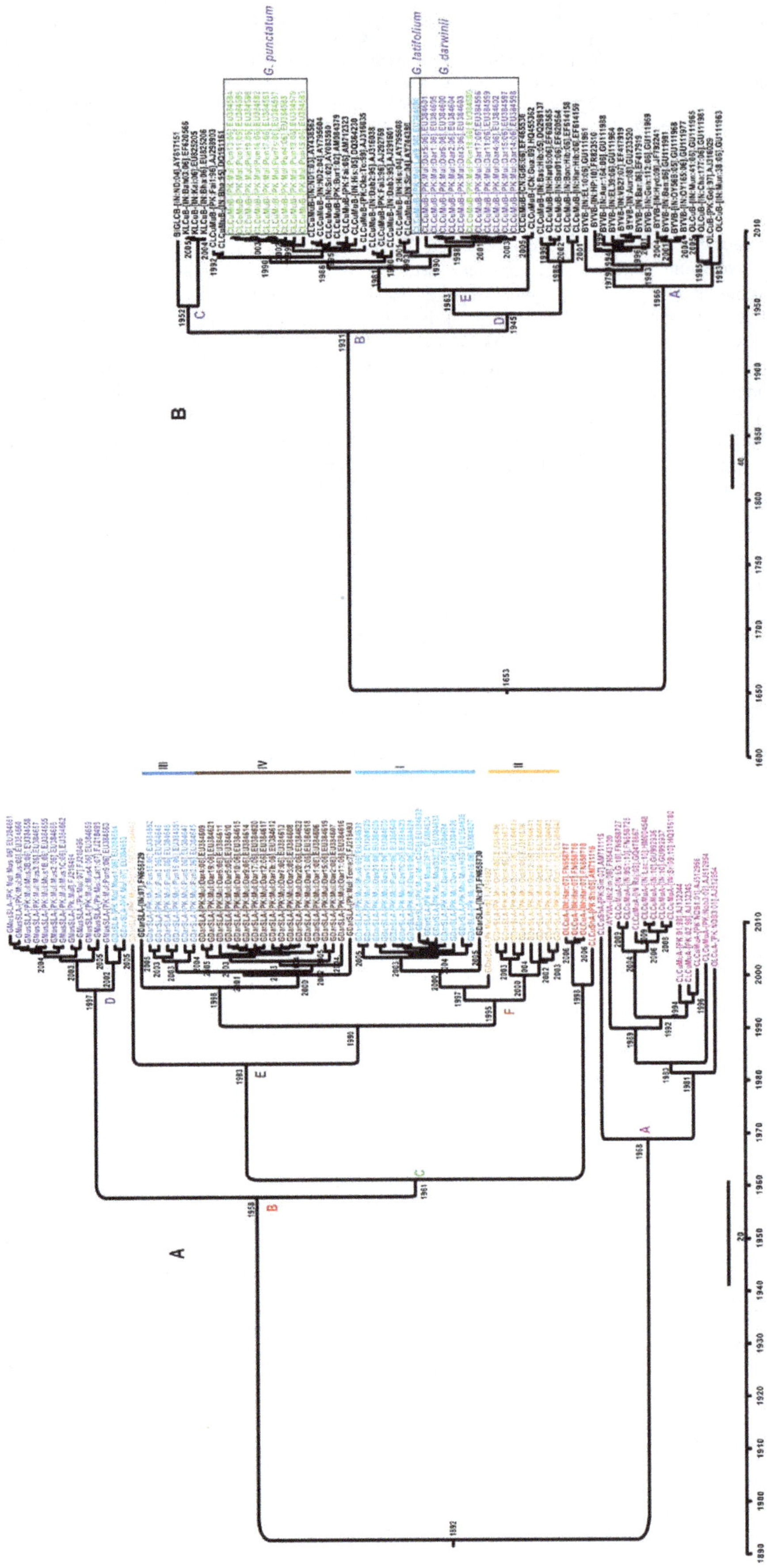
A
B
G. punctatum
G. latifolium
G. darwinii

Figure 8. Dating estimates for the origins of cotton leaf curl begomovirus-associated satellites. Dated phylogeny of the Rep (panel-A) and βC1 (panel-B) proteins of alphasatellites and betasatellites, respectively, associated with CLCuD. The sequences were selected on the basis of prior information from pairwise comparisons and phylogenetic trees presented in Figures 5 and 6. The trees were automatically rooted through uncorrelated relaxed LogNormal clock model in BEAST program (V1.6). The years of divergence are shown on bases of each node. For panel A the groups as sequences labeled I to IV are discussed in the text. For panel B the cotton species from which betasatellites obtained here were isolated are shown on the right.

The complexity with regards to the presence of so many distinct ssDNA molecules makes it difficult to draw any definite conclusions and the suggestion that begomoviruses can infect and induce disease in these species without a betasatellite will need to be confirmed experimentally. There is no information in the literature on the susceptibility of these species to begomoviruses and no reason to believe that they should require a betasatellite for begomovirus infection. In the NW, from where many of these cotton species originate, no betasatellites have been identified and it is bipartite begomoviruses that commonly infect *G. hirsutum* [69].

The orchard in Multan was originally established as an aid to improving *G. hirsutum*, the major component of which has been efforts to identify exotic sources of resistance to CLCuD. It is evident, and consistent with earlier reports, that the A genome cotton species (such as *G. arboreum*) are immune to CLCuD. The fact that no varieties of these species have been shown to be susceptible might indicate that this is a non-host interaction. The remaining species, having D, E and AD genomes, are almost universally susceptible. The one exception here is *G. therburi* which showed no symptoms, nor the presence of any of the geminivirus components. For this species it might be worthwhile examining further varieties to determine whether the lack of support for begomovirus infection constitutes a non-host or a resistance type interaction. If it is a resistance interaction (thus *G. therburi* contains resistance genes for the viruses causing CLCuD) this might be a useful source of resistance genes for incorporation into *G. hirsutum*. If *G. therburi* is a non-host, the fact that all A×D hybrid species are susceptible, might indicate that this species is not useful for improving the CLCuD resistance of *G. hirsutum* since it appears that D genome susceptibility is dominant over A genome non-host resistance (suggesting that the A genome lacks some factor that is essential for infection by begomoviruses).

Since geminiviruses are not seed transmissible, it is unlikely that CGs were introduced with the introduction of cotton species from the NW. The close relationship of CGs to other OW begomoviruses (including their having a V2 gene, which is lacking in NW begomoviruses) is strong evidence that these are viruses native to the sub-Continent which have adapted to infect *G. hirsutum*. Some

149 years after the introduction of *G. hirsutum*, CLCuD was first noted in the vicinity of Multan [49]. TMRCA estimates indicate that for two major groups of CGs (CLCuMuV and CLCuKoV/CLCuRaV: Figure 7A) diversification started in the late 19th century. The exact nature of the begomovirus(es) and satellite(s) involved at the time of the first incidence is unknown. In our analysis for date estimates, it is clear that CLCuMuV was present in the early 20th century. However, CLCuMuB seems to emerge after 1963, consistent with the time of the first emergence of CLCuD on *G. hirsutum*. It is also evident that, at the time of the first epidemic during earlier 1990s all the CGs species were present. However, the fact that CLCuMuV was the most widespread (based on the numbers of sequences isolated) prior to resistance breaking, that CLCuMuV appears to have donated sequences to numerous other CGs by recombination and that only CLCuMuB appears able to induce (in association with CGs) CLCuD, leads to the not unreasonable assumption that the first infection of cotton in the 1980s, that ultimately led to the epidemic, involved CLCuMuV and CLCuMuB. The introduction of resistant cotton varieties in the 1990s was a stringent bottleneck through which only one virus was able to pass. It remains unclear what the molecular basis for resistance breaking is. Both CLCuBuV and CLCuMuB^{Bur} show changes with respect to the viruses and betasatellite occurring pre-resistance breaking. Unfortunately the lack of an infectivity system for cotton prevents a more detailed investigation at this time. However, what happened in 2001 is an indication of the apparent ease with which conventional resistance can be broken and does not bode well for future conventional host-plant resistance, which may be identified

Four distinct alphasatellites (CLCuMuA, CLCuShA, GDarSLA and GMusSLA) have been shown to be associated with CLCuD in Pakistan. The TMCRA estimates suggest that GMusSLA was present before the disease appeared in an epidemic form. However, GDarSLA seems to have evolved more recently, in the early 90s. The role of alphasatellites in begomovirus-betasatellite complexes remains unclear. Several of the *Gossypium* species studied here contain an unusual diversity of satellites and exhibit mild symptoms. GDarSLA, GMusSLA and AYVSGA

Table 5. Mean substitution rates for satellites encoded proteins and CLCuMuV-encoded coat protein.

Model used	Gene	Number of sequences used	Mutation rate Codon position 1	Mutation rate Codon position 2	Mutation rate Codon position 3	Mean rate (substitutions/ nucleotide/year)
Relaxed clock+BSP	CLCuMuB βC1	39	.85	0.73	1.43	3.51×10^{-3}
Relaxed clock+BSP	GDarSLA Rep	63	1.4	0.765	0.831	2.13×10^{-3}
Relaxed clock+BSP	CLCuMuV CP	19	1.64	0.449	0.909	4.24×10^{-4}
Relaxed clock+BSP*	EACMV CP	71	-	-	-	1.37×10^{-3}
Relaxed clock+exponential**	TYLCV CP	54	-	-	-	4.63×10^{-4}

*Duffy and Holmes, 2009.
**Duffy and Holmes, 2008.

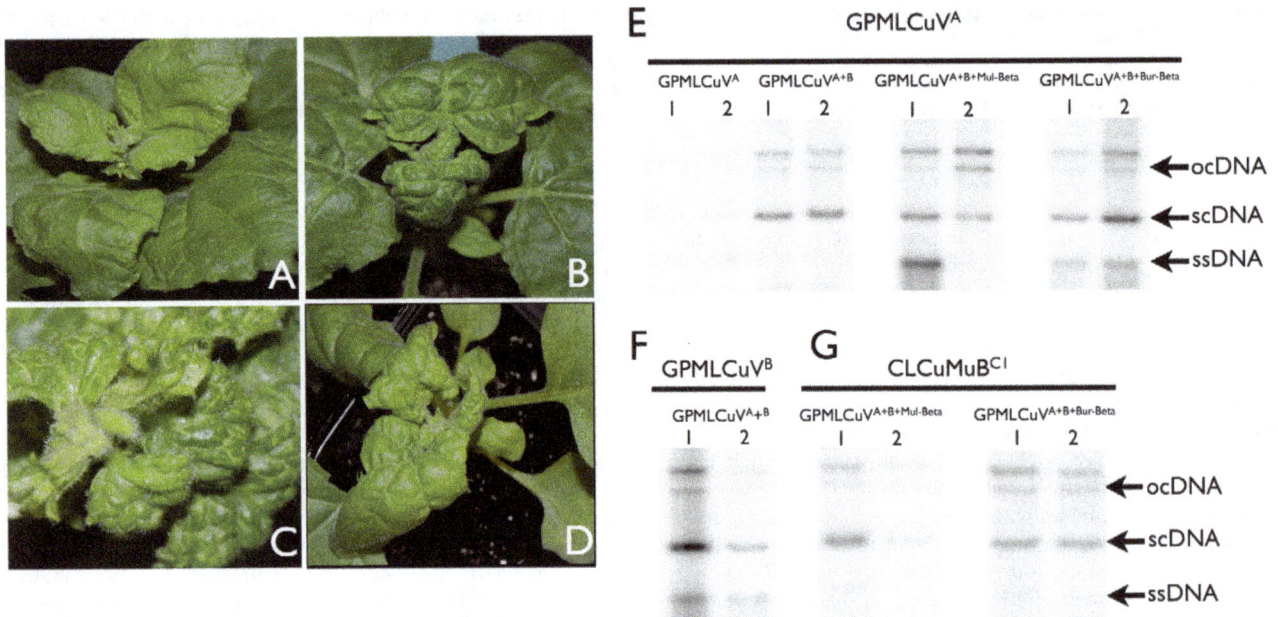

Figure 9. Infectivity of GPMLCuV to *N. benthamiana* in the presence DNA-B and CLCuMuB. *N. benthamiana* plants were inoculated with GPMLCuV (GPMLCuV^A), GPMLCuV with its cognate DNA-B (GPMLCuV^{A+B}), GPMLCuV with its cognate DNA-B and CLCuMuB^{Mul} (GPMLCuV$^{A+B+Mul-Beta}$), and GPMLCuV with its cognate DNA-B and CLCuMuB^{Bur} (GPMLCuV$^{A+B+Bur-Beta}$). Plants were photographed at 21 dpi. Southern hybridization for each combination is shown in panels E, F and G. In each case, approximately, 0.5 ug of DNA isolated from two different plants was loaded (lanes 1 and 2) on agarose gels and transferred to nylon membranes. Panel E was probed with the C4 gene of GPMLCuV. Panels F and G were probed with the BV1 and βC1 genes of DNA-B and CLCuMuB respectively. The replicative DNA forms are indicated as single-stranded (ss), supercoiled (sc) and open-circular (oc).

have the capacity to ameliorate begomovirus symptoms in plants [47,70]. It is thus possible that, in the *Gossypium* spp., this diversity of alphasatellites acts to reduce the pathogenicity of the begomoviruses/betasatellite and could thus promote virus diversification by allowing plants to tolerate infection. This together with the suggestion, from nucleotide substitution rates, that virus and satellites are under different evolutionary pressures and the finding that each *Gossypium* species is selecting for distinct satellites (a possible indication of little exchange of components between *Gossypium* spp.) may be the reasons for the accumulation of such a diversity of viruses and satellites in the orchard.

It is evident that the long term maintenance of the plants in Multan has led to a build-up of a diverse array of begomoviruses and begomovirus-associated satellites and has, more than likely, provided a crucible for the evolution of new species, strains and variants by mutation, recombination and component exchange. It is likely that the diversity identified in the plants in this orchard is a reflection of the diversity of begomoviruses and begomovirus-associated satellites that have been, or are, present in the environment. If this is the case, it is evident that the diversity is much greater than previously estimated by analyzing single isolates from single cultivated host plants. Likely this identification of a wider diversity results not only from the unusual nature of the plants in the orchard (exotic germplasm that has been maintained vegetatively over a very long period of time) but also from the ability of RCA to amplify circular DNA components without prior knowledge of the sequence to be amplified (a requirement that is a definite draw-back of the widely used PCR technique to clone geminivirus molecules). It is possibly coincidental that we have identified CLCuBuV with an intact TrAP gene in the orchard, whereas across the rest of the country it appears that CLCuBuV lacks this gene, possibly suggesting that this virus species could

have originated in the wild species. However, there are other possible sources for this virus such as malvaceous perennials that are maintained long term in an infected state. A pertinent example here is *Hibiscus rosa-sinensis*. *Hibiscus* is a malvaceous perennial that is grown widely as an ornamental and hedge plant across Pakistan and India and, in areas where the environmental conditions are suitable (thus the same areas where cotton is grown), almost universally exhibits symptoms typical of CLCuD and harbors the components that cause CLCuD [26]. Such plants could thus, like the exotic *Gossypium* plants in Multan, be the source of new virus/satellite diversity. The identification, in our study, of a CLCuMuV isolate that contains some sequence derived from ACMV in *G. hirsutum* that did not harbor ACMV, is clear evidence that such new variants are spreading out of the orchard into the crop.

The take-home message from these finding is that this long term maintenance of such virus-infected perennials is unwise, particularly in areas where susceptible crops are grown. The advice must therefore be for the *Gossypium* orchard in Multan, as well as similar plant collections across Pakistan and India, to be removed each year and replanted from seed, since geminiviruses are not seed transmissible. For farmers, the removal of infected perennials, such as *Hibiscus* spp., would be a wise precaution in an effort to reduce the availability of overwintering hosts and potential sources of virus diversity. Farmers are already advised to remove cotton plants at the end of every season and not allow ratooning (regrowth) from the previous year's plants, in an effort to avoid build-up of virus early in the season. Although, with the benefit of hindsight, we can now say that maintaining the *Gossypium* orchard in the middle of the cotton growing belt of Pakistan was/is unwise, it has proven a valuable source of information on the diversity, evolution and adaptation of begomoviruses and their associated satellites.

Materials and Methods

Sample collection and DNA extraction

Fresh young leaves were frozen in liquid nitrogen and total genomic DNA was extracted by the CTAB method [71]. All necessary permits were obtained from USDA/APHIS and USDA/BRS to import the material to the Danforth Center.

Amplification, cloning and sequencing of begomoviruses and their satellites

DNA was quantified using an Ultrospec3000 (Pharmacia Biotech) spectrophotometer and each DNA sample was diluted to 10 ng/μl. RCA reactions were prepared as recommended by manufacturer (TempliPhi[TM], GE Healthcare). Briefly, DNA (~20 ng) was diluted in 5 μl of reaction buffer and denatured for 5 min at 95°C. After cooling at room temperature, 5 μl of sample buffer was added followed by 0.2 μl of φ29 DNA polymerase enzyme mix. The reaction was allowed to proceed for 16 hrs at 28°C and then stopped by heating at 65°C for 10 min. Approximately 1 μg of RCA product was digested with selected restriction endonucleases and bands of approx. 2800 nt and/or 1400 nt were selected for cloning. Selected restriction products were gel isolated and cloned in the plasmid vector pUC19 [72]. Due to the non-specific nature of the amplification of all circular DNA molecules by φ29 DNA polymerase, clones of betasatellites and alphasatellites were difficult to distinguish based only upon size. Colony hybridization [73] was performed to identify recombinant E. coli colonies harboring betasatellites using the βC1 gene of CLCuMuB (EU384588) as a probe. Cloned products were sequenced using a model 377 Automated DNA Sequencer (Perkin Elmer). All clones were sequenced, using the primer walking strategy, in both orientations with no ambiguities remaining.

Sequence assembly, manipulation and analysis

Sequences were assembled and analyzed using the Lasergene (v.8) package (DNASTAR, Madison, Wisconsin). The sequences obtained were compared with sequences available in public databases using BLASTn (NCBI). The coding regions and coding capacities of virus/satellites clones were predicted using the EditSeq module of Lasergene.

Phylogenetic, sequence distance analysis and detection of recombination

Sequences were aligned using Mega5 by applying the Clustal-W algorithm and resulting alignments were used for pairwise comparisons and construction of phylogenetic trees using the neighbor-joining algorithm [74]. Recombination was detected using RDP3 [75] available at http://darwin.uvigo.es/rdp/rdp.html. The default X-over panel was used for recombination analysis in all cases.

Estimation for times of divergence

Divergence times for virus and satellite components associated with CLCuD were estimated by Bayesian Marcov Chain Monte Carlo (MCMC), applied as implemented in BEAST v1.6 [76]. No out-group sequences were used in the BEAST analysis and the relationship based on tree topologies in the preliminary analysis (described earlier) was enforced as prior assumption for the Bayesian analysis. The year wise dates for each sequence were manually added in the tree dates panel of BEAST for TMRCA. The relaxed molecular clock (uncorrelated lognormal) along with exponential population growth models was set for further analysis as described previously for EACMV, TYLCV and influenza virus [50,77,78,79]. Each species of begomovirus and satellite was divided into monophyletic taxa in the taxon-set tab of BEAUti module of BEAST. This was performed to ensure that during the dating estimates each species was kept under a monophyletic group. To ensure an adequate sample size, a chain length of 40,000,000 was used in the MCMC panel. The log and tree files generated by BEAST were combined in the LogCombinar (v.1.6.1) application of BEAST. The LogCombined trees were annotated using TreeAnnotator (v.1.6.1). MCC phylogenetic trees were calculated with 10–15% burn-in followed by visual inspection in TRACER (v.1.5; http://tree.bio.ed.ac.uk/software/tracer/). Phylogenetic trees were analyzed using FigTree (v.1.3.1; http://tree.bio.ed.ac.uk/software/figtree). Since sequence data produced prior to 2010 was used, this year was used as the offset for the time limit in the time scale panel of the FigTree package. The Log and Tree files of the analysis can be obtained from the authors upon request.

Estimation of nucleotide substitution rates by Bayesian Skyline Plot

For all three datasets (CP, βC1 and alpha-Rep) nexus files were generated after aligning sequences in Mega5 as described above. The general time reverse (GTR+E) substitution model was chosen and the nucleotides dataset was partitioned into 3 sets (codon positions 1, 2 and 3). Coalescent Bayesian Skyline was chosen in the tree panel for the tree prior and each dataset was run for a chain length of 4×10^7 to ensure an adequate sample size in the MCMC panel of the BEAUTi module in BEAST.

Inoculation of plants

GPMLCuV (EU365617), DNA-B (EU384577), CLCuMuBMul (FJ607041) and CLCuMuB[Bur] (EU384587) were introduced into N. benthamiana plants by biolistic inoculation. Briefly, the components were excised from their cloning vectors by restriction digestion and then circularized by ligation. Each component was then amplified by RCA, as described earlier [12]. Approximately, 100 ng of multimeric RCA product was coated on 1 μM gold particles and inoculated to N. benthamiana plants using a Helios gene gun (BioRad) as described previously [80]. For co-inoculation of components equal amounts of each component were coated on a single aliquot of gold particles. After inoculation, plants were maintained in a glasshouse and scored for infection at 2 week post-inoculation [81].

Southern blot hybridization

Total DNA was extracted from inoculated plants at 15 dpi using a DNAeasy mini kit (Qiagen). Total DNA (~500 ng) was resolved on 1.5% agarose gels and transferred to nylon membranes (Hybond N+, Amersham). Digoxygenin (DIG) labeled probes for each component were prepared using a PCR DIG Probe Synthesis Kit (Roche). For GPMLCuV primers for amplification of the C4 gene (5'-TCAGGGCCTCTGCTGCTGCATCATT-3' and 5'-ATGGGTCTCTGCATATCCACGC-3') and for the DNA-B component primers for the BV1 gene (5'-ATGAGAA-ATGTTGGTTATACTCTCC-3' and 5'-TTAACCAATA-TATCGCATTACATA-3') were used to PCR-amplify the DIG labeled probes. For betasatellites, the βC1 probe described earlier was used [82].

Supporting Information

Figure S1 Mutations of the virion-sense genes of ACMV DNA-A components isolated from cotton spp. Alignment of the sequence of an isolate of ACMV (ACMV-[CM:Mg:98].AY211884) originating from Africa (Cameroon) with one of the ACMV isolates obtained from cotton spp. (ACMV-[PK:Mul:Dar:06].GQ169505). Conserved nucleotides are highlighted in red. The positions of the AV2 (purple box) and AV1 (CP; orange box) genes of ACMV-[CM:Mg:98] are shown. Gaps (−) were introduced into the sequences to optimise the alignment.

Figure S2 Duplicated stem-loop sequences in the DNA-B components isolated from *Gossypium* species. Alignment of the 3′ sequence, from the hairpin-loop structures (coordinate 1) onwards, of the DNA-B components isolated in this study. Duplicated sequences spanning the right (3′) leg of the stem loop structure (underlined) are highlighted in red. The 3′ repeat is the *bona fide* right (3′) leg of the components. The blue and orange boxes highlight sequences originating from BYVMV and CLCuMuV, respectively.

Figure S3 Phylogenetic analysis of betasatellite sequences. This tree differs from that in Figure 5 only in containing more sequences; sequences representing all so far identified betasatellite species. For other details about the tree see the figure legend of Figure 5.

Table S1 List of virus genomes and genomic components isolated from *Gossypium* species.

Table S2 ACMV DNA-A clones isolated from cotton species.

Table S3 Details of recombination between CGs detected using RDP3.

Table S4 CLCuMuB clones isolated from *Gossypium* species.

Table S5 Alphasatellites isolated from *Gossypium* species.

Table S6 Upper and lower bounds of the 95% highest posterior density (HPD) estimates for divergence dates of CLCuMuV-encoded CP, CLCuMuB-encoded βC1 and GDarSLA-encoded Rep.

Acknowledgments

We are thankful to Mr. Riaz Qureshi (head of cytology section) of the Central Cotton Research Institute (Multan, Pakistan) for his help in collecting samples. The authors thank Dr. Nazia Nahid for critical reading of the manuscript.

Author Contributions

Conceived and designed the experiments: MSNR CMF RWB. Performed the experiments: MSNR. Analyzed the data: MSNR CMF RWB. Contributed reagents/materials/analysis tools: MSNR. Wrote the paper: MSNR CMF RWB.

References

1. Fauquet C, Briddon R, Brown J, Moriones E, Stanley J, et al. (2008) Geminivirus strain demarcation and nomenclature. Arch Virol 153: 783–821.

2. Brown JK, Fauquet CM, Briddon RW, Zerbini M, Moriones E, et al. (2011) Geminiviridae. In: King AMQ, Adams MJ, Carstens EB, Lefkowitz EJ, eds. Virus Taxonomy - Ninth Report of the International Committee on Taxonomy of Viruses. London, Waltham, San Diego: Associated Press, Elsevier Inc. pp. 351–373.

3. Varma A, Malathi VG (2003) Emerging geminivirus problems: A serious threat to crop production. Ann Appl Biol 142: 145–164.

4. Moffat AS (1999) Geminiviruses emerge as serious crop threat. Science 286: 1835.

5. Briddon RW, Stanley J (2006) Subviral agents associated with plant single-stranded DNA viruses. Virology 344: 198–210.

6. Stanley J (1983) Infectivity of the cloned geminivirus genome requires sequences from both DNAs. Nature 305: 643–645.

7. Rojas MR, Hagen C, Lucas WJ, Gilbertson RL (2005) Exploiting chinks in the plant's armor: Evolution and emergence of geminiviruses. Ann Rev Phytopathol 43: 361–394.

8. Stanley J, Gay MR (1983) Nucleotide sequence of cassava latent virus DNA. Nature 301: 260–262.

9. Argüello-Astorga GR, Ruiz-Medrano R (2001) An iteron-related domain is associated to Motif 1 in the replication proteins of geminiviruses: identification of potential interacting amino acid-base pairs by a comparative approach. Arch Virol 146: 1465–1485.

10. Chatterji A, Chatterji U, Beachy RN, Fauquet CM (2000) Sequence parameters that determine specificity of binding of the replication-associated protein to its cognate site in two strains of tomato leaf curl virus-New Delhi. Virology 273: 341–350.

11. Li Z, Xie Y, Zhou X (2005) *Tobacco curly shoot virus* DNA β is not necessary for infection but intensifies symptoms in a host-dependent manner. Phytopathology 95: 902–908.

12. Nawaz-ul-Rehman MS, Mansoor S, Briddon RW, Fauquet CM (2009) Maintenance of an Old World betasatellite by a New World helper begomovirus and possible rapid adaptation of the betasatellite. J Virol 83: 9347–9355.

13. Saunders K, Briddon RW, Stanley J (2008) Replication promiscuity of DNA-β satellites associated with monopartite begomoviruses; deletion mutagenesis of the ageratum yellow vein virus DNA-β satellite localizes sequences involved in replication. J Gen Virol 89: 3165–3172.

14. Saeed M, Zafar Y, Randles JW, Rezaian MA (2007) A monopartite begomovirus-associated DNA β satellite substitutes for the DNA B of a bipartite begomovirus to permit systemic infection. J Gen Virol 88: 2881–2889.

15. Cui X, Li G, Wang D, Hu D, Zhou X (2005) A begomovirus DNAβ-encoded protein binds DNA, functions as a suppressor of RNA silencing, and targets the cell nucleus. J Virol 79: 10764–10775.

16. Saunders K, Norman A, Gucciardo S, Stanley J (2004) The DNA β satellite component associated with ageratum yellow vein disease encodes an essential pathogenicity protein (βC1). Virology 324: 37–47.

17. Briddon RW, Bull SE, Amin I, Mansoor S, Bedford ID, et al. (2004) Diversity of DNA 1: a satellite-like molecule associated with monopartite begomovirus-DNA β complexes. Virology 324: 462–474.

18. Moulherat C, Tengberg M, Haquet J-F, Mille B (2002) First Evidence of cotton at Neolithic Mehrgarh, Pakistan: Analysis of mineralized fibres from a copper bead. J Archaeol Sci 29: 1393–1401.

19. Briddon RW, Mansoor S, Bedford ID, Pinner MS, Saunders K, et al. (2001) Identification of DNA components required for induction of cotton leaf curl disease. Virology 285: 234–243.

20. Rahman M, Hussain D, Malik TA, Zafar Y (2005) Genetics of resistance to cotton leaf curl disease in *Gossypium hirsutum*. Plant Pathol 54: 764–772.

21. Mansoor S, Amin I, Iram S, Hussain M, Zafar Y, et al. (2003) Breakdown of resistance in cotton to cotton leaf curl disease in Pakistan. Plant Pathol 52: 784–784.

22. Kirthi N, Priyadarshini CGP, Sharma P, Maiya SP, Hemalatha V, et al. (2004) Genetic variability of begomoviruses associated with cotton leaf curl disease originating from India. Arch Virol 149: 2047–2057.

23. Mansoor S, Briddon RW, Bull SE, Bedford ID, Bashir A, et al. (2003) Cotton leaf curl disease is associated with multiple monopartite begomoviruses supported by single DNA β. Arch Virol 148: 1969–1986.

24. Zhou X, Liu Y, Robinson D, Harrison B (1998) Four DNA-A variants among Pakistani isolates of cotton leaf curl virus and their affinities to DNA-A of geminivirus isolates from okra. J Gen Virol 79: 915–923.

25. Tahir MN, Amin I, Briddon RW, Mansoor S (2011) The merging of two dynasties-identification of an african cotton leaf curl disease-associated begomovirus with cotton in Pakistan. PLoS ONE 6: e20366.

26. Briddon RW, Bull SE, Amin I, Idris AM, Mansoor S, et al. (2003) Diversity of DNA β, a satellite molecule associated with some monopartite begomoviruses. Virology 312: 106–121.

27. Amrao L, Amin I, Shahid MS, Briddon RW, Mansoor S (2010) Cotton leaf curl disease in resistant cotton is associated with a single begomovirus that lacks an intact transcriptional activator protein. Virus Res 152: 153–163.

28. Amin I, Mansoor S, Amrao L, Hussain M, Irum S, et al. (2006) Mobilisation into cotton and spread of a recombinant cotton leaf curl disease satellite. Arch Virol 151: 2055–2065.

29. Wendel JF, Brubaker C, Alvarez I, Cronn R, Stewart JM (2009) Evolution and natural history of the cotton genus. Genet Genom Cotton: Springer New York. pp. 1–20.

30. Akhtar KP, Haidar S, Khan MKR, Ahmad M, Sarwar N, et al. (2010) Evaluation of Gossypium species for resistance to cotton leaf curl Burewala virus. Ann Appl Biol 157: 135–147.

31. Homs M, Kober S, Kepp G, Jeske H (2008) Mitochondrial plasmids of sugar beet amplified via rolling circle method detected during curtovirus screening. Virus Res 136: 124–129.

32. Fauquet CM, Bisaro DM, Briddon RW, Brown JK, Harrison BD, et al. (2003) Revision of taxonomic criteria for species demarcation in the family Geminiviridae, and an updated list of begomovirus species. Arch Virol 148: 405–421.

33. Lefeuvre P, Lett J-M, Varsani A, Martin DP (2009) Widely conserved recombination patterns amongst single stranded DNA viruses. J Virol 83: 2697–2707.

34. Lefeuvre P, Martin DP, Hoareau M, Naze F, Delatte H, et al. (2007) Begomovirus 'melting pot' in the south-west Indian Ocean islands: molecular diversity and evolution through recombination. J Gen Virol 88: 3458–3468.

35. Gao S, Qu J, Chua N-H, Ye J (2010) A new strain of Indian cassava mosaic virus causes a mosaic disease in the biodiesel crop Jatropha curcas. Arch Virol 155: 607–612.

36. Saunders K, Salim N, Mali VR, Malathi VG, Briddon R, et al. (2002) Characterisation of Sri Lankan cassava mosaic virus and Indian cassava mosaic virus: Evidence for acquisition of a DNA B component by a monopartite begomovirus. Virology 293: 63–74.

37. Briddon R, Brown J, Moriones E, Stanley J, Zerbini M, et al. (2008) Recommendations for the classification and nomenclature of the DNA-β satellites of begomoviruses. Arch Virol 153: 763–781.

38. Tao X, Zhou X (2008) Pathogenicity of a naturally occurring recombinant DNA satellite associated with tomato yellow leaf curl China virus. J Gen Virol 89: 306–311.

39. Saunders K, Bedford ID, Stanley J (2001) Pathogenicity of a natural recombinant associated with ageratum yellow vein disease: Implications for geminivirus evolution and disease aetiology. Virology 282: 38–47.

40. Das S, Roy A, Ghosh R, Paul S, Acharyya S, et al. (2008) Sequence variability and phylogenetic relationship of betasatellite isolates associated with yellow vein mosaic disease of mesta in India. Virus Genes 37: 414–424.

41. Chatterjee A, Ghosh S (2007) Association of a satellite DNA β molecule with mesta yellow vein mosaic disease. Virus Genes 35: 835–844.

42. Paul S, Ghosh R, Roy A, Mir JI, Ghosh SK (2006) Occurrence of a DNA β-containing begomovirus associated with leaf curl disease of kenaf (Hibiscus cannabinus L.) in India. Aust Plant Dis Notes 1: 29–30.

43. Tahir MN, Amin I, Briddon RW, Mansoor S (2011) The merging of two dynasties–identification of an African cotton leaf curl disease-associated begomovirus with cotton in Pakistan. PLoS ONE 6: e20366.

44. Amrao L, Akhter S, Tahir MN, Amin I, Briddon RW, et al. (2010) Cotton leaf curl disease in Sindh province of Pakistan is associated with recombinant begomovirus components. Virus Res 153: 161–165.

45. Mansoor S, Khan SH, Bashir A, Saeed M, Zafar Y, et al. (1999) Identification of a novel circular single-stranded DNA associated with cotton leaf curl disease in Pakistan. Virology 259: 190–199.

46. Saunders K, Bedford ID, Stanley J (2002) Adaptation from whitefly to leafhopper transmission of an autonomously replicating nanovirus-like DNA component associated with ageratum yellow vein disease. J Gen Virol 83: 907–913.

47. Idris AM, Shahid MS, Briddon RW, Khan AJ, Zhu JK, et al. (2011) An unusual alphasatellite associated with monopartite begomoviruses attenuates symptoms and reduces betasatellite accumulation. J Gen Virol 92: 706–717.

48. Paprotka T, Metzler V, Jeske H (2010) The first DNA 1-like α satellites in association with New World begomoviruses in natural infections. Virology 404: 148–157.

49. Hussain T, Tahir M, Mahmood T (1991) Cotton leaf curl virus. Pak J Phytopathol 3: 57–61.

50. Duffy S, Holmes EC (2009) Validation of high rates of nucleotide substitution in geminiviruses: phylogenetic evidence from East African cassava mosaic viruses. J Gen Virol 90: 1539–1547.

51. Zaffalon V, Mukherjee S, Reddy V, Thompson J, Tepfer M (2012) A survey of geminiviruses and associated satellite DNAs in the cotton-growing areas of northwestern India. Arch Virol 157: 483–495.

52. Rajagopalan P, Naik A, Katturi P, Kurulekar M, Kankanallu R, et al. (2012) Dominance of resistance-breaking cotton leaf curl Burewala virus (CLCuBuV) in northwestern India. Arch Virol 157: 855–868.

53. Shahid MS, Mansoor S, Briddon RW (2007) Complete nucleotide sequences of cotton leaf curl Rajasthan virus and its associated DNA β molecule infecting tomato. Arch Virol 152: 2131–2134.

54. Padidam M, Beachy RN, Fauquet CM (1995) Tomato leaf curl geminivirus from India has a bipartite genome and coat protein is not essential for infectivity. J Gen Virol 76: 25–35.

55. Ilyas M, Qazi J, Mansoor S, Briddon RW (2010) Genetic diversity and phylogeography of begomoviruses infecting legumes in Pakistan. J Gen Virol 91: 2091–2101.

56. Tahir M, Haider MS, Briddon RW (2010) First report of Squash leaf curl China virus in Pakistan. Aust Plant Dis Notes 5: 21–24.

57. Lewis JD, Lazarowitz SG (2010) Arabidopsis synaptotagmin SYTA regulates endocytosis and virus movement protein cell-to-cell transport. Proc Natl Acad Sci USA 107: 2491–2496.

58. Carvalho MF, Turgeon R, Lazarowitz SG (2006) The geminivirus nuclear shuttle protein NSP inhibits the activity of AtNSI, a vascular-expressed Arabidopsis acetyltransferase regulated with the sink-to-source transition. Plant Physiol 140: 1317–1330.

59. Carvalho MF, Lazarowitz SG (2004) Interaction of the movement protein NSP and the Arabidopsis acetyltransferase AtNSI is necessary for cabbage leaf curl geminivirus infection and pathogenicity. J Virol 78: 11161–11171.

60. McGarry RC, Barron YD, Carvalho MF, Hill JE, Gold D, et al. (2003) A novel Arabidopsis acetyltransferase interacts with the geminivirus movement protein NSP. Plant Cell 15: 1605–1618.

61. Gilbertson RL, Sudarshana M, Jiang H, Rojas MR, Lucas WJ (2003) Limitations on geminivirus genome size imposed by plasmodesmata and virus-encoded movement protein: Insights into DNA trafficking. Plant Cell 15: 2578–2591.

62. Rojas MR, Noueiry AO, Lucas WJ, Gilbertson RL (1998) Bean dwarf mosaic geminivirus movement proteins recognize DNA in a form- and size-specific manner. Cell 95: 105–113.

63. Alabi O, Ogbe F, Bandyopadhyay R, Lava Kumar P, Dixon A, et al. (2008) Alternate hosts of African cassava mosaic virus and East African cassava mosaic Cameroon virus in Nigeria. Arch Virol 153: 1743–1747.

64. Monde G, Walangululu J, Winter S, Bragard C (2010) Dual infection by cassava begomoviruses in two leguminous species (Fabaceae) in Yangambi, Northeastern Democratic Republic of Congo. Arch Virol 155: 1865–1869.

65. Patil BL, Fauquet CM (2009) Cassava mosaic geminiviruses: actual knowledge and perspectives. Mol Plant Pathol 10: 685–701.

66. Nahid N, Amin I, Mansoor S, Rybicki E, van der Walt E, et al. (2008) Two dicot-infecting mastreviruses (family; Geminiviridae) occur in Pakistan. Arch Virol 153: 1441–1451.

67. Briddon RW, Mansoor S, Bedford ID, Pinner MS, Markham PG (2000) Clones of cotton leaf curl geminivirus induce symptoms atypical of cotton leaf curl disease. Virus Genes 20: 19–26.

68. Qazi J, Amin I, Mansoor S, Iqbal MJ, Briddon RW (2007) Contribution of the satellite encoded gene βC1 to cotton leaf curl disease symptoms. Virus Res 128: 135–139.

69. Idris AM, Brown JK (2004) Cotton leaf crumple virus is a distinct western hemisphere begomovirus species with complex evolutionary relationships indicative of recombination and reassortment. Phytopathology 94: 1068–1074.

70. Nawaz-Ul-Rehman MS, Nahid N, Mansoor S, Briddon RW, Fauquet CM (2010) Post-transcriptional gene silencing suppressor activity of two non-pathogenic alphasatellites associated with a begomovirus. Virology 405: 300–308.

71. Doyle JJ, Doyle JL (1987) A rapid DNA isolation procedure for small quantities of fresh leaf tissue. Phytochem Bull 19: 11–15.

72. Yanisch-Perron C, Vieira J, Messing J (1985) Improved M13 phage cloning vectors and host strains: nucleotide sequences of the M13mpl8 and pUC19 vectors. Gene 33: 103–119.

73. Cami B, Kourilsky P (1978) Screening of cloned recombinant DNA in bacteria by in situ colony hybridization. Nucleic Acids Res 5: 2381–2390.

74. Tamura K, Peterson D, Peterson N, Stecher G, Nei M, et al. (2011) MEGA5: Molecular evolutionary genetics analysis using maximum likelihood, evolutionary distance, and maximum parsimony methods. Mol Bio Evol.

75. Martin DP, Lemey P, Lott M, Moulton V, Posada D, et al. (2010) RDP3: a flexible and fast computer program for analyzing recombination. Bioinformatics 26: 2462–2463.

76. Drummond A, Rambaut A (2007) BEAST: Bayesian evolutionary analysis by sampling trees. BMC Evol Biol 7: 214.

77. Bataille A, van der Meer F, Stegeman A, Koch G (2011) Evolutionary analysis of inter-farm transmission dynamics in a highly pathogenic avian influenza epidemic. PLoS Pathog 7: e1002094.

78. Lefeuvre P, Martin DP, Harkins G, Lemey P, Gray AJA, et al. (2010) The spread of tomato yellow leaf curl virus from the middle East to the world. PLoS Pathog 6: e1001164.

79. Duffy S, Holmes EC (2008) Phylogenetic evidence for rapid rates of molecular evolution in the single-stranded DNA begomovirus Tomato yellow leaf curl virus (TYLCV). J Virol 82: 957–965.

80. Ariyo OA, Atiri GI, Dixon AGO, Winter S (2006) The use of biolistic inoculation of cassava mosaic begomoviruses in screening cassava for resistance to cassava mosaic disease. J Virol Methods 137: 43–50.

81. Fauquet C, Fargette D (1990) African cassava mosaic virus: etiology, epidemiology and control. . Plant Dis 74 404–411.

82. Nawaz-ul-Rehman MS, Nahid N, Mansoor S, Briddon RW, Fauquet CM (2010) Post-transcriptional gene silencing suppressor activity of two non-pathogenic alphasatellites associated with a begomovirus. Virology 405: 300–308.

Aerial Application of Pheromones for Mating Disruption of an Invasive Moth as a Potential Eradication Tool

Eckehard G. Brockerhoff[1]*, David M. Suckling[2], Mark Kimberley[3], Brian Richardson[3], Graham Coker[1], Stefan Gous[3], Jessica L. Kerr[1], David M. Cowan[4], David R. Lance[4], Tara Strand[5,3], Aijun Zhang[6]

1 Scion (New Zealand Forest Research Institute), Christchurch, New Zealand, 2 The New Zealand Institute for Plant & Food Research Ltd, Lincoln, New Zealand, 3 Scion (New Zealand Forest Research Institute), Rotorua, New Zealand, 4 Plant Protection and Quarantine, United States Department of Agriculture, Animal and Plant Health Inspection Service, Buzzards Bay, Massachusetts, United States of America, 5 Pacific Wildland Fire Sciences Laboratory, United States Department of Agriculture, Forest Service, Seattle, Washington, United States of America, 6 Invasive Insect Biocontrol and Behavior Laboratory, United States Department of Agriculture, Agricultural Research Services, Beltsville Agricultural Research Center - West, Beltsville, Maryland, United States of America

Abstract

Biological invasions can cause major ecological and economic impacts. During the early stages of invasions, eradication is desirable but tactics are lacking that are both effective and have minimal non-target effects. Mating disruption, which may meet these criteria, was initially chosen to respond to the incursion of light brown apple moth, *Epiphyas postvittana* (LBAM; Lepidoptera: Tortricidae), in California. The large size and limited accessibility of the infested area favored aerial application. Moth sex pheromone formulations for potential use in California or elsewhere were tested in a pine forest in New Zealand where LBAM is abundant. Formulations were applied by helicopter at a target rate of 40 g pheromone per ha. Trap catch before and after application was used to assess the efficacy and longevity of formulations, in comparison with plots treated with ground-applied pheromone dispensers and untreated control plots. Traps placed at different heights showed LBAM was abundant in the upper canopy of tall trees, which complicates control attempts. A wax formulation and polyethylene dispensers were most effective and provided trap shut-down near ground level for 10 weeks. Only the wax formulation was effective in the upper canopy. As the pheromone blend contained a behavioral antagonist for LBAM, 'false trail following' could be ruled out as a mechanism explaining trap shutdown. Therefore, 'sensory impairment' and 'masking of females' are the main modes of operation. Mating disruption enhances Allee effects which contribute to negative growth of small populations and, therefore, it is highly suitable for area-wide control and eradication of biological invaders.

Editor: Randall P. Niedz, United States Department of Agriculture, United States of America

Funding: E.G.B. and D.M.S. declare having received research funds for this project from the United States Department of Agriculture (USDA) and the New Zealand government (NZ Foundation for Research, Science and Technology). The funders (in particular, USDA scientists who are listed as co-authors) contributed to the study design, aspects of the data collection, and the preparation of the manuscript. The funders had no role in the data analysis or the decision to publish.

Competing Interests: The authors have declared that no competing interests exist.

* E-mail: eckehard.brockerhoff@scionresearch.com

Introduction

Biological invasions resulting from trade and other human activities represent a major threat to biodiversity and the integrity of ecosystems [1]. The economic impact of invasive species can reach billions of dollars annually, resulting from control costs and loss of commodities as well as loss of ecosystem goods and services and other non-market values [2,3]. Some invaders, such as emerald ash borer [4] and the pathogens that cause chestnut blight [5] and Dutch elm disease [6], are particularly harmful and can cause large-scale ongoing damage or even the near disappearance of their host plants. Despite our increasing awareness of biosecurity issues and attempts to mitigate invasion pathways, increasing international trade makes the ongoing arrival of exotic species inevitable. For example, establishments of forest insects in the U.S. continue at a rate of ca. 2.5 species per year [7].

Biological invasions and their impacts are mostly irreversible except when invaders are detected early enough, while their distribution is still limited, when area-wide eradication may be feasible. Successful eradications are difficult to achieve although our increasing understanding of the dynamics of invading

populations [8] and of the integrated use of tools has led to several successful campaigns against insects, mammals, and plants [9,10,11,12]. Despite these successes, eradications that involve the use of pesticides, which may have non-target effects, face increasing opposition [13]. Mating disruption, the use of synthetic sex pheromones which interfere with mate finding and reproduction, is recognized as an environmentally friendly method for the management of pest insects [14,15,16,17]. Mating disruption (MD) uses small amounts of naturally occurring compounds that function as odors rather than toxins, it is highly target-specific, and can be combined with other integrated pest management techniques. Although MD is now widely used for crop protection over smaller areas, its use for area-wide control and eradication has been limited to few species [16]. MD by aerial application of moth pheromones has been widely used in the U.S. for the gypsy moth, *Lymantria dispar*, "slow-the-spread" program; up to 2008 more than 1.4 million ha were treated with MD formulations [10]. This is considered environmentally friendly and more cost-effective than alternatives for suppressing gypsy moth populations [18]. Pink bollworm control has also relied on pheromone application over large areas integrated with other methods [19]. MD may be particularly successful for the management of recent invaders because it can enhance Allee effects that increase extinction pressure [8].

Epiphyas postvittana (Lepidoptera: Tortricidae), the light brown apple moth (LBAM) is an economically important pest, especially in horticulture in its native Australia and in New Zealand where it was introduced over 100 years ago [20]. A detection of the highly polyphagous LBAM in California in 2007 led to a response by federal and state authorities aiming at eradication or containment [20]. LBAM has also become established in Hawaii, the UK and Ireland, proving that it is a successful invader [20]. Although biological control and IPM programs, using insecticides and other methods, contribute to successful management in New Zealand, LBAM remains very abundant in organic orchards [21] and pine forests [22]. However, strict control regimes are required for quarantine reasons as there is zero tolerance on export fruit [23]. California is a major producer and exporter of a wide range of horticultural and other crops, and the establishment and potential spread of LBAM was expected to cause considerable crop damage and impediments to international and domestic trade [24], as well as direct and indirect environmental damage.

Given the scale of the LBAM infestation evident across several hundred square kilometers in 2007, aerial treatment was considered the only effective way of covering such a large area, using pheromone formulations as the most appropriate eradication technology [20]. Use of conventional insecticides in the mostly urban and peri-urban areas was ruled out. MD for LBAM using ground-applied dispensers was first developed in New Zealand in pine plantations and orchards [25]. Today, MD is used with success in Australia and New Zealand for LBAM control in citrus orchards [26] and organic apple orchards [21,23]. But our experience with aerial application of formulations for MD is limited. In California, two initial aerial applications of micro-encapsulated LBAM pheromone (using an incomplete blend without a 'sticker') were made in 2007 over ca. 20,000 ha in Monterey and Santa Cruz counties [20]. The primary means of assessment of efficacy was disruption of pheromone trap catch, however traps operated within the treated and untreated zones provided unclear results [20]. The use of MD for the incursion response in California has been criticized and its effectiveness has been questioned [27]. Before further consideration of pheromone applications, it was necessary to test the efficacy and longevity of different formulations experimentally in areas where LBAM was abundant.

Here we report the results of a substantial study involving aerial application by helicopter of four different pheromone formulations as well as hand-applied pheromone polyethylene tubing dispensers (hereafter referred to as 'twist-ties') in New Zealand, in a pine forest with abundant LBAM [22]. Formulations were tested in 5 ha plots (Fig. S1, Fig. S2), replicated five times. A total of 802 traps baited with either synthetic lures or female moths were used before and after application to assess populations and treatment effects (i.e., whether pheromone trap catch was inhibited, which is indicative of females not being able to attract males for mating). Vertical trap transects were installed from near the ground up to a height of 17 m to monitor the vertical distribution of LBAM and to assess treatment effects in stands of taller trees, in the canopy of trees as well as near ground level. We hypothesized that successful MD was more difficult to achieve in the upper canopy than at ground level because of the lower surface area available for deposition of formulations and dilution of airborne pheromone associated with greater wind speeds. We also examined the longevity of the different formulations by quantifying pheromone evaporation over time, which is expected to affect the resulting

aerial concentrations and effectiveness of MD. This study is important not only for the response to the incursion of LBAM in California but also for similar uses against other species where 'greener' yet effective alternative incursion response and management methods are required.

Results

The results are presented in four section addressing (1) mating disruption treatment effects near ground level and relationships with pheromone release characteristics of formulations, (2) effects of horizontal trap position (i.e., edge effects from the center to beyond the treated area), (3) distribution of catches in vertical transects across the canopy, and (4) mating disruption effects at different heights across the canopy.

LBAM Trap Catches and Treatment Effects Near Ground Level

Over 28,700 male LBAM were trapped during this trial, of which over 25,000 catches occurred near ground level (i.e., at ca. 1.5 m) where mating disruption is usually assessed, and where the majority of our traps were located. The covariate-adjusted mean catch and percent presence following the application of treatments (log-transformed trap catch for each plot in the 2 weeks prior to treatment was used as a covariate) showed that treatment effects were significant for traps at 1.5 meters across the entire treatment period (see Table S1 for ANCOVA results) except near the end of the flight period when catches generally declined. In the first few weeks following treatment, catches were considerably reduced in all treatments, compared with the controls. All pheromone formulations provided better than 90% disruption in the first week post treatment and better than 80% in the second week (Fig. 1). Traps in several treatments started catching considerable numbers of LBAM from week 5 and catches were recorded in over 10% of the traps in plots treated with CheckMate and NoMate (Fig. S3), providing less than 65% disruption in terms of absence of catches (Fig. 1). Traps in plots treated with Splat and twist-ties remained suppressed, with the Disrupt flakes providing an intermediate effect. Splat and twist-tie treatments maintained between 95% and near 100% disruption during the 10 weeks following application while moth activity was sufficient for evaluation (Fig. 1). Summed over the entire 13 weeks post treatment, all formulations except CheckMate reduced catches significantly from the controls (Table 1). Plots treated with Splat and twist-ties showed the greatest degree of suppression (ca. 98%), although the values were only marginally significantly different from those for Disrupt and NoMate, probably due to variation between plots.

Suppression of trap catch largely mirrored the pheromone release profiles of these formulations. The small droplets of the micro-encapsulated formulations (NoMate and CheckMate) initially had the greatest rate of loss of pheromone (Fig. S4). This resulted in the highest release rates compared with the two other formulations, during the first and second weeks post-application (Fig. 2). However, by week 5, NoMate and CheckMate had already lost ca. 92% and 74%, respectively, of the main pheromone component (Fig. S4), resulting in a decreased release rate (Fig. 2). Splat and especially Disrupt had a slower and longer-lasting release, although in the case of Disrupt, this caused a considerably lower release rate already during the first few weeks following the application (Fig. 2). The actual pheromone release rate per ha (Fig. 2) also varied among treatments due to application differences among formulations. The post-trial analysis

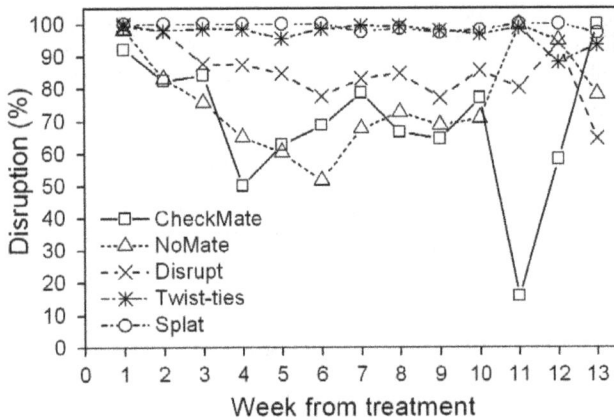

Figure 1. Effects of application of pheromone formulations on trap catch (percent disruption) of light brown apple moth. Note: Percent disruption is the difference in presence between 'treated' and 'controls' expressed as a percentage of the 'control' (values are based on back-transformed covariate-adjusted percent moth presence in traps – see Fig. S3). Results shown are for traps at 1.5 m above ground. Data for weeks 11–13 are based on very low catches at the end of the flight period and percent disruption in those weeks should be interpreted with caution.

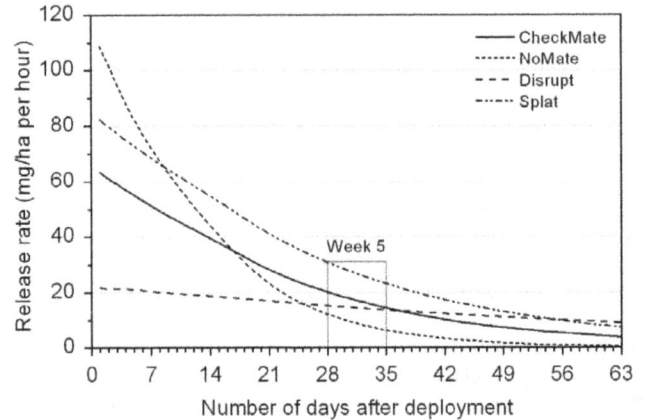

Figure 2. Pheromone release rates (mg ha^{-1} hr^{-1}) of the main component (E-11-tetradecen-1-yl acetate) of four formulations applied for mating disruption. Release rates were calculated using actual application rates (mass/area) and the change in mass over time from decay rate curves (Fig. S4). 'Week 5' is highlighted as this is mentioned specifically in the results section. Values for Disrupt and CheckMate are based on the deployment target of 40 g pheromone ha^{-1}. Slight under-application of NoMate and over-application of Splat, by 50%, were taken into consideration. See methods for details.

revealed that the Splat formulation exceeded the targeted application rate (discussed below).

Effects of Trap Position (Edge Effects) and Lure Type

Edge effects were assessed by comparing trap catch and disruption relative to horizontal position within or outside treated plots, for traps near ground level (1.5 m) (Fig. S2). ANCOVAs revealed a gradient of disruption from the 'Outside' towards the plot centre (Table S2). This was particularly the case when traps baited with females were considered separately from traps with rubber septa lures (Table S2). A disruption gradient appeared to occur also in control plots, although it was much less pronounced (Table S2). Therefore, treatments were clearly most effective in the centre of plots, where pheromone concentrations would be expected to be higher than near or beyond the edge of the treated areas.

It is of interest to know whether *trap shutdown* is representative of *mating disruption* of females. To examine this we compared trap catch and catch suppression between traps baited with three caged freshly-emerged virgin females and those with synthetic lures

(rubber septa) containing different doses of pheromone. Field survival of females was satisfactory with an average of 1.68 (± 0.05 S.E.) females alive when they were replaced after one week. Traps baited with females ('Centre - females') had significantly higher catches, on average, than nearby traps baited with 3 mg rubber septa ('Centre') (Table S2). This difference was partly due to the fact that a single rubber septum was compared with several females (between 1.7 and 3.0, on average), and therefore this comparison needs to be viewed with caution. Despite this apparently greater attractiveness of female-baited traps, even these were significantly disrupted to a high degree (Table S2), indicating that the overall results are likely to be representative of successful MD of actual female moths. A dose response in trap catch or suppression relative to lure loading was not consistently apparent, although traps baited at the lowest dose appeared to have the lowest catches (Table S2).

Vertical Transect Trap Catches and Recapture Results

Additional traps were installed along vertical transects in taller stands at ca. 1.5 m, 5 m, 9 m, 13 m and 17 m above ground to

Table 1. Trap catch of light brown apple moth and percent trap disruption following application of pheromone formulations for mating disruption over 13 weeks following treatment, expressed as covariate-adjusted, back-transformed mean summed counts, percent presence, and percent disruption (for traps at 1.5 m above ground).

Treatment	Mean count		Presence of LBAM (%)		Percent Disruption
Control	14.44	a	32.5	a	n/a
CheckMate (Suterra)	2.83	ab	10.7	ab	67.1
NoMate (Scentry)	1.25	bc	8.7	bc	73.2
Disrupt (Hercon)	0.96	bc	5.1	bc	84.3
Twist-ties (Shin-Etsu)	0.20	c	0.7	c	97.8
Splat (ISCA)	0.18	c	0.6	c	98.2

Values not sharing lower case letters are significantly different at $\alpha = 0.05$ according to least significant difference tests.

assess the vertical distribution of moths across the canopy and to determine disruption efficacy at different heights. In taller stands moths were found to be considerably more abundant in the upper canopy (at 13–17 m) than at mid-canopy and near ground level (Fig. S5). In younger stands with trees ca. 6 m tall, average trap catch at 4 m was also much higher than at 1.5 m. Analysis of covariance of vertical transects (based on 3,797 catches) indicated that traps in the upper canopy (i.e., 13–17 m in older stands; 4 m in younger stands) had significantly higher counts and presence of LBAM than those at middle or lower heights (P<0.01). In older stands, there were no significant differences in counts or presence of LBAM between the lowest height class (1.5 m) and the middle height class (5–9 m). Additional evidence for the preference of male LBAM moths for the upper canopy comes from recaptures of males that were released in plots in older stands. Of 19,190 laboratory-reared males marked with fluorescent powder that were released, 0.8% were recaptured. Ca. 68% of all recaptures were in traps at either 13 m or 17 m even though these traps represented only 19% of all the traps in these plots. Therefore, the recapture rate at 13 m and 17 m was ca. ten-fold of that recorded for the traps located at lower heights.

Disruption of Vertical Transect Traps

For the comparison of disruption efficacy at different heights across the canopy, no pre-treatment catches were available, and there were fewer traps in total in the vertical transects than for assessments of treatment effects at 1.5 m. This provided less statistical power and the results were less clear than for traps located at 1.5 m. However, the results generally indicate that control of LBAM in the upper canopy was less successful than near ground level and lasted for a shorter period of time. In weeks 1–5 post-treatment, only the Splat treatment (count, P = 0.0025; % presence, P = 0.0033) showed clear evidence of a reduction in LBAM in the upper canopy compared with the Control treatment (Fig. 3A) and, to a lesser extent, the No Mate treatment (count, P = 0.066; % presence, P = 0.098). In weeks 6–10 post-treatment, there was no evidence of any treatment differences in the upper canopy (Fig. 3B), indicating that treatments were no longer effective except near ground level. However, the presence of moths in the canopy of plots treated with Splat during weeks 6–10 (Fig. 3B) reinforces the successful disruption in the upper canopy of these plots during weeks 1–5 (Fig. 3A).

Discussion

Our study demonstrated that aerial application of sex phero-mone formulations can successfully disrupt trap catch of LBAM near ground level at a scale of 5 ha plots. Although the trap-shutdown effect was relatively short-lived for microencapsulated formulations, other formulations with a longer-lasting pheromone release provided better than 95% shutdown for at least 10 weeks near ground level. This effective period lies within the ranges reported by other studies on MD of LBAM [25,26,28]. Disrupt flakes, with the slowest pheromone release among the aerially applied formulations, provided an intermediate effect. The slow release of Disrupt probably resulted in lower aerial concentrations of pheromone that were insufficient for lasting disruption, but this could be improved by increasing the application rate. In addition, a rain event in the second week of this trial may have carried the Disrupt flakes to the forest floor, reducing the effectiveness of the Disrupt formulation in the upper part of the canopy.

Catches of traps baited with three females were slightly greater than catches to synthetic lures used to assess trap shutdown. The difference is probably explained by the fact that

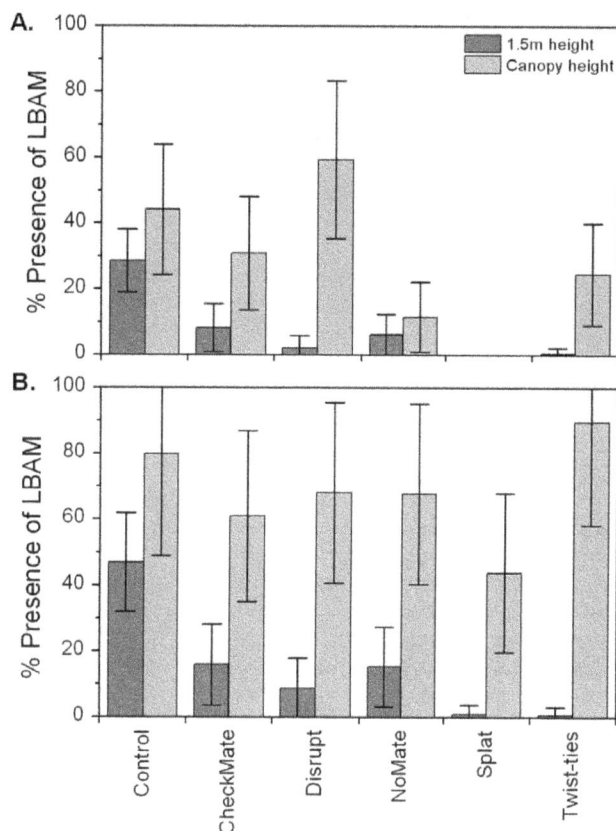

Figure 3. Covariate-adjusted percent presence (mean ± S.E.) of light brown apple moth in traps near ground level and at canopy height. Data shown are for weeks 1–5 (A) and for weeks 6–10 following pheromone application (B).

live female traps contained three females (vs. single lures). Furthermore, "centre" traps, which had the lowest catches, were surrounded by more traps than female-baited and other traps at other locations (Fig. S2). This additional competition probably contributed to the relatively lower catch rate of centre-traps. Recently, a further minor pheromone compound has been identified in the natural sex pheromone of LBAM [29], which was not included in the synthetic lures. However, even female-baited traps were effectively disrupted, indicating that the overall results are representative of successful MD of actual female moths. This was achieved despite the fact that the pheromone formulations contained at least 15% Z11-14Ac, a behavioral antagonist for LBAM. which renders release points practically unattractive if more than ca. 10% is present, relative to E-11-14Ac, the main pheromone component of LBAM [30,31]. The lack of attraction of all formulations clearly ruled out 'false trail following' as a mechanism explaining trap shutdown. False trail following and competition between pheromone release devices and female moths were considered a primary mechanism according to results of a recent field cage study of another tortricid (Cydia pomonella) [32]. We contend that sensory impairment resulting from receptor adaptation and habituation of the central nervous system of male LBAM as well as 'camouflage' of calling females are more likely to be the main mechanisms [16], at least in the case of LBAM with formulations that are not attractive due to the high content of the antagonist, Z11-14Ac. A decrease in male wing fanning responsiveness to pheromone was evident with an increase in atmospheric pheromone which included the antagonist [33], suggesting that sensory adaptation was operating. This agrees with studies documenting reduced male responsiveness to the female sex pheromone, following pre-exposure of males [16,32,34].

The extent of trap catches in the upper canopy, demonstrated here for the first time for LBAM, strongly suggests that in arboreal situations it is critical to achieve successful MD across the vertical extent of the canopy. In taller woody vegetation, particularly when fresh growth in the understorey is limited, female LBAM are likely to occur primarily in the upper canopy, where more suitable oviposition sites are located. Achieving pheromone concentrations sufficient for successful MD at the heights of taller trees is challenging. Placing sufficient amount of formulation in the upper canopy may prove difficult due to the limited surface area available for deposition. Exposure to sunlight may accelerate degradation of formulations, and rain may drive formulations from the upper canopy. Furthermore, higher wind velocities and increased exposure to coherent turbulent structures due to the proximity of the canopy top to the atmospheric roughness sub-layer will dilute the aerial pheromone concentration [35,36,37]. These effects may be even more important in some LBAM-infested areas in California where trees may be taller and forest canopies more sparse compared to those in the present study. By contrast, in lower vegetation, these issues would influence MD to a lesser degree. Surprisingly few previous studies of MD in forests have examined the possible implications of the vertical dimension on pheromone concentrations.

Factors influencing plume structures from polyethylene tubing dispensers were studied in apple orchards to visualize an insect's detection of the pheromone plume [38]. A modeling study has shown a change in biological efficacy and predicted concentration with respect to the height of orchard trees, depending on pheromone dispenser release height, canopy distribution and other factors [35]. The use of field electroantennograms was combined with a dispenser release rate model [39] and a Lagrangian model [35] to estimate atmospheric concentrations and plume structures required for disruption for LBAM in an apple orchard. Trap shutdown was achieved at an aerial concentration of >10 ng/m^3, or emission of >5 mg per ha per hour [33]. According to our estimates based on the release rates measured in the forest, all formulations initially exceeded emissions of 5 mg per ha per hour, but only the two most effective aerially applied formulations maintained pheromone emissions at or near this level at the beginning of week 10. However, as discussed above, a greater release rate is probably required to achieve an effective aerial concentration in a forest environment. Concentrations are also influenced by the deposition of aerially-applied formulations at different heights, and this can be examined using deposition models that are suitable for forest environments [40]. With pheromone-based MD it is essential to obtain the highest aerial concentrations in the part of the canopy where the target insect is most abundant, and using modeling tools may help to achieve this objective.

Despite these complex factors, our results nevertheless show that it is possible to achieve successful MD with aerially applied formulations, and one of the formulations we tested provided suppression for many weeks at upper canopy heights and near the ground. Furthermore, the results are likely to be better near ground level *and* at greater heights when pheromone has been applied forest-wide, reducing the influence of untreated adjacent areas. However, re-application may be necessary in tall canopies and for species with an extended mating period or overlapping

generations (such as LBAM in warmer climates). Evidently, to achieve successful suppression of populations in taller vegetation will require the use of tactics that can reach LBAM at these heights, and aerial application of pheromone is one of the most promising area-wide tactics available when considering environmental and human health issues.

MD is likely to be particularly successful as a tactic for eradication of recent invaders which are still of low or moderate local abundance. Such low-density populations are highly susceptible to mate-location failure causing Allee effects, a positive relationship between the size and growth rate of a population. MD raises the Allee threshold below which the population growth rate becomes negative, thereby enhancing the effect of driving the population to extinction [8,41]. Modeling simulations showed that MD will be particularly effective when mechanisms operate like sensory impairment of males and camouflage of calling females [41], which are implicated in the case of LBAM. Therefore, MD is promising as an eradication tactic to stem the invasion of insects, although this is limited to species where a suitable pheromone is known and potentially available in large quantities.

Materials and Methods

Permits

All approvals and permits necessary for this work were obtained from the New Zealand Environmental Risk Management Authority (Approval numbers HSC000309– HSC000314, which included consultation of the Food Safety Authority, the Canterbury Regional Council, and the Department of Conservation). Ngai Tahu Properties (land owner) and Matariki Forest (forest manager) kindly allowed this trial to take place.

Study Design and Plot Layout

The study took place in Eyrewell Forest (between 43°26′10.81″S and 43°24′20.14″S, and 172°27′14.05″E and 172°20′30.59″E), a *Pinus radiata* plantation forest northwest of Christchurch, New Zealand. Disruption efficacy of four aerially-applied formulations was compared with ground-applied 'twist-tie' polyethylene tubing dispensers (positive control) and untreated (blank) controls in a field trial using a single application over 5 ha plots (225 m×225 m) (Fig. S1, Fig. S2), replicated five times. Additional 'external control plots' were added as a sixth block at a greater distance from treated areas. The trap layout within each plot consisted of transects from the plot center to the plot edge and 75 m outside the plot perimeter. Traps were at least 16 m apart. Buffers of >200 m were maintained between plots and between plots, forest boundaries and open water. Plots were arranged in five blocks of spatially clustered stands of trees. A detailed canopy characterization was undertaken (Table S3). Tree heights ranged from 3–28 m.

Pheromone Formulations

CheckMate® LBAM-F (Suterra LLC, Bend, OR) is a micro-encapsulated suspension (in water) with an average capsule size of 120 μm, with 17.6% active ingredient (i.e., LBAM pheromone). NoMate® LBAM MEC (Scentry Biologicals, Inc., Billings, MT) is a micro-encapsulated suspension (in water) with an average capsule size of 40–60 μm, with 20.3% active ingredient. Disrupt Bio-Flake® LBAM (Hercon Environmental, Emigsville, PA) is a biodegradable solid flake, each measuring approximately 3.0 mm×2.5 mm×1.9 mm and containing 13.6% of active ingredient. This formulation was applied in slurry with X3221 Micro-Tac II sticker (Lock n' Pop, Everett, WA) with 2.5% guar gum as a suspension agent. Splat LBAM^{TM} (ISCA Technologies,

Inc., Riverside, CA) is an amorphous polymer (wax) carrier containing 10% active ingredient. CheckMate, NoMate, Disrupt, and Splat were applied aerially (details below). The pheromone composition of all these formulations was 81% E-11-tetradecen-1-yl acetate (E11-14Ac), 15% Z-11-tetradecen-1-yl acetate (Z11-14Ac), and 4% (E,E)-9,11-tetradecadien-1-yl acetate (EE9,11-14Ac) (according to supplier, Bedoukian Research, Inc., Danbury, CT).

The ground-applied ISOMATE®-LBAM PLUS is a 'twist-tie' polyethylene tubing dispenser (Shin-Etsu Chemical Co., Ltd., Tokyo, Japan) containing 125 mg active ingredient (68% E11-14Ac, 29% Z11-14Ac, 3% EE9,11-14Ac). These were applied to trees by hand on Feb. 19, 2008 at ca. 1.5 m above ground, at a density of ca. 600 per ha (a total of 15,000 dispensers across five plots) as a positive control [25].

Aerial Application Details and Climatic Conditions during Application

Each of the four formulations was applied with a single aerial application using Hughes MD 500D helicopters at a target rate of 40 g LBAM pheromone per ha. The microencapsulated formulations CheckMate and NoMate were applied with rear-mounted boom systems resulting in relatively large droplets as per the target size of 390 µm. The Disrupt flake formulation was applied with a modified fertilizer bucket suspended under the helicopter. Splat was applied using a pressurized supply tank, an internal piston pump, and three oscillating solenoid valves on a rear mounted boom to give a target droplet size of 3 mm (see Table S4 for further application details). Applications occurred on Feb. 20–21, 2008. GPS tracking data from helicopters confirmed that applications of CheckMate, NoMate, and Splat were achieved with considerable accuracy in terms of targeting the plots (see also Table S4). Applications of Disrupt flakes to the specified plot area were not as accurate because of an unpredicted delay between the pump and release from the spinner both at switch on and off times. We took these delays into account as much as possible during the application by adjusting the timing of switching on and off. This ensured good plot coverage, particularly of the central area where trap disruption was assessed. Disrupt flakes initially had the slowest rate of pheromone release, until about week four (Fig. 2). The Splat application was accurately targeted to the plot area but the application rate exceeded the target rate by ca. 50% despite previously successful calibration. This apparent bias was partly compensated by the slower pheromone release compared with NoMate and CheckMate (Fig. S4). The resulting slower decay rate suggests that the aerial pheromone concentration, which is the measure ultimately relevant for successful MD, remained at an effective concentration for MD for longer. In addition, the slower decay is likely to have extended the longevity of the Splat treatment. The plot coverage of NoMate was ca. 20% below the target area but this was partly compensated by a ca. 14% greater application rate. Furthermore, the central plot area, where the main assessment of treatment effects occurred, was treated as intended.

Trapping of LBAM

Red delta pheromone traps (Plant & Food Research, hereafter 'PFR'), baited with 3 mg PFR lures, were installed to assess treatment effects. Lures contained a 95:5 ratio of E11-14Ac (99.7% purity) and E,E9,11-14Ac (>99% purity) (both supplied by Pherobank, The Netherlands) loaded onto red rubber septa (Thomas Scientific Inc., Philadelphia, PA). Additional traps with 0.1 mg–3 mg PFR lures were deployed to assess dose responses and compare catches with those to three live caged female moths.

We also used 3 mg and 0.1 mg rubber septa lures by Suterra (Suterra LLC, Bend, OR) for comparison (3 mg Suterra lures are used in California). Suterra lures contained a 96:4 ratio of E11-14Ac (99.1% purity) and E,E9,11-14Ac (96.8% purity) (both supplied by Bedoukian Research). Female-baited traps contained three laboratory-reared, freshly-emerged virgin females placed in plastic vials with a mesh bottom and top to allow for pheromone dispersal. LBAM females were replaced weekly from a colony maintained by PFR.

All traps at 1.5 m were activated on Feb. 4, 2008 prior to the application of MD treatments. Additional traps in vertical transects up to 17 m high on a pulley system were set up in older stands from Feb. 17 to Mar. 4, 2008, using a motorized cherry picker. A further replicate of vertical transects was activated on Mar. 26, 2008. A total of 802 traps were used consisting of 330 transect traps placed at 1.5 m (baited with 3 mg PFR lures), 180 dose response traps at 1.5 m (30 each of 0.1 mg, 0.3 mg, 1 mg, 3 mg PFR lures and 0.1 mg and 3 mg Suterra lures), 120 female-baited traps at 1.5 m (four per plot), 154 additional traps in tall vertical transects (up to 17 m height, in tall stands, 3 transects per plot, 2 replicates; all baited with 3 mg PFR lures), and 18 additional traps in short vertical transects (up to 4 m height, in young stands, 3 transects per plot, all baited with 3 mg PFR lures). Traps were checked and cleared on the day when treatments were applied and then at weekly intervals until May 15–16, 2008. A final collection was made on June 12–13, 2008.

Releases of Male LBAM

Between Mar. 6 and May 2, 2008, a total of 19,190 laboratory-reared male LBAM, marked with fluorescent powder, were released to supplement populations in older stands to assist with the detection of MD effects. Weekly releases of between 100 and 480 moths per plot were made at plot centers. Numbers varied between weeks, depending on moth availability. From March onwards, all trapped moths were viewed under UV light to detect recaptures of dyed moths. Additional releases and recaptures after 2 May were excluded from the recapture analysis because few catches occurred after the onset of the southern hemisphere winter.

Longevity of Formulations

Pheromone release over time was assessed in an untreated part of the forest. Twenty-five 3 µl droplets of CheckMate, NoMate and Splat, and 25 Disrupt flakes, were applied to 5×10 cm Strathmore canvas paper and placed in a shaded area, protected by wire mesh. Three replicates were removed for analysis on days 0 (Feb. 23, 2008), 1, 3, 7, and then in weekly intervals until day 63. Each card was placed in a borosilicate glass vial and held in a freezer or under dry ice. LBAM pheromone was extracted with 25 ml of acetone/hexane (1/1) and analyzed using an Agilent 6890N Network GC system (Agilent Technologies 7683 B Series auto injector; Agilent 19091 J-413, HP-5 5% Phenyl Methyl Siloxane column (30 m×320 µm×0.25 µm; 100DISP1, splitless, 100°C (2 min), 15°C/min to 280°C (10 min); Hydrogen carrier). Loss of mass of the main active ingredient (E11-14Ac) over time was determined using these GC measurements. For liquid and waxy formulations, mass for the initial day was calculated using the specific gravity of E11-14Ac. Using the day zero and subsequent E11-14Ac mass measurements, an exponential decay curve was fitted to the data. To compare treatment decay rates, mass from the sampling days was normalized to the day zero mass. Decay curves were also used to estimate the resulting pheromone release rates per ha for given application rates of formulations.

Data Analysis

Two response variables were analyzed, (i) trap catch ('count') per hundred trap days (log transformed using $\ln(N+0.1)$) and (ii) the percentage of traps catching LBAM within each plot (analyzed using angular or arcsine transformation). The main analysis was for data from traps at 1.5 m height. Analyses of covariance using SAS Version 9.1 PROC MIXED were used to test for treatment differences. The log-transformed trap count for each plot in the 2 weeks prior to treatment was used as a covariate. Separate analyses of covariance were performed for each response variable using plot means for each assessment (weekly), for five-weekly averages (Weeks 1–5 and 6–10), and for the average across all 13 weeks following treatment, along with least significant difference post-hoc tests comparing all treatments. Traps located 'outside' plot areas (75 m from the edge of the treated area) were excluded from this analysis. All control traps were included in the analysis. Percentage disruption was calculated as 100-(100*(treatment catch/control catch)).

Analyses comparing trap position (Outside, Edge, Centre/Edge, Centre) and lure type (rubber septa vs. females) within each plot were performed with ANOVAs separately for control plots and for treated plots combined. These were carried out using PROC MIXED with plot included in the model as a random effect and position × type as a fixed interaction effect.

For vertical transects, no pre-treatment catches were available above 1.5 m height, and pre-treatment log count at 1.5 m height was therefore used as a covariate. Count and presence data was classified into 3 height classes, (1) 1.5 m, (2) middle (5–9 m in older stands only), (3) upper canopy (13–17 m in older stands or 4 m in younger stands). Average count and presence by height class and plot (excluding outside traps) were then subjected to analysis of covariance (ANCOVA) using PROC MIXED with fixed-effect model terms for treatment, height class, and the interaction between these two factors, and random plot effects.

Supporting Information

Figure S1 Aerial view of mating disruption plot locations in Eyrewell Forest (New Zealand).

Figure S2 Plot layout showing pheromone aerial treatment area and core area, trap locations and lures used. Most lures were standard Plant&Food Research ('PFR') lures loaded with 3 mg LBAM pheromone. Other lures were used to examine dose responses and to compare PFR and Suterra ('S') lures which are used in California. See text for more details on plot design and lure types.

Figure S3 Effects of application of pheromone formulations on trap catch (percent presence) of light brown apple moth. Note: Results are for traps at 1.5 m above ground shown as back-transformed covariate-adjusted percent moth presence in traps. Data for weeks 11–13 are based on very low catches at the end of the flight period.

Figure S4 Loss of pheromone in the field expressed as a percentage of the main component (E-11-tetradecen-1-yl acetate) remaining over time for each formulation, with 95% confidence intervals, based on analysis by gas chromatography of extracts from 3 µl droplets or Disrupt 'flakes' applied to canvas paper cards placed in the forest where the trial took place. See methods for details. Note that the actual aerial application of Splat resulted in a range of droplet sizes, with 3 µl being at the lower range of Splat droplets. A proportion of Splat droplets were considerably larger than 3 µl, and the actual longevity of the Splat treatment was therefore somewhat longer than shown here.

Figure S5 Vertical distribution of catches of light brown apple moth (mean ± S.E.) in tall stands based on traps in control plots and untreated areas (n = 22 transects for all heights except for 17 m where n = 11).

Table S1 Analysis of covariance (ANCOVA) table showing significance of covariate and treatment effects in the analysis of log-transformed counts and angular-transformed percent presence of light brown apple moth for the average of the 13 weeks following treatment, and for weeks 1–5 and 6–10 following treatment, and each of the 13 weeks following treatment.

Table S2 Trap catch of male light brown apple moth at different locations within and outside plots and for different lure types following application of pheromone formulations. Covariate-adjusted, back-transformed mean summed counts and percent presence are shown for control plots and for all treated plots combined over 13 weeks following application, for traps at 1.5 m above ground. Values (within columns) not sharing lower case letters are significantly different at $\alpha = 0.05$ according to least significant difference tests.

Table S3 Mean tree height, canopy thickness (i.e., green crown height) and tree density for replicated plot areas (n = 5). Block 6 contained five extra control plots at a greater distance from treated areas.

Table S4 Target swath width, speed, altitude of application (above ground) and droplet size for the four helicopter-applied formulations. See footnote [§] for further application details.

Acknowledgments

We thank the many participants, especially Kirsty Trotter, David Henley, Tom Sullivan, Nicola White and Salah Araj, for outstanding technical assistance, Gerry Bohmfalk, Don Kauffman, Tom Larsen, Reg Coler, and Agenor Mafra-Neto and staff of Amuri Helicopters and Mt. Hutt Helicopters for their cooperation. Matariki Forest/Rayonier and Ngai Tahu Properties kindly allowed this trial to take place and provided GIS support. Ring Cardé, Miriam Cooperband, Andrew Liebhold, Vic Mastro, Mike Stefan and anonymous reviewers provided helpful comments on earlier versions of the manuscript.

Author Contributions

Conceived and designed the experiments: EGB DMS MK BR DRL. Performed the experiments: EGB DMS GC SG JLK DMC AZ. Analyzed the data: EGB MK GC TS. Wrote the paper: EGB DMS MK.

References

1. Mack RN, Simberloff D, Lonsdale WM, Evans H, Clout M, et al. (2000) Biotic invasions: Causes, epidemiology, global consequences, and control. Ecol Applic 10: 689–710.

2. Holmes TP, Aukema JE, Von Holle B, Liebhold A, Sills E (2009) Economic impacts of invasive species in forests: Past, present, and future. Annals of the New York Academy of Sciences 1162: 18–38.

3. Aukema JE, Leung B, Kovacs K, Chivers C, Britton K, et al. (2011) Economic impacts of non-native forest insects in the continental United States. PLoS ONE 6: e24587.

4. Poland TM, McCullough DG (2006) Emerald ash borer: Invasion of the urban forest and the threat to North America's ash resource. J For 104: 118–124.

5. Liebhold AM, Macdonald WL, Bergdahl D, Maestro VC (1995) Invasion by exotic forest pests: A threat to forest ecosystems. For Sci Monogr 30: 1–49.

6. Brasier CM (1991) *Ophiostoma novo-ulmi* sp. nov., causative agent of current Dutch elm disease pandemics. Mycopathologia 115: 151–161.

7. Aukema JE, McCullough DG, Holle BV, Liebhold AM, Britton K, et al. (2010) Historical accumulation of nonindigenous forest pests in the continental United States. BioScience 60: 886–897.

8. Liebhold AM, Tobin PC (2008) Population ecology of insect invasions and their management. Annu Rev Entomol 53: 387–408.

9. Simberloff D (2003) Eradication - Preventing invasions at the outset. Weed Science 51: 247–253.

10. Brockerhoff EG, Liebhold A, Richardson B, Suckling DM (2010) Eradication of invasive forest insects: Concepts, methods, costs and benefits. New Zeal J For Sci 40 suppl: S117–S135.

11. Myers JH, Savoie A, Van Randen E (1998) Eradication and pest management. Annu Rev Entomol 43: 471–491.

12. Veitch CR, Clout MN, editors (2002) Turning the tide: The eradication of invasive species. Gland, Switzerland, and Cambridge, UK: International Union for Conservation of Nature.

13. Barinaga M (1991) A new buzz in the medfly debate. Science 253: 1351.

14. Roelofs WL, editor (1979) Establishing efficacy of sex attractants and disruptants for insect control. Lanham, MD: Entomological Society of America.

15. Carde RT, Minks AK (1995) Control of moth pests by mating disruption: Successes and constraints. Annu Rev Entomol 40: 559–585.

16. Cardé RT (2007) Using pheromones to disrupt mating of moth pests. In: Kogan M, Jepson P, editors. Perspectives in ecological theory and integrated pest management. Cambridge: Cambridge University Press. 122–169.

17. Suckling DM, Brockerhoff EG, Stringer LD, Butler RC, Campbell DM, et al. (2012) Communication disruption of *Epiphyas postvittana* (Lepidoptera: Tortricidae) using two formulations at four point source densities in vineyards. Journal of Economic Entomology (in press).

18. Sharov AA, Leonard D, Liebhold AM, Clemens NS (2002) Evaluation of preventive treatments in low-density gypsy moth populations using pheromone traps. J Econ Entomol 95: 1205–1215.

19. Walters ML, Staten RT, Roberson RC, Tan KH (2000) Pink bollworm integrated management using sterile insects under field trial conditions, Imperial Valley, California. In: Tan KH, editor. Joint Proc Intl Conference on Area-Wide Control of Insect Pests and the 5th Intl Symp on Fruit Flies of Economic Importance; 28 May-2 June 1998. Penang, Malaysia: Penerbit Universiti Sains Malaysia. 201–206.

20. Suckling DM, Brockerhoff EG (2010) Invasion biology, ecology, and management of the light brown apple moth (Tortricidae). Annu Rev Entomol 55: 285–306.

21. Rogers DJ, Walker JTS, Moen IC, Weibel F, Lo PL, et al. (2003) Understory influence on leafroller populations in Hawke's Bay organic apple orchards. New Zeal Plant Prot 56: 168–173.

22. Brockerhoff EG, Jactel H, Leckie AC, Suckling DM (2002) Species composition and abundance of leafrollers in a Canterbury pine plantation. New Zeal Plant Prot 55: 85–89.

23. Varela LG, Walker JTS, Lo PL, Rogers DJ (2010) New Zealand lessons may aid efforts to control light brown apple moth in California. California Agric 64: 6–12.

24. Fowler G, Garrett L, Neeley A, Magarey R, Borchert D, et al. (2009) Economic analysis: Risk to U.S. apple, grape, orange and pear production from the light brown apple moth, *Epiphyas postvittana* (Walker). United States Department of Agriculture: Raleigh, North Carolina.

25. Suckling DM, Shaw PW (1995) Large-scale trials of mating disruption of lightbrown apple moth in Nelson, New Zealand. New Zeal J Crop Hort Sci 23: 127–137.

26. Mo J, Glover M, Munro S, Beattie GA (2006) Evaluation of mating disruption for control of lightbrown apple moth (Lepidoptera: Tortricidae) in citrus. J Econ Entomol 99: 421–426.

27. Carey JR (2008) Invasion biology of the light brown apple moth: Presentation to Assembly California Legislature Committee on Agriculture.

28. Suckling DM, Shaw PW (1992) Conditions that favor mating disruption of *Epiphyas postvittana* (Lepidoptera: Tortricidae). Environ Entomol 21: 949–956.

29. El-Sayed AM, Mitchell VJ, Manning L-AM, Suckling DM (2011) New sex pheromone blend for the lightbrown apple moth, *Epiphyas postvittana*. Journal of Chemical Ecology 37: 640–646.

30. Rumbo ER, Deacon SM, Regan LP (1993) Spatial discrimination between sources of pheromone and an inhibitor by the light-brown apple moth *Epiphyas postvittana* (Walker) (Lepidoptera: Tortricidae). J Chem Ecol 19: 953–962.

31. Stephens AEA, Suckling DM, El-Sayed AM (2008) Odour quality discrimination for behavioural antagonist compounds in three tortricid species. Entomol Exp Appl 127: 176–183.

32. Miller JR, McGhee PS, Siegert PY, Adams CG, Huang J, et al. (2010) General principles of attraction and competitive attraction as revealed by large-cage studies of moths responding to sex pheromone. Proc Natl Acad Sci USA 107: 22–27.

33. Suckling DM (1999) Atmospheric pheromone concentrations affect the behaviour of lightbrown apple moth (Lepidoptera: Tortricidae) in the orchard. J Chem Ecol 25: 2011–2025.

34. Bartell RJ, Lawrence LA (1973) Reduction in responsiveness of males of *Epiphyas postvittana* (Lepidoptera) to sex pheromone following previous brief pheromonal exposure. J Insect Physiol 19: 845–855.

35. Suckling DM, Green SR, Gibb AR, Karg G (1999) Predicting atmospheric concentration of pheromone in treated apple orchards. J Chem Ecol 25: 117–139.

36. Thistle HW, Peterson H, Allwine G, Lamb B, Strand T, et al. (2004) Surrogate pheromone plumes in three forest trunk spaces: Composite statistics and case studies. For Sci 50: 610–625.

37. Strand T, Lamb B, Thistle H, Allwine E, Peterson H (2009) A simple model for simulation of insect pheromone dispersion within forest canopies. Ecol Model 220: 640–656.

38. Suckling DM, Angerilli NPD (1996) Point source distribution affects pheromone spike frequency and communication disruption of *Epiphyas postvittana* (Lepidoptera: Tortricidae). Environ Entomol 25: 101–108.

39. Bradley SJ, Suckling DM, McNaughton KG, Wearing CH, Karg G (1995) A temperature-dependent predictive model for polyethylene tubing pheromone dispenser release rates. J Chem Ecol 21: 745–760.

40. Richardson B, Thistle HW (2006) Measured and predicted aerial spray interception by a young *Pinus radiata* canopy. Transactions ASABE 49: 15–23.

41. Yamanaka T, Liebhold AM (2009) Spatially implicit approaches to understand the manipulation of mating success for insect invasion management. Popul Ecol 51: 427–444.

Migrating Giant Honey Bees (*Apis dorsata*) Congregate Annually at Stopover Site in Thailand

Willard S. Robinson*

Biology Department, Casper College, Casper, Wyoming, United States of America

Abstract

Giant honey bees (*Apis dorsata*) of southern Asia are vital honey producers and pollinators of cultivated crops and wild plants. They are known to migrate seasonally up to 200 km. It has been assumed their migrations occur stepwise, with stops for rest and foraging, but bivouacking bees have rarely been seen by scientists. Here I report discovery of a site in northern Thailand where bivouacs appeared in large congregations during the wet seasons of 2009 and 2010. The bivouac congregation stopover site is a small mango orchard along the Pai River. Bivouacs rested in branches of mango and other tree species in the immediate vicinity. Departures were preceded by dances indicating approximate direction and apparently, distance of flights. Such consistent stopover sites likely occur throughout southern Asia and may support critical, vulnerable stages in the life history of giant honey bees that must be conserved for populations of the species to survive.

Editor: Anna Dornhaus, University of Arizona, United States of America

Funding: A private citizen, Gail Zimmerman, contributed $6,000 for travel expenses for this work. The funders had no role in study design, data collection and analysis, decision to publish, or preparation of the manuscript.

Competing Interests: The author has declared that no competing interests exist.

* E-mail: wrobins@caspercollege.edu

Introduction

The giant honey bee, *Apis dorsata*, is an important pollinator [1,2] and honey producer [3] that ranges through southern Asia from Pakistan to Indonesia [4]. Colonies migrate seasonally. Though details vary with region, colonies typically live in lofty communal nest sites of $\approx$20–100 colonies at relatively high elevation (>1,000 m) during the dry season. There they build $\approx$1.5-m-diameter single wax combs. The colonies grow quickly and reproduce by fission (swarming). As forage decreases toward the end of the season, colonies abandon their combs and migrate to lower elevations, establishing new nests there for the mass flowering of the monsoon season [5–9]. Microsatellite DNA fingerprinting studies indicate that returning colonies are faithful to original nesting sites after many months or even years away [10,11].

Migration distances have been inferred by the presence of colonies on combs in different places at different times of year. Swarms reportedly cross the 50-km-wide Strait of Malacca between Sumatra and Malaysia [12]. Colonies in Sri Lanka travel for a month or so, up to 200 km in each direction, resting in trees along the way [13]. At stopovers the bees form combless clusters, or bivouacs, and accumulate food reserves for flight and for comb construction upon arrival at nest sites [7,13,14]. Shorter, altitudinal migrations of the closely related montane giant honey bee *Apis laboriosa* are well documented [15–17]. Widely dispersed bivouacs of this species survive the coldest months of Himalayan winter huddling near the ground, "…no swarm…visible from the site of another" [15].

Bivouacking *A. dorsata* have received no systematic study. Rare, opportunistic discoveries of bivouacs allowed researchers to describe dances of scout bees and subsequent flights of widely scattered Sri Lankan swarms, which were assumed to be near the end of migrations and seeking nearby nest sites [13].

Although I recently described congregated bivouacs [18], no one has reported annual congregations at traditional sites. Here I report timing and phenology of arrival and departure of 52 *A. dorsata* bivouacs found over 2 consecutive monsoon seasons, congregated in and around a riverside mango grove in northern Thailand. As the same site was occupied in consecutive years, and knowledgeable field workers report that groups of giant honey bees gather there annually, I call it a "bivouac congregation stopover site" (BCSS). I describe the aggregated nature of the bivouacs, some of their general features, dances the bees performed in preparation for departure flights, and ensuing flight directions.

Finally I stress the possible implications of BCSS's for conservation of this spectacular honey bee, in light of worldwide pollinator decline, Southeast Asia's rapid deforestation, dwindling numbers of giant honey bees and other insect species' vulnerability to ecological bottlenecks.

Study Site

Timing of Bivouac Searches

I began searching for bees on 25 August 2009 at the Mae Hong Son Agricultural Research and Development Center. The Center is 4 km southwest of Mae Hong Son, Thailand, elevation 200 m, 19°16′09 N, 97°56′39 E. Searches were approximately weekly until 7 September, when I encountered 2 *A. dorsata* bivouacs in mango (*Mangifera indica*) trees in an orchard at the Center. Other swarms soon began to arrive at the study site. That season's work concluded 31 October, when the last of 16 swarms departed.

Deciduous trees were then shedding their leaves as the dry season commenced.

In 2010 I began searching for bivouacs at the Agricultural Center 17 August. Observations concluded 19 November, thus bracketing the 2009 study period. Although the study ended 3 weeks later than in the previous year, the dry season was not nearly as advanced.

Physical Setting of Study Site

Fig. 1 depicts the mango orchard, its surroundings and bivouac locations for the two study years. The orchard is narrow, 4–8 trees in width, stretching north-south along the Pai River. In 2009, it was ≈ 350 m by ≈30 m. Tree removal reduced its length in 2010 to ≈270 m. It is bordered to the west by a narrow road partly lined by *Eucalyptus* and then a 150-m-wide strip comprising macadamia (*Macadamia integrifolia*) and pomelo (*Citrus maxima*) orchards, and mixed vegetable plantings. ≈250 m west of the river is another large mango grove. Directly east of the mango orchard is the Pai River, ≈70 m wide, with trees and undergrowth of various species along its bank. Dense moist broadleaf forest lies to the south; to the north is Research Center land cultivated with such plants as macadamia, scattered mango, vegetables, rice and ornamental flowers. Surrounding the Center on all sides, including the east side of the river, is steeply rising, relatively undisturbed mountain forest of diverse deciduous species.

The rainy season in northern Thailand is April through September, peaking in August. The dry season is November to February. The summer and fall of 2009 were abnormally dry, while in 2010 a strong monsoon caused widespread flooding. From July through October, more than twice the amount of rain fell in 2010 as in 2009 (Fig. 2).

Results

Features of Bivouacs

Dates of arrival and departure of bivouacs are presented in Fig. 3. Arrival of bivouacs was later and more protracted in the wet autumn of 2010 than in the dry 2009. In 2009, bivouacs remained in the study area for 11.1 ± s.d. 8.8 days; 2010 mean stay was 12.5 ± s.d. 12.1 days. The difference of 1.4 days was not significant according to an unpaired t-test (2-tailed P = 0.68; df = 50). Median duration of residence for 2009 and 2010 was 11 and 8.5 days respectively. Stays ranged from 2–31 days in 2009; 2 hours to ≥57 days in 2010. Concentration of bivouacs in the core area of the study site was such that I frequently saw and filmed swarms landing nearby as I studied the bivouacs; at one point in 2010, 12 bivouacs were simultaneously crowded within an area of 2,100 m^2, all on mango. If not for the tree foliage all would have been easily visible from the center of the occupied area.

Bivouacs favored mango trees more in 2010 and were also more densely concentrated than in 2009 (Fig. 1). Bivouacs all hung from tree branches, comprising 10 tree species. Of the 52 total bivouacs, 28 rested in mango, 7 in *Acacia* sp., 6 in macadamia, with lesser numbers in other species. In 2009, 6 of 16 were in *Acacia*, 5 in mango, 5 others distributed among several species. In 2010, 23 of the 36 bivouacs rested in mango, 5 in macadamia, with the 8 others distributed among various species. Miscellaneous trees utilized included native species of teak (*Tectona grandis*), cluster fig (*Ficus racemosa*), *Dipterocarpus* sp. and 4 unidentified species.

Mean size of 2009 bivouacs was 1,807 ± s.d. 683 cm^2; in 2010 mean size was 2,562 ± s.d. 3,301 cm^2. This difference was not significant according to an unpaired t-test (2-tailed P = 0.37; df = 50). However, size range and variability were markedly different in the 2 years, as reflected in disparate coefficients of

variation. In 2009 bivouacs ranged in size from 226 to 2,743 cm^2; in 2010 the range was 130 to 12,993 cm^2. The coefficient of variation [19] in 2010 was 3.4 times that of 2009. Six of the 2010 bivouacs were smaller than 500 cm^2 and 3 were larger than 11,000 cm^2 (Figs. 4, 5). Rough estimates of the numbers of individual bees, calculated from photographs, indicated that the smallest bivouacs contained a few hundred bees, the largest well over 100,000 bees.

Bivouacs generally rested at lower heights in 2010 than in 2009. Resting height varied from 1.8 to 18.0 m (mean = 10.1±5.0 m) in 2009; in 2010 resting height ranged from 1.0 to 15.0 m (mean = 5.6±4.3 m). This 4.5-m difference in mean resting height was statistically significant (2-tailed P = 0.002; df = 50).

None of the 52 bivouacs built comb. Departing swarms left numerous small beads of white wax behind on the bark at their resting places. In 2010, 2 transitory colonies, not included among the bivouacs in this study, established small (≈30 cm-diameter) combs near the study site during the first week, only to abscond within a few days.

Departure Dances and Flights

All swarm flights were preceded in the 30–60 min before liftoff by a varying number of simultaneously dancing bees, spread over the outer curtain of bees (Video S1, Video S2). Number of dancers increased with bivouac size. Dancers indicated unanimously (r = 1, Rayleigh test) [20] and fairly precisely the direction, and apparently the relative distance of the upcoming flight. For longer dances (waggle runs >3 sec) these were distinctive, staggering runs [13,18,21]. I describe details of these dances elsewhere (W.S. Robinson, in preparation).

I watched 4 swarms depart in 2009 (Video S3). In each, dancers indicated impending flights ranging from ENE to SE, and flights were generally easterly (Fig. 6A). I thus hypothesized [18] that all bivouacs might be *en route* to a known annual nesting aggregation near the village of Pai, about 50 km E. However, in 2010 the dances and flights of 11 witnessed departures were oriented to widely varying compass directions (Fig. 6B, C). I traced the flights of 2 of these bivouacs that took flights of ≈150 m, both preceded by dances with waggle circuits of 2 s (Fig.6b, 6c). Both bivouacs made these short flights in the afternoon, settled, and had departed the study site unobserved by the next morning. All other flights were impossible to trace, mostly disappearing from view after ≈ 150–200 m.

Flight directions of the departing bivouacs quite accurately reflected preceding waggle dance orientations (Fig. 6). For each of the 15 witnessed flights, the probability that the direction indicated by unanimously dancing bees would by chance alone be so close to the departure direction is extremely low (p<0.001, Rayleigh V-test) [20].

Discussion

Functions of Stopover Site

It is possible that BCSS's occur as migration stopover grounds throughout southern Asia, likely in geographical locations with landmarks that are easily recognizable to bees and could be found instinctively, just as drone congregation areas are found by both instinct and environmental cues [22]. This is especially likely considering evidence that *A. dorsata* migration has a genetic basis [13]. At this BCSS, the Pai River is a pronounced navigational landmark and a water source to traveling bees. Flowering native teak and assorted understory herbs provide nectar, as do plentiful non-native *Eucalyptus* trees (Fig. 7). In addition, numerous *A. cerana*

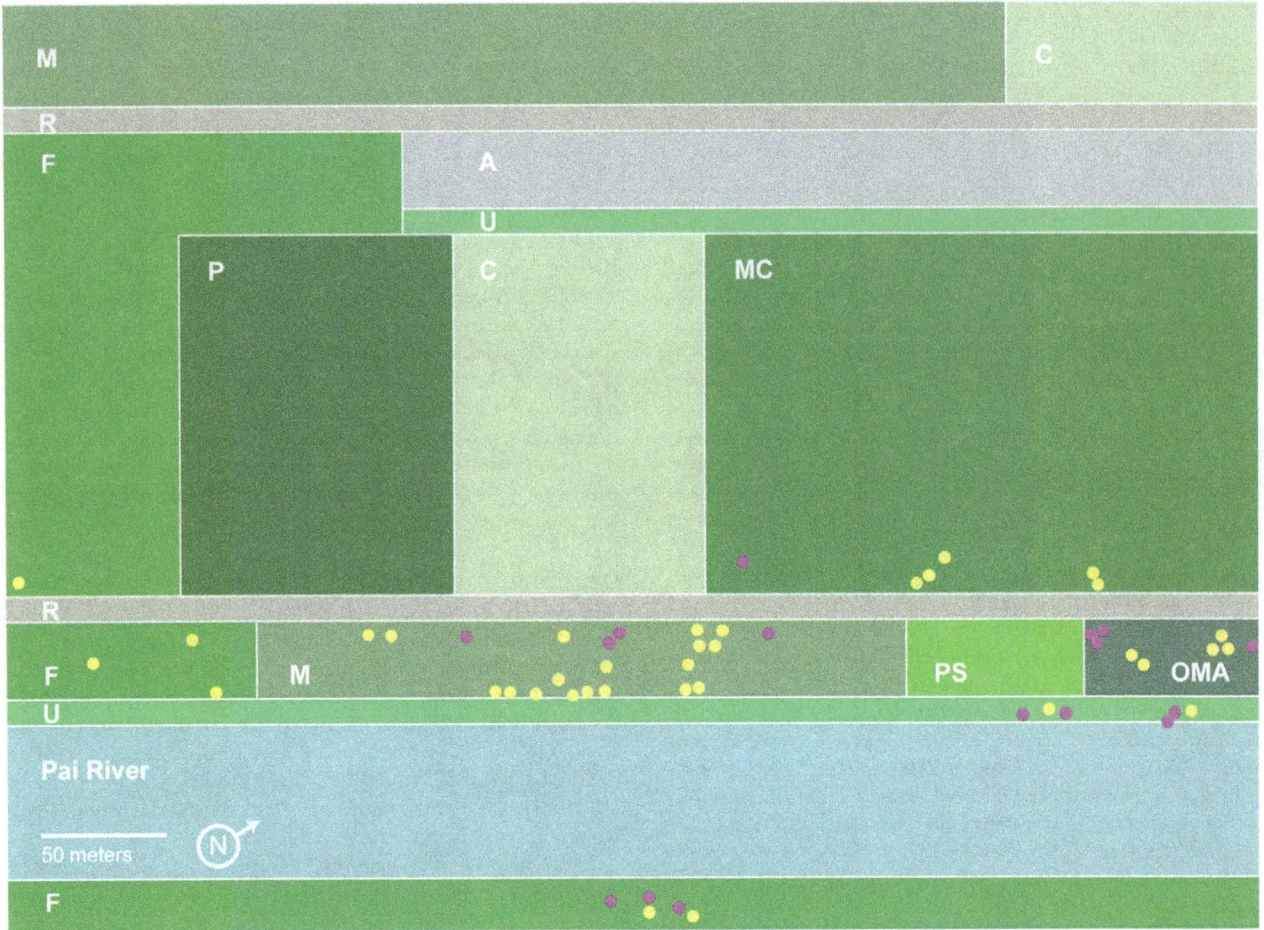

Figure 1. Study site map to scale, with bivouac locations marked for 2009 (purple dots) and 2010 (yellow dots). Note concentration of bivouacs in riverside mango trees. In 2010 one other bivouac was seen, 1 km NE of study site, along transect road. A = airstrip; C = assorted vegetable crops; F = forest; M = mango orchards; Ma = macadamia orchard; OMA = old mango trees; P = pomelo orchard; PS = passion fruit; R = road; U = undergrowth.

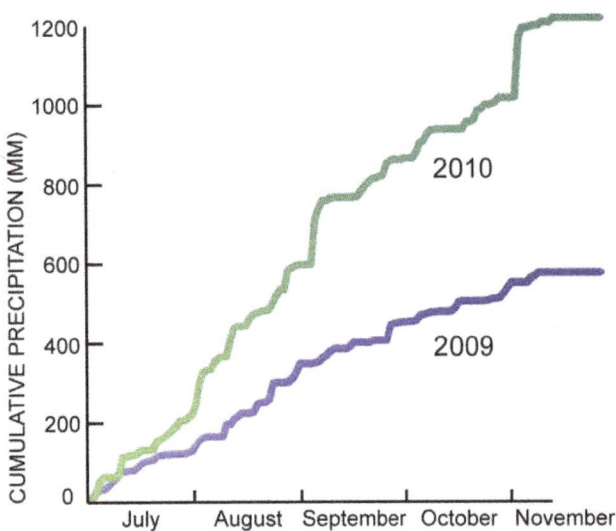

Figure 2. Cumulative rainfall, Mae Hong Son, July-November 2009 and 2010. Source: Air Force Datsav3 Weather Station 483000, call sign VTCH.

swarms utilize the site as a refuge in defense against hornet attacks (W. S. Robinson, submitted for publication).

Migration Origin and Destination Questions

Without marking or radio telemetry studies, it is impossible to know either the origin(s) or ultimate destination(s) of the migrating bees. Given the variety of flight directions from this valley surrounded by higher ground, it appears the bees are making altitude shifts of varying distance similar to what occurs in Nepal [15], but related more to a search for forage as the dry season commences [7,14]. This would explain the later colony arrivals and departures in the intense, prolonged 2010 wet season.

An alternative hypothesis to account for the BCSS is that bees were considering, but discarding, the mango orchard as a possible nesting site. This alternative is supported by the presence in 2010 of 2 small transitory colonies that built comb and quickly absconded, but seems unlikely in that 52 swarms built no comb.

Bivouac Features

The 2010 bivouacs, larger in number, were also more densely crowded than in 2009. Unlike those in Sri Lanka [13], bivouacs did not generally rest in exposed branches above surrounding vegetation. Many rested low in shady, protected places. This may

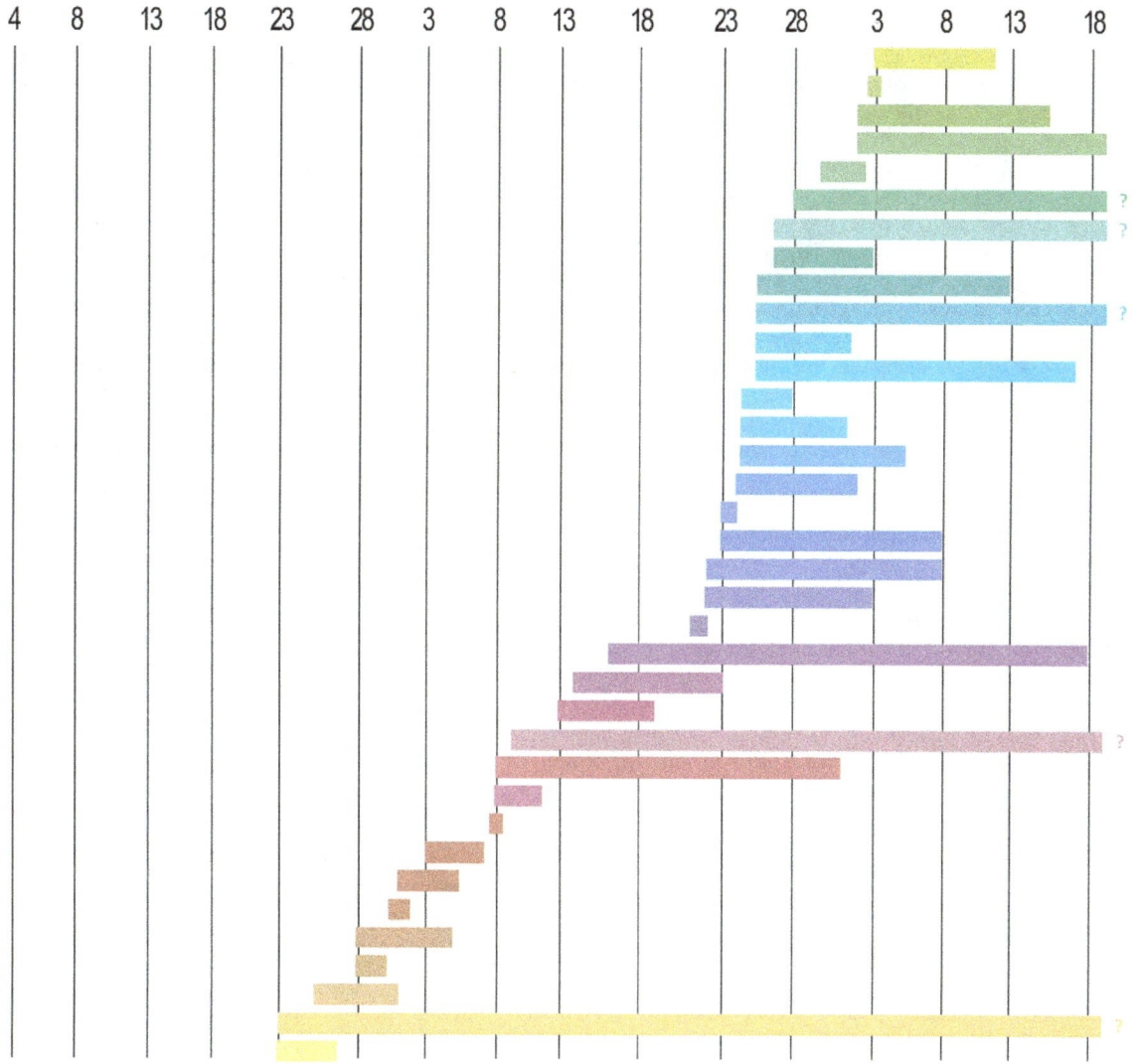

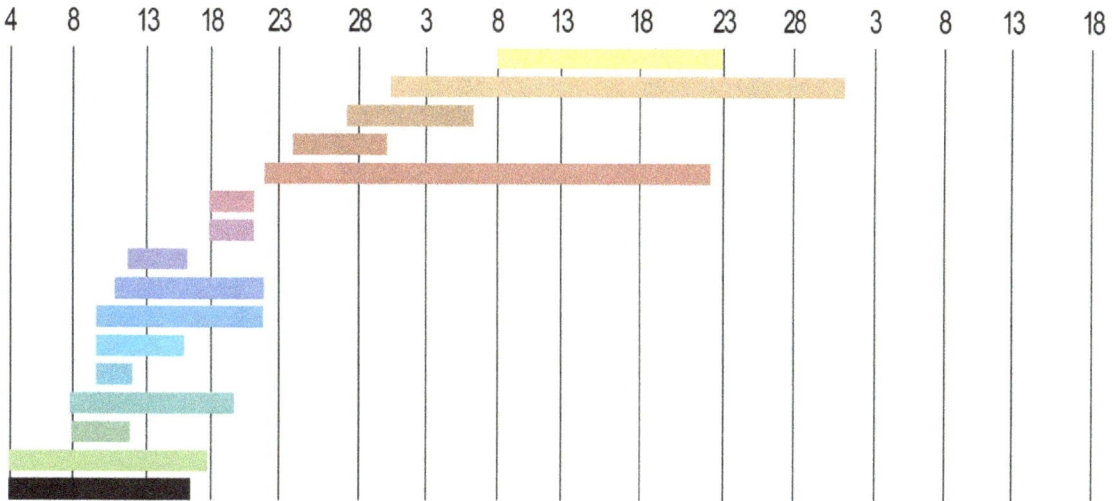

Figure 3. Timeline chart of bivouac arrival and departure, 2009 and 2010. Bars followed by a question mark represent bivouacs still on site at conclusion of 2010 study.

be a function of the mango trees themselves, most of which were pruned to <5 m in height. The greater concentration in the mangoes in 2010 thus also accounts for those bivouacs resting at significantly lower heights than those of 2009.

The range in size of the bivouacs was remarkable, given the typical *A. dorsata* colony population of 35,000–40,000 bees [8]. The smallest, 130 cm^2 surface area on one side, (approximately equal to a circle with a diameter of 13 cm) was about the size of a grapefruit and contained a few hundred bees; the 3 largest, >11,000 cm^2 (approximately equal to a circle with diameter 118 cm), were as large as average-sized adult humans and comprised >100,000 bees. Yet all functioned normally, with foragers arriving and departing, ventilators fanning in the heat, defensive shimmering to deter hornets, frequent mass defecation flights, normal dancing and eventual flight. Also notable was the range in duration of stay (2 hr to >57 days). Assuming simply that the bees are resting and foraging until they have adequate food or fat reserves in their bodies to fuel further flight does not seem to account for this huge variation.

Departure Dances and Flights

Though the sample size is small, the 2 observations of short flights preceded by short-duration waggle runs, with longer flights portended by longer dances, may contradict earlier conclusions for *A. dorsata*, *Apis mellifera* and *A. m. scutellata* [7,23,24] that migration dances preceding departure indicate direction but not distance. Correlation between dance duration and flight distance for the 2 observed short flights approximates earlier findings for foraging flights [25]. Of course, it may also be that dancing and departure behavior of bivouacking bees, or those selecting a nest site at the end of migration, simply differs from the behavior of bees initiating migration, i.e., departing from comb. This hypothesis is supported by the bivouacking dancers' showing unanimity in the duration of their waggle dances in this study, compared to the great variation witnessed among dancers at the initiation of migration [7,23,24]. The witnessed flights and dances, short and long, indicate that migration may consist of multiple steps of variable length.

Conservation Implications of Stopover Sites

Pollinators are in decline worldwide [2,26–28]. Giant honey bees are not only a splendid spectacle but are indispensable pollinators of cultivated crops [2] and wild flowering plants [1]. They manufacture honey and other hive products, and are culturally revered in Asia [8]. Their current population decline and vulnerability to human impacts [8] and "local extinctions across extensive areas" [29] make them deserving of "the sort of conservation attention that is normally reserved for charismatic vertebrates" [1].

Though attention has been paid to disappearance of giant honey bee nesting sites in large trees [8,30], none has been given to heretofore-unknown BCSS's. *A. laboriosa* may be in serious decline in Nepal due to threats to wintering clusters from forest clearing and grazing [31]. Migrating *A. dorsata* need tree branches in which to rest; none of the bivouacs from this or other studies have rested elsewhere. I speculate that they have used trees on the site of the Mae Hong Son BCSS for millennia. Doubtless some tree species are more desirable than others. For example, though the BCSS was lined with tall *Eucalyptus* trees, I never saw a bivouac in this smooth-barked species, to which it may be difficult for bivouacs to cling. No clusters utilized nearby, numerous pomelo trees. Factors such as shade [18] and angle of tree branches may be important, as they are in nest site selection [32]. Plentiful nectar from flowering plants, to fuel further flight, would also be critical to bivouacking bees, as would water for evaporative cooling.

Figure 4. The smallest bivouac at the study site.

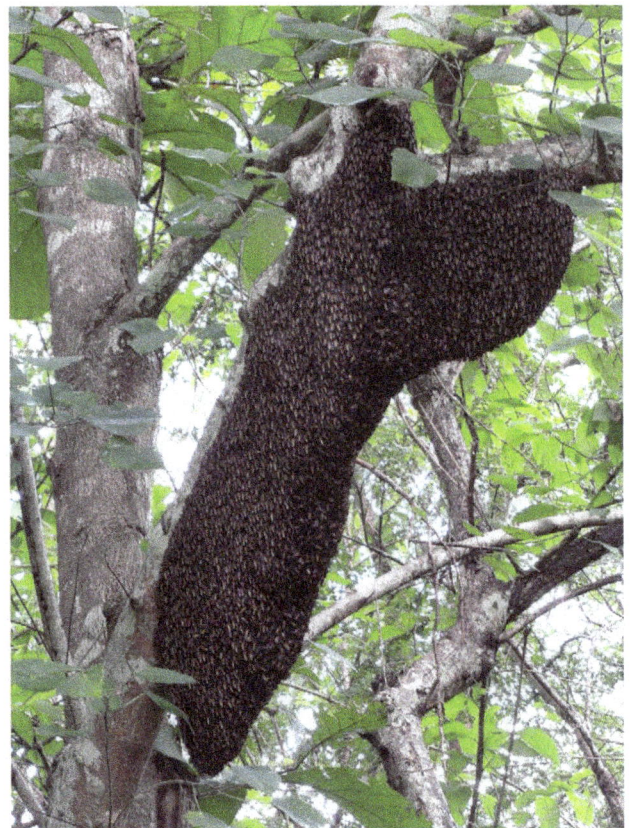

Figure 5. One of the largest bivouacs at the study site.

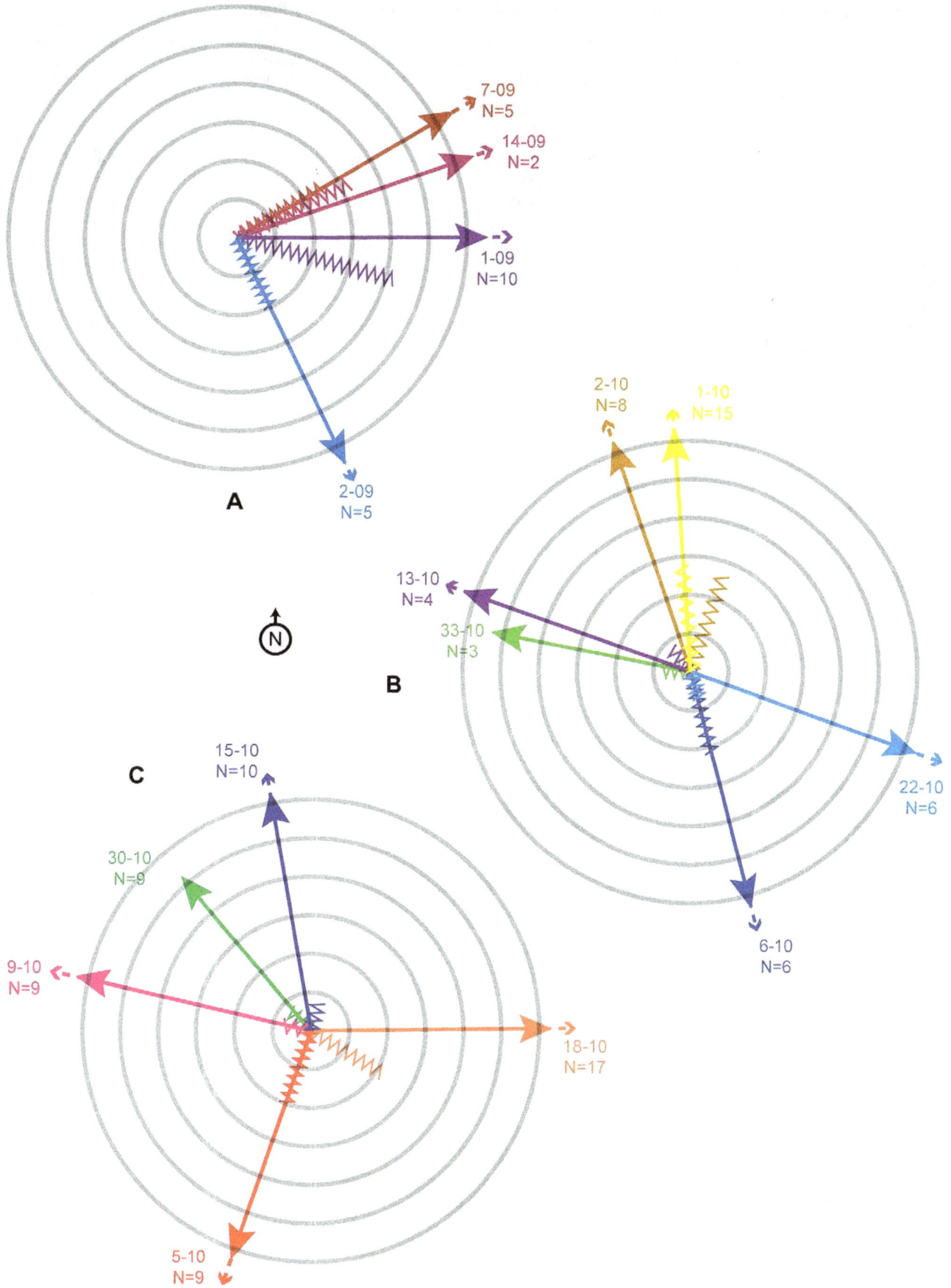

Figure 6. Compass plots depicting flight directions indicated by dancers and following flights. (A) observations from 2009, (B, C) observations from 2010. Dances of each worker were numerous; directions and distances indicated were unanimous for all dancers in each bivouac. Each color represents dances and flight of a single bivouac. Each concentric ring represents either 3 s of dancing (zigzag lines)–waggle runs only for dances of >3 s duration, complete circuits for dances of <3 s–or 30 m of flight (straight lines). Most flights exceeded 150 m and disappeared from view, as indicated by broken lines. At periphery are identification numbers of bivouacs. N = maximum number of workers simultaneously dancing within 30 min before flight.

Populations of many migratory animals, and the migrations themselves, have been damaged or destroyed by loss of specific habitat crucial to vulnerable stages of the animals' life histories [33–35]. Insects are not immune to such loss [36,37]; monarch butterflies (*Danaus plexippus*) [38] and Rocky Mountain locusts (*Melanoplus spretus*) [39] provide vivid examples. With ongoing rapid destruction of Southeast Asian forests [40–43], BCSS's may represent ecological or population bottlenecks [44–46] whose disruption could be responsible for a substantial portion of giant honey bees' decline and/or loss of their phenomenal migrations. There is a pressing need for further study; for example, to discover other migration waypoints, and to determine the bees' flexibility in accepting altered habitat, different tree species and alternative kinds of resting substrates in established BCSS's.

Methods

Ethics Statement

Permission for access to the study site was granted by the director of the Mae Hong Son Agricultural Research and Development Center. No permits were required for very limited insect collecting. Field studies did not involve endangered or protected species.

Survey Methods and Measurements

Commencing Sept 7, at the beginning and end of every day I surveyed the area for newly arrived and departed bivouacs, by walking or bicycling slowly for 2.5 km along the road flanking the west side of the mango orchard. This twice-daily transect included

close inspection of the orchard, and of trees on both sides of the road for ≈1 km north and south. I also patrolled the site during lulls in colony activity, and sometimes found new swarms as I made rounds studying bivouacs already present. Field workers occasionally told me of swarms and their arrival times.

At least twice daily I surveyed the macadamia and pomelo orchards to the west. No bivouacs were encountered in the pomelos. I found several bivouacs in the macadamias. Approximately weekly I searched the other mango orchard, west ≈250 m from the river, but never encountered bees there. I also regularly traveled through northern Thailand both near the study area and well outside it, constantly looking for more bivouacs without success. This included, for example, a daily bicycle ride of ~10 km to and from the village of Mae Hong Son, through human settlements, agricultural and forested land. At the height of the 2009 bivouac season I rafted the Pai River from Pai to Mae Hong Son (~70 km), looking unsuccessfully along the shores for bivouacs.

Bivouacs were distinguishable by their widely varying sizes and shapes. Upon finding a new bivouac I attached an identifying tag to a nearby branch. With the exception of 2 that took short flights that I was able to track, bivouacs did not move around the study site before leaving it completely, so there was no potential for confusing one with another.

I recorded arrival time as the date I first saw the swarm. Often this was very precise, as I actually saw the swarm land. In rare cases that date could be late by as much as several days, if a newly arrived swarm went unnoticed. I recorded departure time as the date at which I first noted the bivouac's absence. Again, in many

Figure 7. *A. dorsata* **folragers in high density on** *Eucalyptus* **flowers.**

cases I saw swarms depart the study site; in other instances the swarm may have left the previous afternoon but its absence was not noticed until morning. In calculating the mean length of stay, I recorded a stay as 0.5 day if it departed the same day it arrived, 1 day if its absence was noticed the day after arrival, 2 days if it was gone 2 days after its arrival, etc.

I was unsure of the precise arrival time of the first swarms seen in 2009. They were not present during a preliminary search on 1 September, but I did not survey the orchard again until 7 Sept, when I found the first bivouacs. I have thus interpolated 4 Sept as their arrival date. In 2010 I left Thailand on 19 November, when 5 bivouacs were still present, and was unable to record their departure dates. In calculating duration of stay, I used 19 November as the date of departure, so these calculations are overly conservative.

I used a tape measure to measure colony size and height at which swarms rested. For occasional lofty colonies, I estimated height by sighting against a 2-m reference pole, and colony dimensions using the known 17-mm length of an A. dorsata worker as a reference. Because I could not always see all sides of distant or inaccessible colonies, I did not attempt to estimate swarm volume. Rather, I recorded maximum length and width of the colonies. I present these measurements as approximate surface area of one side of the bivouac. Their depth was approximately the diameter of the branch on which they rested, normally ≈10–20 cm.

I employed unpaired t-tests to test the hypotheses that mean swarm size, mean bivouac resting height and mean length of stay varied between the 2 years.

I patrolled the swarms on a 30–60 min circuit, observing any dancing worker bees, and apparent preparations for swarm flight. Bivouacs were often low in the trees and easily viewed from 1–3 m with the unaided eye. I used Leupold™ 10×42 binoculars to view higher swarms. I photographed and video-recorded the bees' behavior using Canon Power Shot™ S3 IS or SX10 IS digital cameras.

Departure Dances and Flights

I measured duration of waggle dance circuits [47,48] to the nearest second with a digital stopwatch. For bees dancing repeated circuits a mean time was calculated, again to the nearest second, though individual bees were remarkably consistent in their dances. Note that for longer dances circuit integrity breaks down; a dancing bee performs a waggle run, but instead of returning to the site where the run began, the dancer often turns and does another waggle run, or moves to another site on the mantle and does its waggle run [13,25]. For dances lasting >3 s, I therefore measured duration just of waggle runs, not entire circuits. I estimated by eye

the angle of waggle runs to the nearest 30°. For example, a dance to 12 o'clock was labeled 0°, a dance in the 1 o'clock direction was 30°, 2 o'clock was 60°, etc.

When a swarm flew, I used a compass to determine its flight direction to the nearest 5° by sighting a line from point of takeoff to point of disappearance or observed landing. I used the NOAA Sun Location Calculator [49] to find the sun's azimuth at flight time. To test the hypothesis that dance direction and duration are correlated with flight direction and distance, I then compared the direction and, when possible, distance of the flight to that indicated by the dances [25,47,48,50]. For presentation here, I have converted directions of observed dances to their indicated directions of flight, by adding the angle of dancing to the sun's azimuth at the time [25].

Statistical Analysis and Voucher Specimens

Statistical tests were employed following Mead et al. [19] and Batchelet [20]. Voucher specimens are housed in the University of Wyoming Insect Museum, Laramie, WY, USA.

Supporting Information

Video S1   Several bees dance 8-s waggle runs as a large bivouac prepares for a long flight out of the study area. Note discreet waggle runs of individuals separated by a looping movement; no complete circuits are performed.

Video S2   Several bees dance 2-s complete circuits as a medium-sized bivouac prepares for a short flight of 150 m.

Video S3   An agitated bivouac departs its resting place. Elapsed time from liftoff until branch was vacated was approximately 60 s.

Acknowledgments

I thank Maria Katherman for her many valuable insights. Jeffrey Lockwood and Quinn Robinson critiqued the manuscript. Jesseda Weera, Yotin Bootchalaey and Jaroon Kumnuanta provided logistical help in Thailand. Phanida Ponya helped monitor bivouacs. Jane Ifland prepared the figures.

Author Contributions

Conceived and designed the experiments: WSR. Performed the experiments: WSR. Analyzed the data: WSR. Contributed reagents/materials/analysis tools: WSR. Wrote the paper: WSR.

References

1. Corlett RT (2011) Honeybees in natural ecosystems. In: Hepburn R, Radloff, SE, editors. Honeybees of Asia. Berlin: Springer-Verlag. 215–226. DOI 10.1007/978-3-642-16422-4 (2011).

2. Partap U (2011) The pollination role of honeybees. In: Hepburn R, Radloff, SE, editors. Honeybees of Asia. Berlin: Springer-Verlag. 227–256. DOI 10.1007/978-3-642-16422-4 (2011).

3. Crane E (1999) The world history of beekeeping and honey hunting. New York: Routledge. 720 pp.

4. Hepburn HR, Radloff SE (2011) Biogeography. In: Hepburn R, Radloff, SE, editors. Honeybees of Asia. Berlin: Springer-Verlag. 51–68. DOI 10.1007/978-3-642-16422-4 (2011).

5. Ahmad R (1989) A note on the migration of Apis dorsata in the Andaman and Nicobar Islands. Bee World 70: 62–65.

6. Deodikar GB, Ghatge AL, Phadke RP, Mahindre DB, Kshirsagar KK, et al. (1977) Nesting behaviour of Indian honey bees III. Nesting behaviour of Apis dorsata Fab. Indian Bee J. 39: 1–12.

7. Dyer FC, Seeley TD (1994) Colony migration in the tropical honeybee Apis dorsata F. (Hymenoptera: Apidae). Insect Soc 41: 129–140.

8. Oldroyd BP, Wongsiri S (2006) Asian honey bees: biology, conservation and human interactions. Cambridge: Harvard University Press. 340 pp.

9. Venkatesh G, Reddy CC (1989) Rates of swarming and absconding in the giant honeybee, Apis dorsata F. Proc Indian Acad Sci 98: 425–430.

10. Neumann P, Koeniger N, Koeniger G, Tingek S, Kryger P, et al.(2000) Home-site fidelity in migratory bees. Nature 406: 474–475.

11. Paar J, Oldroyd BP, Kastberger G (2000) Giant honeybees return to their nest sites. Nature 406: 475.

12. Mardan MB (1989) Thermoregulation in the Asiatic giant honeybee Apis dorsata (Hymenoptera: Apidae). Thesis, University of Guelph.

13. Koeniger N, Koeniger G (1980) Observations and experiments on migration and dance communication of Apis dorsata in Sri Lanka. J Apic Res 19: 21–34.

14. Hepburn HR (2011) Absconding, migration and swarming. In: Hepburn R, Radloff, SE, editors. Honeybees of Asia. Berlin: Springer-Verlag. 133–158. DOI 10.1007/978-3-642-16422-4 (2011).

15. Underwood BA (1990) Seasonal nesting cycle and migration patterns of the Himalayan honeybee Apis laboriosa. Nat Geogr Res 6: 276–290.

16. Thapa R, Wongsiri S, Oldroyd BP, Prawan S (1998) Migration of Apis dorsata in northern Thailand. In: Matsuka M, Verma LR, Wongsiri S, Shrestha K K,

Partap U, editors. Asian bees and beekeeping: progress of research and development. Proc 4th Asian apic assoc conf, Kathmandu, March 23–28, 1998. Enfield, New Hampshire: Science Publishers. 39–43.

17. Woyke J, Wilde J, Wilde M (2001) A scientific note on *Apis laboriosa* winter nesting and brood rearing in the warm zone of the Himalayas. Apidologie 32: 601–602.

18. Robinson WS (2010) Studies on *Apis dorsata* bivouacs congregating in the north of Thailand. In: Matsuka M, editor. Proc 9th international conf apitherapy health care, Chiang Rai, Thailand, 2009. Tokyo: Asian Apiculture Association. 111–124.

19. Mead R, Curnow RN, Hasted AM (1993) Statistical methods in agriculture and experimental biology. 2nd ed. London: Chapman & Hall. 415 pp.

20. Batchelet E (1981) Circular statistics in biology. New York: Academic Press. 371 pp.

21. Duangphakdee O, Hepburn HR, Tautz J (2011) The dance language. In: Hepburn R, Radloff, SE, editors. Honeybees of Asia. Berlin: Springer-Verlag. 313–332. DOI 10.1007/978–3–642–16422–4 (2011).

22. Koeniger G, Koeniger N, Phiancharoen M (2011) Comparative reproductive biology of honeybees. In: Hepburn R, Radloff, SE, editors. Honeybees of Asia. Berlin: Springer-Verlag. 159–206. DOI 10.1007/978–3–642–16422–4 (2011).

23. Schneider SS, McNally LC (1994) Waggle dance behavior associated with seasonal absconding in colonies of the African honey bee, *Apis mellifera scutellata*. Insect Soc 41: 115–127.

24. Lewis LA, Schneider SS (2008) "Migration dances" in swarming colonies of the honey bee, *Apis mellifera*. Apidologie 39: 354–361. DOI: 10.1051/apido:2008018.

25. Dyer FC, Seeley TD (1991) Dance dialects and foraging range in three Asian honey bee species. Behav Ecol Sociobiol 28: 227–233.

26. Allen-Wardell GP, Bernhardt R, Bitner A (1998) The potential consequences of pollinator declines on conservation of biodiversity and stability of food crop yields. Cons Biol 12: 8–17.

27. National Research Council (2007) Status of pollinators in North America Washington: National Academies Press. 322 pp.

28. Buchmann SL, Nabhan GP (1997) The forgotten pollinators. Washington: Island Press. 312 pp.

29. Oldroyd BP, Nanork P (2009) Conservation of Asian honey bees. Apidologie 40: 296–312. DOI 10.1051/apido/2009021.

30. Basavarajappa S, Raghunandan KS, Hegde SN (2009) Studies on the nest host trees of Asian giant honey bee, Apis dorsata F. in southern Karnataka, India. Bioscan 4, 647–652.

31. Underwood BA (1992) Impact of human activities on the Himalayan honeybee, *Apis laboriosa*. In: Verma, LR, editor. Honey bees in mountain agriculture (Boulder, Colorado: Westview Press. 51–57.

32. Tan NQ, Chinh PH, Thai PH, Mulder V (1997) Rafter beekeeping with *Apis dorsata*: some factors affecting the occupation of rafters by bees. J Apic Res 36: 49–54.

33. Brower LP, Malcolm SP (1991) Animal migrations: endangered phenomena. Amer Zool 31: 265–276.

34. Wilcove DS (2008) Animal migration: an endangered phenomenon? Issues in Sci. and Tech. 24: 71–78.

35. Wilcove DS (2010) No way home: the decline of the world's great animal migrations. Washington: Island Press. 256 pp.

36. Dybas HS, Lloyd M (1974) The habits of 17-year periodical cicadas (Homoptera: Cicadidae: *Magicicada* spp. Evol 44: 279–324.

37. Janzen DH (1986) The external threat. In: Soule ME, editor. Conservation Biology. Sunderland, Massachusetts: Sinauer & Associates. 286–303.

38. Brower LP, Castilleja G, Peralta A, Lopez-Garcia J, Bojorquez-Tapia L, et al. (2002) Quantitative changes in forest quality in a principal overwintering area of the monarch butterfly in Mexico, 1971–1999. Cons Biol 16: 346–359.

39. Lockwood JA, DeBrey LD (1990) A solution for the sudden and unexpected extinction of the Rocky Mountain locust, *Melanoplus spretus*. Environ Entomol 19: 1194–1205.

40. Food and Agriculture Organization of the United Nations (2011) State of the World's Forests, 2011. Rome: FAO. 164 pp.

41. Miettinen J, Shi C, Liew SC (2011) Deforestation rates in insular Southeast Asia between 2000 and 2010. Global Change Biol 17: 2261–2270. DOI: 10.1111/j.1365–2486.2011.02398.x

42. Sodhi NS, Koh LP, Brook BW, Ng PKL (2004) Southeast Asian biodiversity: an impending disaster. Trends Ecol Evol 19: 654–660. DOI: 10.1016/j.tree.2004.09.006.

43. Laurance WF (2007) Forest destruction in tropical Asia. Current Sci 93: 1544–1550.

44. Lockwood JA (2005) Locust: the devastating rise and mysterious disappearance of the insect that shaped the American frontier New York: Basic Books. 320 pp.

45. Lovejoy TE (1983) Tropical deforestation and North American birds. Bird Cons 1: 126–128.

46. Matthews GVT (1981) The conservation of migratory birds. In: Aidley DJ, editor. Animal migration. Cambridge: University Press. 231–239.

47. Frisch K von (1967) The dance language and orientation of bees. Cambridge: Belknap Press of Harvard University Press. 566 pp.

48. Seeley TD (2010) Honeybee democracy. Princeton, New Jersey: Princeton University Press. 280 pp.

49. NOAA Solar Position Calculator. Available: http://www.srrb.noaa.gov/highlights/sunrise/azel Accessed 2011 November 11.

50. Punchihewa RKW, Koeniger N, Kevan PG, Gadawaski RM (1985) Observations on the dance communication and natural foraging ranges of *Apis cerana*, *Apis dorsata* and *Apis florea* in Sri Lanka. J Apic Res 24: 168–175.

PERMISSIONS

The contributors of this book come from diverse backgrounds, making this book a truly international effort. This book will bring forth new frontiers with its revolutionizing research information and detailed analysis of the nascent developments around the world.

We would like to thank all the contributing authors for lending their expertise to make the book truly unique. They have played a crucial role in the development of this book. Without their invaluable contributions this book wouldn't have been possible. They have made vital efforts to compile up to date information on the varied aspects of this subject to make this book a valuable addition to the collection of many professionals and students.

This book was conceptualized with the vision of imparting up-to-date information and advanced data in this field. To ensure the same, a matchless editorial board was set up. Every individual on the board went through rigorous rounds of assessment to prove their worth. After which they invested a large part of their time researching and compiling the most relevant data for our readers.

The editorial board has been involved in producing this book since its inception. They have spent rigorous hours researching and exploring the diverse topics which have resulted in the successful publishing of this book. They have passed on their knowledge of decades through this book. To expedite this challenging task, the publisher supported the team at every step. A small team of assistant editors was also appointed to further simplify the editing procedure and attain best results for the readers.

Apart from the editorial board, the designing team has also invested a significant amount of their time in understanding the subject and creating the most relevant covers. They scrutinized every image to scout for the most suitable representation of the subject and create an appropriate cover for the book.

The publishing team has been an ardent support to the editorial, designing and production team. Their endless efforts to recruit the best for this project, has resulted in the accomplishment of this book. They are a veteran in the field of academics and their pool of knowledge is as vast as their experience in printing. Their expertise and guidance has proved useful at every step. Their uncompromising quality standards have made this book an exceptional effort. Their encouragement from time to time has been an inspiration for everyone.

The publisher and the editorial board hope that this book will prove to be a valuable piece of knowledge for researchers, students, practitioners and scholars across the globe.

LIST OF CONTRIBUTORS

Yousry A. El-Kassaby, Cherdsak Liewlaksaneeyanawin and Jaroslav Klápštš
Department of Forest Sciences, Faculty of Forestry, University of British Columbia, Vancouver, British Columbia, Canada

Eduardo P. Cappa
Instituto Nacional de Tecnología Agropecuaria (INTA), Instituto de Recursos Biológicos, Hurlingham, Buenos Aires, Argentina

Milan Lstibůrek
Department of Dendrology and Forest Tree Breeding, Faculty of Forestry and Wood Sciences, Czech University of Life Sciences Prague, Praha, Czech Republic

Pierre Franck, Myriam Siegwart, Jerome Olivares, Jean-François Toubon and Claire Lavigne
INRA, UR1115 Plantes et Systèmes de culture Horticoles, Avignon, France

Theresa M. Nogeire and Jennifer M. Duggan
School of Environmental and Forest Sciences, University of Washington, Seattle, Washington United States of America

Frank W. Davis
Bren School of Environmental Science and Management, University of California Santa Barbara, Santa Barbara, California, United States of America

Kevin R. Crooks
Department of Fish, Wildlife, and Conservation Biology, Colorado State University, Fort Collins, Colorado, United States of America

Erin E. Boydston
U.S. Geological Survey, Western Ecological Research Center, Thousand Oaks, California, United States of America

Grzegorz Buczkowski and Douglas S. Richmond
Department of Entomology, Purdue University, West Lafayette, Indiana, United States of America

Kevin Cloonan and Walter Leal
Honorary Maeda-Duffey Laboratory, University of California Davis, Davis, California, United States of America

Robert H. Bedoukian
Bedoukian Research Inc., Danbury, Connecticut, United States of America

David J. Biddinger and Neelendra K. Joshi
Fruit Research and Extension Center, Pennsylvania State University, Biglerville, Pennsylvania, United States of America
Department of Entomology, Pennsylvania State University, University Park, Pennsylvania, United States of America

Jacqueline L. Robertson
USDA Forest Service PSW Station, Albany, California, United States of America

Chris Mullin, James Frazier, Sara A. Ashcraft and Edwin G. Rajotte
Department of Entomology, Pennsylvania State University, University Park, Pennsylvania, United States of America

Mace Vaughn
The Xerces Society, Portland, Oregon, United States of America

Mo-Qian Zhang, Hai-Qiang Guo, Xiao Xie, Ting-Ting Zhang, Zu-Tao Ouyang and Bin Zhao
Coastal Ecosystems Research Station of the Yangtze River Estuary, Ministry of Education Key Laboratory for Biodiversity Science and Ecological Engineering, Institute of Biodiversity Science, Fudan University, Shanghai, P.R. China

Valeria Malagnini, Federico Pedrazzoli, Christian Cainelli, Rosaly Zasso, Valeria Gualandri and Claudio Ioriatti
Centre for Technology Transfer, FEM-IASMA, San Michele all'Adige (TN), Italy

Chiara Papetti
Department of Biology, University of Padua, Padova, Italy

Alberto Pozzebon
Department of Agronomy, Food, Natural Resources, Animals and Environment, University of Padua, AGRIPOLIS, Legnaro (PD), Italy

Miguel Montes-Borrego and Blanca B. Landa
Department of Crop Protection, Institute for Sustainable Agriculture (IAS-CSIC), Cordoba, Spain

Madis Metsis
Tallinn University, Institute of Mathematics and Natural Sciences, Tallinn, Estonia

Qiang Zhang, Jian Sun, Songzhong Liu and Qinping Wei
Institute of Forestry and Pomology, Beijing Academy of Agriculture & Forestry Sciences, Beijing, China

Daniele Bevacqua, Michel Gènard and Françoise Lescourret
INRA, UR1115 PSH, Avignon, France

Xuan Fang and Guoan Tang
Key Laboratory of Virtual Geographic Environment, Ministry of Education, School of Geography Science, Nanjing Normal University, Nanjing, China

Bicheng Li
Research Center of Soil and Water Conservation and Ecological Environment, Chinese Academy of Sciences, Yangling, Shaanxi, China

Ruiming Han
School of Geography Science, Nanjing Normal University, Nanjing, China

Elisa González-Domínguez and Josep Armengol
Instituto Agroforestal Mediterráneo, Universidad Politécnica de Valencia, Valencia, Spain

Vittorio Rossi
Istituto di Entomologia e Patologia vegetale, Universitá Cattolica del Sacro Cuore, Piacenza, Italy

Ashutosh K. Srivastava, Satyajeet Rai, M. K. Srivastava, M. K. R. Mudiam and L. P. Srivastava
Pesticide Toxicology Laboratory, CSIR-Indian Institute of Toxicology Research, (Council of Scientific and Industrial Research, Govt. of India), Mahatma Gandhi Marg, Lucknow, India

M. Lohani
Department of Biotechnology, Integral University, Lucknow, India

Hai Ren, Linjun Li, Lianlian Yuan, Shuguang Jian, Jun Wang, Xubing Chen, Hongfang Lu, Guoyi Zhou, Xuli Tang and Qianmei Zhang
Key Laboratory of Vegetation Restoration and Management of Degraded Ecosystems, South China Botanical Garden, Chinese Academy of Sciences, Guangzhou, China

Qiang Liu
College of Life Science, Hainan Normal University, Haikou, China

Xu Wang
College of Environment and Plant Protection, Hainan University, Haikou, China

Yide Li and Huai Yang
Research Institute of Tropical Forestry, Chinese Academy of Forestry, Guangzhou, China

Dafeng Hui
Department of Biological Sciences, Tennessee State University, Nashville, TN, United States of America

Dong Wang
School of Life Sciences, Central China Normal University, Wuhan, China

Muhammad Shah Nawaz-ul-Rehman and Claude M. Fauquet
Danforth Plant Science Center, St. Louis, Missouri, United States of America

Rob W. Briddon
Agricultural Biotechnology Division, National Institute for Biotechnology and Genetic Engineering, Jhang Road, Faisalabad, Pakistan

Eckehard G. Brockerhoff, Jessica L. Kerr and Graham Coker
Scion (New Zealand Forest Research Institute), Christchurch, New Zealand

David M. Suckling
The New Zealand Institute for Plant & Food Research Ltd, Lincoln, New Zealand

Mark Kimberley, Brian Richardson and Stefan Gous
Scion (New Zealand Forest Research Institute), Rotorua, New Zealand

David M. Cowan and David R. Lance
Plant Protection and Quarantine, United States Department of Agriculture, Animal and Plant Health
Inspection Service, Buzzards Bay, Massachusetts, United States of America

Tara Strand
Pacific Wildland Fire Sciences Laboratory, United States Department of Agriculture, Forest Service, Seattle, Washington, United States of America

Aijun Zhang
Invasive Insect Biocontrol and Behavior Laboratory, United States Department of Agriculture, Agricultural Research Services, Beltsville Agricultural Research Center - West, Beltsville, Maryland, United States of America

Willard S. Robinson
Biology Department, Casper College, Casper, Wyoming, United States of America

Index

A

A Melting Pot, 141, 143, 145, 147, 149, 151, 153, 155, 157, 159, 161, 163

Aerial Application Of Pheromones, 164-165, 167, 169, 171

Agricultural-wildland Gradient, 22-23, 25, 27

Andalusian Landscape, 69, 71, 73, 75, 77, 79

Ant Abundance And Diversity, 28-31, 33-35

Arbuscular Mycorhizal Fungi, 69, 71, 73, 75, 77, 79

B

Biodiversity Loss, 28-29

Biological Invasions, 36, 164

Breeding Concept, 1

Breeding Without Breeding (bwb), 2

Buffer Natural Areas, 22

C

Camera Stations, 22-24

Carbon Storage, 106, 130-131, 133, 135, 137, 139-140

Carnivore Use Of Avocado Orchards, 22-23, 25, 27

Codling Moth Populations, 12-13, 15-21

Collection Of Gossypium Species, 141, 143, 145, 147, 149, 151, 153, 155, 157, 159, 161, 163

Community Differentiation, 69, 78, 80

Comparative Toxicities, 43-45, 47

Complete Pedigree Information, 1-2

Cotton Leaf Curl Multan Betasatellite (clcumub), 141

Cultivated Crops, 172, 176

D

Detoxification Activity, 13

Dna Fingerprinting, 1, 9-10

Dna Variability, 12

Dna-based Pyrosequencing, 81

Dose-mortality Regressions, 43-44

E

Ecological And Genetic Differences, 60-61, 63, 65, 67

Ecosystem Service Values, 95-99, 101, 103-105, 107

Efficient Breeding, 1-3, 5, 7, 9, 11

Electroantennogram (eag) Recordings, 38

F

Factors Affecting Biodiversity, 28-29, 31, 33, 35

Features Of Bivouacs, 173

Field Oviposition Assay, 38

Forest Ecosystems, 11, 58, 130-131, 133-139, 171

Fungicide Applications, 108, 117

Fusicladium Eriobotryae, 108-110, 112, 118

G

Gc-ms Confirmation, 126, 128

Geminiviruses, 20, 141-142, 144, 149, 154, 159-160, 162-163

Genotyping Microsatellite Markers, 60

H

Hazard Index Estimation, 120, 126

Honey Producers And Pollinators, 172

Host-switching Experiments, 60-62

I

Invasive Moth, 164-165, 167, 169, 171

L

Land Use Pattern, 95-103, 105-107

Land-cover Characteristics, 49-53, 55, 57-59

Land-cover Classification, 23, 49, 53-54, 58

Land-use Change, 22, 140

Loess Plateau Of China, 95, 107

Loss Of Ant Biodiversity, 28

M

Mammalian Carnivores, 22, 27

Manure Refinement, 81-83, 85, 87

Mating Disruption, 37, 42, 164-167, 169-171

Migrating Giant Honey Bees, 172-173, 175, 177, 179

Modified Quechers Extraction Method, 120-121, 123, 125, 127, 129

Modis Time Series Data, 49, 51, 53, 55, 57, 59

N

Navel Orangeworm Oviposition, 37, 39, 41

O

Old World Begomoviruses, 141, 143, 145, 147, 149, 151, 153, 155, 157, 159, 161, 163

Olive Oil Producer, 69
Organophosphates (ops), 120

P
Plant Breeding, 1, 11
Pollinator Species, 43, 45, 47-48
Potential Eradication Tool, 164-165, 167, 169, 171
Predicting Infection Of Loquat Fruit, 108-109, 111, 113, 115, 117, 119
Proof-of-concept Study, 37
Pyrethroid Resistance, 12, 18, 20-21

Q
Quasi-double-blind Screening, 37, 39, 41

R
Resistance To Insecticides, 12, 20
Rhizosphere Sampling, 70

S
Sandy Soil, 81-83, 85, 87
Semiochemicals, 37, 39, 41
Sodium Channel Kdr Mutations, 12-13, 15, 17, 19, 21
Spatial And Temporal Patterns, 130-131, 133, 135, 137, 139

Spatial And Temporal Variations, 95, 97, 99, 101, 103, 105, 107
Species Host Plant Preference, 60
Species Richness, 23, 28-30, 34-36
Spore Dispersal, 108
Stopover Site, 172-173, 175, 177, 179
Synergism Of Apple Orchard Pesticides, 43, 45, 47

T
Taxonomic Composition Analysis, 86
Temporal Examination, 28-29, 31, 33, 35
Terrestrial C Budget, 130
The Effect Of Urbanization, 28-29, 31, 33, 35
The Proportion Of Forestland, 95, 106
Timing Of Bivouac Searches, 172

W
Weather-based Model, 108-109, 111, 113, 115, 117, 119
Wide-ranging Species, 22
Winter Pruning Intensity, 89-90, 92

Y
Yangtze River Estuary, 49

www.ingramcontent.com/pod-product-compliance
Lightning Source LLC
Chambersburg PA
CBHW082011190326

41458CB00010B/3152